Quantum Optics

T0178356

Miguel Orszag

Quantum Optics

Including Noise Reduction, Trapped Ions, Quantum Trajectories, and Decoherence

Second Edition

With 90 Figures and 92 Problems with Hints for Solutions

 Springer

Miguel Orszag
Pontificia Universidad Católica de Chile
Facultad de Física
Av. Vicuña Mackenna 4860
Macul, Santiago
Chile
morszag@fis.puc.cl

ISBN 978-3-642-09180-3   e-ISBN 978-3-540-72707-1

Springer is a part of Springer Science+Business Media
springer.com
© Springer-Verlag Berlin Heidelberg 2010

Cover design: eStudio Calamar S.L., F. Steinen-Broo, Pau/Girona, Spain

Preface to the Second Edition

Seven years have passed since the publication of '*Quantum Optics*', so I used the opportunity of this Second Edition to correct several errors and misprints, as well as to include some new material.

The *Chapter 20* was enlarged as to include a section on '**Decoherence Free Subspaces**'. This is an interesting topic for those readers interested in Quantum Computation, Teleportation and Cryptography, because tools have been developed to find 'a quiet corner in Hilbert space', that is a region free from the damaging interaction with the environment. Several examples and applications are also given in this section.

Also, we include in this present version *Chapter 21*, dealing with the basics of quantum computation as well as the study of quantum entanglement.

Quantum entanglement is one of the most relevant quantum mechanical properties and crucial to understand a whole family of phenomena as well as applications such as Quantum Teleportation, Quantum Cryptography, etc. We describe first, the entanglement for pure states, developing the **Schmidt decomposition** and then deal with the more complicated mixed state entanglement. For the mixed case, we cover the well-known **Peres–Horodecki criteria** for positive partial transposition.

At the end of the chapter, and as an exciting application, we cover the **Quantum teleportation Protocol**, as originally described by Bennet et al.

Chapter 22 deals with topics such as the **no-cloning theorem**, the **Universal Quantum Copying Machine** as a concept and how to implement it using a logical circuit made of quantum gates. Both the fidelity and the entanglement of the copies are calculated for the duplicator and triplicator. The last topic of this chapter includes the discussion of a simple model of a **stochastic quantum processor** that rotates a qubit and how to improve its efficency, including additional program qubits.

I thank Mr Paul Blackburn for helping me in the drawings, Ms Maritza Hernandez for reading the new chapters and also editing some figures, and Mr Sergio Dagach for re-reading the entire manuscript, Mr Juan Acuña for helping me with the figures and editing the book. Finally, I thank Dr Thorsten Schneider and Ms Jacqueline Lenz, from Springer and MS Sunayana Jain

(Integra Software, India) for their friendly collaboration in the preparation of this second edition.

Santiago, Chile.   *Miguel Orszag*
January, 2007

Preface to the First Edition

This Graduate textbook originated from lectures given by the author at the Universidad Católica de Chile in Santiago, as well as at the University of New Mexico. Also, material has been drawn from short summer courses given in Rio de Janeiro and Caracas.

Chapter 1 is devoted to some basic ideas of interaction of radiation and matter, starting from Einstein's ideas of emission and absorption and ending with an elementary laser theory.

Quantum mechanical description of the atom–radiation interaction is dealt with in Chap. 2, including Rabi's oscillations and Bloch's equations.

Chapter 3 contains the basic quantization of the electromagnetic field, while Chaps. 4, 5 and 6 study special states of the electromagnetic field and quantum theory of coherence.

The Jaynes–Cummings model, which describes in a fully quantized manner the atom–radiation interaction, is studied in the Chap. 8, along with the phenomena of collapse and revival. We also introduce the dressed state description that is useful when studying resonance fluorescence (Chap. 10).

Real physical systems are open, that is, one must always consider dissipative mechanisms, including the electromagnetic losses in the cavity walls or atomic decay. All these effects can be considered in great detail, studying system–reservoir interactions, leading to Master and Fokker–Planck equations. These reservoirs can also be phase dependent, effect that can modify the decay rate of an atom (Chap. 9). As we mention before, Chap. 10 is entirely devoted to resonance fluorescence, and the study and observation, for the first time, of photon antibunching.

The invention of the laser, in the decade of the 1960s, opened up a new area of research, baptized as Quantum Optics. This discovery allowed the growth of new fields as non-linear optics and non-linear spectroscopy. The semiclassical theory, first, and the Quantum Theory of the Laser were well developed by the late 1960s. The quantum theory of the laser, from the master equation and Langevin equation approach is extensively treated in the Chaps. 11 and 12, respectively. We have also added some more recent material, including the micromaser and the effect of the pump statistics, as a form of noise reduction scheme. Although, pump statistics did not play any role in the original laser theory, recent experiments and theoretical calculations showed that one could reduce considerably the photon number fluctuations if one is careful enough in pumping the atoms in an orderly way.

We further study the quantum noise reduction in correlated emission lasers and generation of squeezed states, typically, from a parametric oscillator. These subjects are studied in the Chaps. 13 and 14, respectively. In Chap. 14, we also introduce the Input–Output theory, which is very appropriate to describe the parametric oscillator and other non-linear optical systems.

Quantum phase is a controversial subject, even today and which started with Dirac (Chap. 15). Optical experiments most of the time deal with direct or indirect measurement of a phase. For this reason, I felt it was important to include it in this textbook, even if it may not be a closed subject.

The last five chapters deal with more recent topics in quantum and atom optics. The Montecarlo method and the stochastic Schrödinger equation (Chap. 16) are recent tools to attack optical problems with losses. Theoretically, it shows a different point of view from the more traditional way through master or Fokker–Planck equations, and it is convenient for practical simulations.

Measurements in optics, and physics in general, play a central role. This was recognized early in the history of quantum mechanics.

We introduce the reader the notions of quantum standard limits and quantum non-demolition measurements (Chap. 18). A detailed example is studied, in connection with the quantum non-demolition (QND) measurement of the photon number in a cavity. Also continuous measurements are studied. A somewhat related subject, decoherence (Chap. 20) is quite relevant for quantum computing. This intriguing phenomenon is connected to dissipation and measurement.

Finally, a little outside the scope of Quantum Optics, we have included the topics of atom optics (Chap. 17) and trapped ions (Chap. 19). These are fast growing areas of research.

Throughout the years, I have collaborated with many colleagues and students, who directly or indirectly contributed to this work. In particular, G.S. Agarwal, Claus Benkert, Janos Bergou, Wilhelm Becker, Luiz Davidovich, Mary Fuka, Mark Hillery, María Loreto Ladron de Guevara, Jack K. McIver, Douglas Mundarain, Ricardo Ramírez, Juan Carlos Retamal, Luis Roa, Jaime Röessler, Bernd Rohwedder, Carlos Saavedra, Wolfgang Schleich, Marlan O. Scully, Herbert Walther, K. Wodkiewicz, Nicim Zagury, F.X. Zhao, Sh.Y. Zhu. I thank them all.

I want to thank Prof. Juan Carlos Retamal for reading and correcting the whole manuscript, Dr H. Lotsch for the encouragement of writing this book, and Prof. Hernan Chuaqui and Mr Jaime Fernandez for the invaluable help with the computer-generated figures and photography.

Finally, last but not least as they say, I would like to thank my wife Marta Montoya (Martita) for her love and constant support in this project.

Santiago, November 1998   *Miguel Orszag*

No te escapes
Ahora
Me ayudarás. Un dedo,
una palabra,
un signo
tuyo
y cuando
dedos, signos, palabras
caminen y trabajen
algo
aparecerá en el aire inmóvil,
un
solidario sonido en la ventana,
una estrella en la terrible paz nocturna,
entonces
tu dormirás tranquilo,
tu vivirás tranquilo:
será parte
del sonido qur acude a tu ventana,
de la luz que rompió la soledad.

From: Odas Elementales, Pablo Neruda[1]

[1] Neruda, P.: Antología Fundamental.   (PEHUEN Editores, Santiago, Chile (1988)

Contents

1. Einstein's Theory of Atom–Radiation Interaction

In this chapter, we study spontaneous and stimulated emission phenomenologically, as well as an elementary laser theory.

In 1917, **Einstein** [1] formulated a theory of spontaneous, stimulated emission and absorption, based on purely phenomenological considerations. His results allow to understand in a qualitative way the basic ingredients of the atom-radiation interaction, and could be useful to describe the processes of absorption, light scattering by atoms, stimulated emission in a variety of laser and maser systems, etc. This happened after **Planck** found that the spectral distribution of the blackbody radiation could be explained by quantizing the energy [2], and **Einstein** had explained the photoelectric effect [3] assuming energy packets that were later called **photons** (See Fig. 1.1, 1921.).

Einstein's Theory is based on reasonable postulates which will be justified more rigorously later when we will treat the same problem using Quantum Mechanics. The present arguments are of a heuristic nature [1].

> Recently I found a derivation of Planck's radiation formula which is based upon the basic assumption of quantum theory and which is related to Wien's original consideration: in this derivation, the relationship between the Maxwell distribution and the chromatic black-body distribution plays a role. The derivation is of interest not only because it is simple, but especially because it seems to clarify somewhat the at present unexplained phenomena of emission and absorption of radiation by matter. I have shown, on the basis of a few assumptions, about emission and absorption of radiation by molecules, which are closely related to quantum theory, that molecules distributed in temperature equilibrium over states, in a way which is compatible with quantum theory, are in dynamic equilibrium with Planck's radiation. In this way, I deduced in a remarkably simple and general manner, Planck's formula. [4]

1.1 The A and B Coefficients

We assume a closed cavity with N identical atoms, with two relevant bound state energy levels that we shall label by E_b and E_a quasi resonant with the

Fig. 1.1. Albert Einstein 1921

thermal radiation produced by the cavity walls at a temperature. Let

$$\hbar\omega = E_a - E_b \,. \tag{1.1}$$

We will also assume that there is an external source of electromagnetic energy, for example, a light beam crossing the cavity, and we may be interested in the scattering losses of such a beam.

The average energy density (over a cycle) can be written as

$$U(\omega) = U_T(\omega) + U_E(\omega) \,. \tag{1.2}$$

The total energy density, will, in general be a function of position and frequency, however, for the sake of simplicity, we consider it to be *only a slow varying function of frequency.* The labels T and E refer to thermal and external sources.

Energy-conserving processes are spontaneous emission of a photon, stimulated emission and absorption, as described in Fig. 1.2.

Let A_{ab} be the probability/time for the atom to spontaneously decay from level a to b, emitting a photon of energy $\hbar\omega$. On the other hand, if the atom is in state b, there will be a probability/time for absorption that will

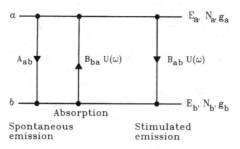

Fig. 1.2. The three processes in the atom-field interaction. Absorption, stimulated emission and spontaneous emission

be *proportional* to the electromagnetic energy present in the cavity, that is the absorption rate will be $B_{ba}U(\omega)$. The two processes described above are quite reasonable.

Now, **Einstein** proposed, that in order to re-discover Planck's radiation law, **it is absolutely necessary to assume a third type of process called stimulated emission,** and the corresponding rate is defined as $B_{ab}U(\omega)$.

So far, we will consider A_{ab}, B_{ba}, B_{ab}, as phenomenological constants.

We can now write a Rate Equation. Calling N_b and N_a the populations of the lower and upper levels respectively, and $N_b + N_a = N$, N being the total number of atoms, then

$$\frac{dN_b}{dt} = -\frac{dN_a}{dt} = \underbrace{A_{ab}N_a + B_{ab}U(\omega)N_a} - \underbrace{N_b B_{ba}U(\omega)} \qquad (1.3)$$

where the terms grouped in the first curly bracket represent the emission and in the second, the absorption.

1.2 Thermal Equilibrium

We will consider the equilibrium condition on (1.3), that is

$$\frac{dN_b}{dt} = 0 , \qquad (1.4)$$

so, that if we have only thermal electromagnetic energy ($U_E = 0$), then:

$$U_T(\omega) = \frac{A_{ab}}{B_{ab}\left(\frac{B_{ba}g_b}{B_{ab}g_a}\exp(\beta\hbar\omega) - 1\right)} , \qquad (1.5)$$

where we have assumed a Boltzmann distribution for the level populations, weighted with their respective degeneracies g_b and g_a

$$\frac{N_b}{N_a} = \frac{g_b \exp(-\beta E_b)}{g_a \exp(-\beta E_a)} = \frac{g_b}{g_a} \exp(\beta \hbar \omega) . \tag{1.6}$$

A comparison of 1.5 with Planck's blackbody energy distribution :

$$U_T(\omega) \,|_{\text{Planck}} = \frac{\hbar \omega^3}{\pi^2 c^3 (\exp(\beta \hbar \omega) - 1)} \tag{1.7}$$

gives us

$$\frac{B_{ba} \, g_b}{B_{ab} \, g_a} = 1 , \tag{1.8}$$

$$\frac{\hbar \omega^3}{\pi^2 c^3} = \frac{A_{ab}}{B_{ab}} . \tag{1.9}$$

From (1.8, 1.9), we can see that there are two equations for three coefficients, that is only one of them is independent.

1.3 Photon Distribution and Fluctuations

If we quantize the thermal photons as

$$E_n = n \hbar \omega , \tag{1.10}$$

then the probability for n photons at Temperature $T \left[\beta = (kT)^{-1} \right]$ is given by the usual Boltzmann factor

$$P_n = \frac{\exp(-\beta E_n)}{\sum_n \exp(-\beta E_n)} = \exp(-\beta(n+1)\hbar \omega) \left[\exp(\beta \hbar \omega) - 1 \right] , \tag{1.11}$$

but as

$$\langle n \rangle = \frac{1}{\exp(\beta \hbar \omega) - 1} , \tag{1.12}$$

we finally get

$$P_n = \frac{\langle n \rangle^n}{(\langle n \rangle + 1)^{n+1}} , \tag{1.13}$$

which is the Bose–Einstein distribution for thermal photons.

We can also express P_n in terms of the level populations

$$P_n = \left(\frac{g_b N_a}{g_a N_b} \right)^n \left(1 - \frac{g_b N_a}{g_a N_b} \right) \tag{1.14}$$

thus showing that, if $g_b = g_a$, one cannot achieve population inversion by only thermal excitation of the atoms. ($N_a \leq N_b$ in order to have $P_n \leq 1$).

It is left to the reader to prove that

$$(\Delta n)^2 \equiv \langle n^2 \rangle - \langle n \rangle^2 = \langle n \rangle + \langle n \rangle^2 . \tag{1.15}$$

In the next two sections, we will apply these simple ideas to the atomic excitation by an external light source and an elementary laser theory .

1.4 Light Beam Incident on Atoms

We assume here that we have a cavity filled with atoms interacting with an external light beam whose frequency is resonant with a pair of atomic levels, consistent with Einstein's description of the atom–radiation interaction. As we saw in a previous section, three important processes take place. One important fact is that stimulated emission from the excited atom tends to remain in the same electromagnetic mode, thus tending to amplify the incident radiation. On the other hand, spontaneous emission is isotropic in all spatial directions and independent of the direction of the incident beam [5].

If we neglect the spatial dependence of the radiation in the cavity (thin cavity), then U = constant and one easily finds a solution for (1.3)

$$N_b(t) = \left(N_b^0 - \frac{N(A+BU)}{A+2BU} \right) \exp\left[-(A+2BU)t \right] + \frac{N(A+BU)}{A+2BU} \quad (1.16)$$

where we have assumed $g_b = g_a = 1$, thus $B_{ba} = B_{ab} = B$, and N_b^0 is the initial population of the lower level.

In the particular case $N_b^0 = N$, that is all the atoms are initially in the lower energy level, then

$$\frac{N_a}{N} = \chi(1 - \exp(-\tau)) \quad (1.17)$$

with $\chi \equiv \frac{BU}{(A+2BU)}$ and $\tau \equiv t(A+2BU)$.

In steady state $(\tau \longrightarrow \infty)$, we have

$$\frac{N_a}{N} = \frac{1}{2 + \frac{A}{BU}} . \quad (1.18)$$

We notice again that even when the stimulated emission term (BU) is much stronger than the spontaneous emission one by, for example having a large energy on the external light beam, one can at best achieve equal population in the two levels.

If we now disconnect the external light source, the (1.3) becomes

$$\frac{dN_a}{dt} = -N_a A \quad (1.19)$$

thus $N_a(t) = N_a^0 \exp(-At)$, describing the exponential decay of the upper level population or spontaneous emission. The average lifetime of the upper level is $\tau_R = A^{-1}$.

1.5 An Elementary Laser Theory

A simplified view of the process of amplification of light can be formulated based on **Einstein**'s ideas on the fundamental processes in the atom–radiation interaction [6].

As in the previous sections, we consider two level atoms, resonant with the electromagnetic field. Also, for the moment, we will neglect the effects of the spontaneous emission. The corresponding rate equations are

$$\frac{\mathrm{d}N_b}{\mathrm{d}t} = -W_{ba}nN_b + W_{ab}nN_a , \qquad (1.20)$$

$$\frac{\mathrm{d}N_a}{\mathrm{d}t} = W_{ba}nN_b - W_{ab}nN_a , \qquad (1.21)$$

where n is the number of photons in the cavity and W_{ij} is the transition rate from level i to j.

We can also define the population difference $D = N_a - N_b$ that obeys

$$\frac{\mathrm{d}D}{\mathrm{d}t} = -2WnD \qquad (1.22)$$

where we have set $W_{ba} = W_{ab} = W$, consistent with the previous arguments.

On the other hand, one also has a rate equation for the photons, namely,

$$\frac{\mathrm{d}n}{\mathrm{d}t} = WnD - \frac{n}{T_c} , \qquad (1.23)$$

where we have included the term $-\frac{n}{T_c}$ to account for the photons coming out of the cavity.

The two coupled equations read

$$\frac{\mathrm{d}D}{\mathrm{d}t} = -2WnD - \frac{1}{T_1}(D - D_0) , \qquad (1.24)$$

$$\frac{\mathrm{d}n}{\mathrm{d}t} = WnD - \frac{n}{T_c}.$$

In the first equation of (1.24), we have included a phenomenological term

$$-\frac{1}{T_1}(D - D_0)$$

that accounts for the spontaneous decay and pump action. T_1 being a characteristic lifetime associated with the decay of the population. Also, D_0 is the equilibrium population in the absence of photons.

The equations given by (1.24) are called **laser rate equations**. These equations, although, as mentioned before, are a simplified version of the fully quantum mechanical laser theory, allow us to study some basic characteristics of the laser action, such as steady state, it's stability and the laser threshold.

1.5.1 Threshold and Population Inversion

Assuming initially a low photon number (say 1), amplification of the number of photons will occur only if

$$\frac{\mathrm{d}n}{\mathrm{d}t} = (WD(0) - \frac{1}{T_c}) > 0 .$$

or, equivalently,

$$D(0) = D_0 > D_{\text{thresh}} = \frac{1}{WT_c} . \tag{1.25}$$

From the above analysis two conclusions can be drawn:

1. There is a laser threshold condition given in the inequalities (1.25). Laser action only takes place if the initial inversion is above the threshold value. Usually, the equilibrium population is equal to the initial one, so the inequality can be referred to the equilibrium value.
2. Clearly, D_0 is positive, which implies $N_a > N_b$, that is population inversion is required, and it has to be enough to compensate for the cavity losses. One would like to have D_{thresh} as small as possible, which implies either T_c or W large, or both. A large T_c means that we need a high-quality optical cavity. On the other hand, high W implies to choose a pair of levels with a large dipole moment.

1.5.2 Steady State

The steady-state condition reads

$$2WnD + \frac{1}{T_1}(D - D_0) = 0 , \tag{1.26}$$

$$n(WD - \frac{1}{T_c}) = 0 . \tag{1.27}$$

There are two possible solutions to the steady-state equations

i) $n = 0$, $D = D_0$,

which is the case of trivial equilibrium with the pump and no photons in the cavity.

ii)

$$D = \frac{1}{T_c W} = D_{\text{thresh}} , \tag{1.28}$$

$$n = \frac{D_0 - D_{\text{thresh}}}{2W D_{\text{thresh}} T_1} = (D_0 - D_{\text{thresh}})\frac{T_c}{2T_1} . \tag{1.29}$$

A serious limitation of this model is that if $n(0) = 0$, then $n(t) = 0$, for all times. The laser does not get started even if $D_0 > D_{\text{thresh}}$. The reason for this limitation is the absence of the quantum noise because of spontaneous emission, which was introduced in the present treatment in a ad hoc manner. A quantum laser theory does not have this limitation, because the quantum noise and the spontaneous emission appear in a natural way.

As we can see from (1.29), the only possibility to have a positive steady-state photon number is if $D_0 > D_{\text{thresh}}$.

1.5.3 Linear Stability Analysis

If we call n_∞ and D_∞ the steady-state values, then we write these quantities, as their steady-state value plus a small deviation, that is

$$n(t) = n_\infty + \varepsilon_n(t) \,, \tag{1.30}$$

$$D(t) = D_\infty + \varepsilon_D(t) \,. \tag{1.31}$$

Linearizing (1.24), up to order ε, we easily get

$$\frac{d\varepsilon_D}{dt} = -2W(n_\infty \varepsilon_D + D_\infty \varepsilon_n) - \frac{1}{T_1}\varepsilon_D \,, \tag{1.32}$$

$$\frac{d\varepsilon_n}{dt} = W(n_\infty \varepsilon_D + D_\infty \varepsilon_n) - \frac{\varepsilon_n}{T_c}.$$

We look now for a solution of the form

$$\begin{pmatrix} \varepsilon_D \\ \varepsilon_n \end{pmatrix} = \exp(\lambda t) \begin{bmatrix} \varepsilon_D(0) \\ \varepsilon_n(0) \end{bmatrix} \,, \tag{1.33}$$

and get two algebraic coupled equations

$$(\lambda + 2W n_\infty + \frac{1}{T_1})\varepsilon_D(0) + 2W D_\infty \varepsilon_n(0) = 0 \,, \tag{1.34}$$

$$-W n_\infty \varepsilon_D(0) + (\lambda - W D_\infty + \frac{1}{T_c})\varepsilon_n(0) = 0 \,. \tag{1.35}$$

From (1.34, 1.35), we get the secular equation

$$(\lambda + 2W n_\infty + \frac{1}{T_1})(\lambda - W D_\infty + \frac{1}{T_c}) + 2W^2 n_\infty D_\infty = 0 \,. \tag{1.36}$$

In order to have a stable solution, one must have

$$\mathrm{Re}\,\{\lambda_j\} < 0 \text{ for } j = 1, 2 \,.$$

For the case i

$(D = D_0, n = 0)$

$$\lambda_1 = -\frac{1}{T_1} \,, \tag{1.37}$$

$$\lambda_2 = W D_0 - \frac{1}{T_c} \,, \tag{1.38}$$

therefore, the above solution is stable only if $W D_0 - \frac{1}{T_c} < 0$, that is for the laser below threshold. On the other hand, for $W D_0 - \frac{1}{T_c} > 0$, the $n = 0$ solution becomes unstable.

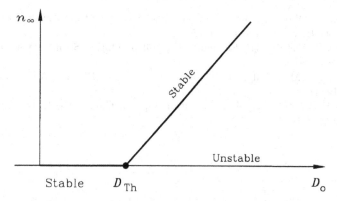

Fig. 1.3. Behaviour of the steady-state photon number versus population

For the case ii

The two roots are

$$\lambda_{1,2} = \frac{1}{2} \left\{ -\frac{D_0}{D_{\text{thresh}}T_1} \pm \left[\left(\frac{D_0}{D_{\text{thresh}}T_1}\right)^2 - \frac{4W(D_0 - D_{\text{thresh}})}{T_1} \right]^{\frac{1}{2}} \right\} \quad (1.39)$$

It is evident that the solution is only stable if $D_0 > D_{\text{thresh}}$.
Finally, this stability analysis is depicted in Fig. 1.3.

Problems

1.1. Prove the thermal photon distribution given by (1.13).
1.2. Prove that, for a thermal distribution, the correct expression for fluctuations of the photon number is given by (1.15).
1.3. Verify that for U = constant, the solution of (1.3) is given by (1.16).

References

1. Einstein, A.: Phys. Z., **18**, 121 (1917)
2. Planck, M.: Verh. Deutsch. Phys. Ges, **2**, 202 (1900)
3. Einstein, A.: Ann. Phys., **17**, 132 (1905)
4. Einstein, A.: The Old Quantum Theory. Pergamon, London (1967)
5. Loudon, R.: The Quantum Theory of Light. Clarendon Press, Oxford (1983)
6. Narducci, L.M., Abraham, N.B.: Laser Physics and Laser Instabilities. World Scientific, Singapore (1988)

Further Reading

- Haken, H.: Light Vol. 1. North Holland, Amsterdam (1981)
- Haken, H.: Light Vol. 2. North Holland, Amsterdam (1985)

- Mandel, L., Wolf, E.: Optical Coherence and Quantum Optics. Cambridge University Press, Cambridge (1995)
- Meystre, P., Sargent, III: Elements of Quantum Optics. Springer Verlag, Berlin (1993)
- Sargent, III, M., Scully, M.O., Lamb, W.L.: Laser Physics. Addison Wesley, USA (1974)
- Scully, M.O., Zubairy, M.S.: Quantum Optics. Cambridge University Press, Cambridge (1997)

2. Atom–Field Interaction: Semiclassical Approach

In this chapter, we study the resonant interaction between atoms and light. The Bloch´s equations are derived, and by adding a relaxation term, various decay effects are included.

We describe here various aspects that appear in the interaction between a collection of atoms or molecules with light.

The basic phenomena may be understood using exactly soluble models and they often are excellent approximations of the real experiments.

The so-called **semiclassical models**, such as the ones presented in this chapter, describe a classical field interacting with quantum mechanical atoms.

A fully quantum mechanical model is described in Chap. 8.

Furthermore, here, we will deal with quasiresonant phenomena, where the electromagnetic field frequency almost coincides with the energy difference between a pair of atomic levels. In this context, we will often use the concept of "two-level atom".

This situation is described in the Fig. 2.1.

The two-level atom is characterized by the ground state $| b \rangle$ and an excited state $| a \rangle$ with energies $\hbar \omega_b$ and $\hbar \omega_a$. The detuning δ is defined by

$$\delta \equiv (\omega_a - \omega_b) - \omega = \omega_{ab} - \omega . \tag{2.1}$$

The strength of the interaction is usually measured by the so-called Rabi frequency that depends on resonance on the square root of the number of photons.

For the calculation of the transition rates, consider an atom interacting with a sinusoidal field

$$\mathbf{E}(t) = \mathbf{e}E_0 \cos \omega t , \tag{2.2}$$

which is switched on in $t = 0$. $\mathbf{e}$ represents the unit vector along the polarization of the field. We will show that the presence of higher excited states can be neglected, when the transition frequencies are very different from ω.

We start with the time-dependent Schrödinger equation

$$i\hbar \frac{\partial \psi}{\partial t} = (H_0 + H')\psi . \tag{2.3}$$

H_0 represents the time-independent Hamiltonian of the free atom and H' is the time-dependent atom field interaction, that in the electric dipole approximation can be written as (e being negative for the electron)

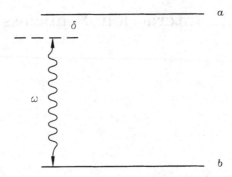

Fig. 2.1. Quasi-resonant interaction between a two-level atom and the electromagnetic radiation, with a detuning δ

$$H' = -e\mathbf{E} \cdot \mathbf{r} = -eE_0(\mathbf{e} \cdot \mathbf{r}) \cos \omega t \ . \tag{2.4}$$

Because H_o is the Hamiltonian of the free atom, one can write

$$i\hbar \frac{\partial \Psi}{\partial t} = (H_0)\Psi \ , \tag{2.5}$$

with solutions of the form

$$\Psi = \psi_n(\mathbf{r}) \exp\left(\frac{-iE_n t}{\hbar}\right) \ , \tag{2.6}$$

where $\psi_n(\mathbf{r})$ and E_n are the eigenfuctions and eigenvalues of H_0

$$H_0\psi_n(\mathbf{r}) = E_n\psi_n(\mathbf{r}) \ . \tag{2.7}$$

The functions $\psi_n(\mathbf{r})$ satisfy the usual orthonormality conditions

$$\int d\mathbf{r}\psi_n^*(\mathbf{r})\psi_m(\mathbf{r}) = \delta_{nm} \ . \tag{2.8}$$

The above wavefunctions will serve as a convenient basis to expand the wavefunction of the time-dependent problem

$$\psi(\mathbf{r}, t) = \sum_n C_n(t)\psi_n(\mathbf{r}) \exp(-i\omega_n t) \ , \tag{2.9}$$

with

$$\omega_n \equiv \frac{E_n}{\hbar} \ .$$

Substituting (2.9) in (2.3) we get

$$i\hbar \sum_n \left(\frac{dC_n}{dt} - i\omega_n C_n\right)\psi_n(\mathbf{r}) \exp(-i\omega_n t) = \sum_n E_n C_n \psi_n(\mathbf{r}) \exp(-i\omega_n t) \tag{2.10}$$

$$-eE_o(\mathbf{e} \cdot \mathbf{r}) \sum_n C_n \psi_n(\mathbf{r}) \exp(-i\omega_n t) \cos \omega t.$$

The second term on the left-hand side cancels the first term on the right-hand side, thus getting, after multiplication by ψ_m^* from the left and integration

$$i\hbar \frac{dC_m}{dt} = \frac{E_o}{2} \sum_n d_{mn} C_n(t) \left[\exp i(\omega_{mn} + \omega)t + \exp i(\omega_{mn} - \omega)t\right] , \quad (2.11)$$

where

$$\omega_{mn} = \omega_m - \omega_n = \frac{1}{\hbar}(E_m - E_n) , \quad (2.12)$$

$$d_{mn} = |e| \int d\mathbf{r} \psi_m^*(\mathbf{r})(\mathbf{e} \cdot \mathbf{r}) \psi_n(\mathbf{r}) .$$

Now, let us assume that initially the atom is at the state $\psi_k(\mathbf{r})$, in other words:

$$C_k(t = 0) = 1, \quad (2.13)$$
$$C_n(t = 0) = 0, n \neq k .$$

As a first approximation, we replace $C_n(t)$ by $C_n(0)$ in (2.11), getting

$$i\hbar \frac{dC_m}{dt} = \frac{E_o}{2} d_{mk} \left[\exp i(\omega_{mk} + \omega)t + \exp i(\omega_{mk} - \omega)t\right] , \quad (2.14)$$

and integrating

$$C_m(t) - C_m(0) = -\frac{E_o}{2\hbar} d_{mk} \left[\frac{\exp i(\omega_{mk} + \omega)t - 1}{(\omega_{mk} + \omega)} + \frac{\exp i(\omega_{mk} - \omega)t - 1}{(\omega_{mk} - \omega)}\right] , \quad (2.15)$$

and for $m \neq k$ we get

$$C_m(t) = -i\frac{E_o}{\hbar} d_{mk} \left\{ \exp\left[i(\omega_{mk} + \omega)\frac{t}{2}\right] \frac{\sin \frac{(\omega_{mk} + \omega)t}{2}}{(\omega_{mk} + \omega)} \right.$$
$$\left. + \exp\left[i(\omega_{mk} - \omega)\frac{t}{2}\right] \frac{\sin \frac{(\omega_{mk} - \omega)t}{2}}{(\omega_{mk} - \omega)} \right\} . \quad (2.16)$$

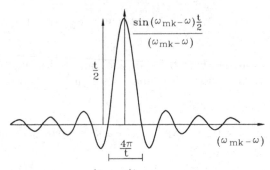

Fig. 2.2. The function $\frac{\sin\frac{(\omega_{mk}-\omega)t}{2}}{(\omega_{mk}-\omega)}$ is sharply peaked around $\omega = \omega_{mk}$

As we can easily see from the Fig. 2.2 , the function:

$$\frac{1}{(\omega_{mk}-\omega)}\sin\frac{(\omega_{mk}-\omega)t}{2} \tag{2.17}$$

is sharply peaked around $\omega_{mk} \simeq \omega$, for large t.

As we can see from the above discussion, for those states with ω_{mk} very different from ω, the transition probability is indeed very small, thus justifying the two-level approximation [1].

In (2.16), ω_{mk} is negative for emission and positive for absorption, thus the main contribution comes from the first term in the first case and from the second term for absorption.

Now, we return to the two-level model from Fig. 2.1 and assume that initially the atom is in its lower level b. The transition probability to a is

$$|C_a(t)|^2 = \frac{|d_{ab}|^2 E_o^2}{4\hbar^2}\left(\frac{\sin\left(\frac{\omega_{ab}-\omega}{2}\right)t}{\left(\frac{\omega_{ab}-\omega}{2}\right)}\right)^2 . \tag{2.18}$$

The result given in (2.18) is approximate and represents for $|C_a(t)|^2 \ll 1$ the probability for stimulated absorption.

2.1 Broad-Band Radiation Spectrum

Now, we modify our previous analysis by considering a broad-band light, rather than a monochromatic one. Because the average electromagnetic energy density per unit volume is $\frac{\epsilon_o E_o^2}{2}$, and because $U(\omega)$ is the spectral energy, such that $U(\omega)d\omega$ represents the field energy per unit volume and in the frequency interval ω and $\omega + d\omega$, we just replace E_o^2 by $\frac{2U(\omega)d\omega}{\epsilon_o}$ and integrate over the spectrum, getting

$$|C_a(t)|^2 = \frac{|d_{ab}|^2}{2\epsilon_o\hbar^2}\int d\omega U(\omega)\left(\frac{\sin\left(\frac{\omega_{ab}-\omega}{2}\right)t}{\left(\frac{\omega_{ab}-\omega}{2}\right)}\right)^2 \tag{2.19}$$

and assuming $U(\omega)$ to be slowly varying with ω, we get the approximate result:

$$| C_2(t) |^2 \simeq \frac{| d_{ab} |^2}{2\epsilon_o \hbar^2} U(\omega_{ab}) \int d\omega \left(\frac{\sin \left(\frac{\omega_{ab}-\omega}{2} \right) t}{\left(\frac{\omega_{ab}-\omega}{2} \right)} \right)^2 , \qquad (2.20)$$

$$= \frac{\pi | d_{ab} |^2}{\epsilon_o \hbar^2} U(\omega_{ab}) t .$$

From (2.20), we see that the absorption probability per unit time is proportional to the energy density, as in Einstein's theory. In order to get the B coefficient, we have to average over all the directions, because

$$\overline{| d_{ab} |^2} \sim || e | \int dr \psi_a^*(\mathbf{r}) \overline{(\mathbf{e} \cdot \mathbf{r})} \psi_b(\mathbf{r}) |^2$$

$$\sim || e | \int dr \psi_a^*(\mathbf{r}) \mathbf{r} \psi_b(\mathbf{r}) |^2 \overline{\cos^2 \theta}$$

$$\sim | \left[| e | \int dr \psi_a^*(\mathbf{r}) \mathbf{r} \psi_b(\mathbf{r}) \right] |^2 \frac{1}{3} ,$$

so, finally

$$\frac{| C_a(t) |^2}{t} = \frac{\pi p_{ab}^2}{3\epsilon_o \hbar^2} U(\omega_{ab}) \equiv B_{ba} U(\omega_{ab}) , \qquad (2.21)$$

$$p_{ab}^2 = || e | \int dr \psi_a^*(\mathbf{r}) \mathbf{r} \psi_b(\mathbf{r}) |^2 , \qquad (2.22)$$

so, we may write

$$B_{ba} = \frac{\pi p_{ab}^2}{3\epsilon_o \hbar^2} . \qquad (2.23)$$

2.2 Rabi Oscillations

Instead of the general expansion given in (2.9), consider two levels only

$$\psi(\mathbf{r}, t) = C_b'(t) \psi_b(\mathbf{r}) \exp(-i\omega_b t) + C_a' \psi_a(\mathbf{r}) \exp(-i\omega_a t) . \qquad (2.24)$$

Making use of (2.11) for this particular case, we get

$$i\hbar \frac{dC_b'}{dt} = \frac{E_o d_{ba}}{2} C_a'(t) \{ \exp[-i(\omega_{ab} - \omega)t] + \exp[-i(\omega_{ab} + \omega)t] \} , \qquad (2.25)$$

$$i\hbar \frac{dC_a'}{dt} = \frac{E_o d_{ab}}{2} C_b'(t) [\exp i(\omega_{ab} - \omega)t + \exp i(\omega_{ab} + \omega)t] .$$

Defining

$$C_{b,a} = \exp \left(\pm i \frac{\delta t}{2} \right) C_{b,a}'$$

and performing the rotating wave approximation to neglect the rapidly varying term $\exp i(\omega_{ab} + \omega)t$ versus the $\exp i(\omega_{ab} - \omega)t$ term, we write

$$\frac{dC_b}{dt} = -\frac{i}{2}\left(-\delta C_b + \frac{E_o d_{ba}}{\hbar}C_a\right) , \tag{2.26}$$

$$\frac{dC_a}{dt} = -\frac{i}{2}\left(\delta C_a + \frac{E_o d_{ab}}{\hbar}C_b\right) . \tag{2.27}$$

We can write (2.26, 2.27) in a matrix form

$$\frac{d}{dt}\begin{bmatrix} C_b(t) \\ C_a(t) \end{bmatrix} = \frac{-i}{2}\begin{bmatrix} -\delta & \frac{E_o d_{ba}}{\hbar} \\ \frac{E_o d_{ab}}{\hbar} & \delta \end{bmatrix}\begin{bmatrix} C_b(t) \\ C_a(t) \end{bmatrix} . \tag{2.28}$$

The eingenvalues of (2.28) are $\mp R$, where

$$R \equiv \sqrt{\delta^2 + R_o^2}, \tag{2.29}$$

$$R_o \equiv \left|\frac{E_o d_{ba}}{\hbar}\right| .$$

Normally, R_o is referred to as the Rabi frequency. [2, 3, 4]
The solution of (2.28) is

$$\begin{bmatrix} C_b(t) \\ C_a(t) \end{bmatrix} = \begin{bmatrix} \cos\frac{Rt}{2} + \frac{i\delta}{R}\sin\frac{Rt}{2} & -i\frac{E_o d_{ba}}{R\hbar}\sin\frac{Rt}{2} \\ -i\frac{E_o d_{ab}}{R\hbar}\sin\frac{Rt}{2} & \cos\frac{Rt}{2} - \frac{i\delta}{R}\sin\frac{Rt}{2} \end{bmatrix}\begin{bmatrix} C_b(0) \\ C_a(0) \end{bmatrix} . \tag{2.30}$$

To take a simple example, if we start from the lower state $[C_b(0) = 1]$, the transition probability for absorption is

$$|C_a(t)|^2 = \left|\frac{R_o}{R}\right|^2 \sin^2\frac{Rt}{2} . \tag{2.31}$$

2.3 Bloch's Equations

Equation (2.25) describes exactly the interaction between a two-level atom and the radiation field

$$i\frac{dC_b'}{dt} = \nu C_a'(t)\exp(-i\omega_{ab}t)\cos\omega t ,$$

$$i\frac{dC_a'}{dt} = \nu^* C_b'(t)\exp(i\omega_{ab}t)\cos\omega t ,$$

where $\nu \equiv \frac{E_o d_{ba}}{\hbar}$.
    A general treatment involves the density matrix.
    Define

$$\rho_{bb} = |C_b'|^2 , \tag{2.32}$$

$$\rho_{aa} = |\, C'_a \,|^2\,,$$
$$\rho_{ba} = C'_b C'^*_a = \rho^*_{ab}\,.$$

Of course, the property $Tr\rho = 1$ is automatically satisfied because

$$\rho_{bb} + \rho_{aa} = |\, C'_b \,|^2 + |\, C'_a \,|^2 = 1\,. \tag{2.33}$$

We differentiate ρ_{ij} with respect the time, getting

$$\frac{d\rho_{ij}}{dt} = C'_i \frac{dC'^*_j}{dt} + C^{*\prime}_j \frac{dC'_i}{dt}\,, \tag{2.34}$$

and replacing (2.32) into (2.34), we get (making use of the rotating-wave approximation)

$$\frac{d\rho_{aa}}{dt} = -\frac{d\rho_{bb}}{dt} = -\frac{i}{2}\nu^* \exp\left[i(\omega_{ba} - \omega)t\right]\rho_{ba} + \frac{i}{2}\nu \exp\left[-i(\omega_{ba} - \omega)t\right]\rho_{ab}\,,$$

$$\frac{d\rho_{ba}}{dt} = \frac{d\rho^*_{ab}}{dt} = \frac{i}{2}\nu \exp\left[-i(\omega_{ba} - \omega)t\right](\rho_{bb} - \rho_{aa})\,. \tag{2.35}$$

Equation (2.35) is the **optical Bloch Equation**.

In order to introduce dissipative effects, we can modify (2.11)

$$i\hbar\frac{dC_m}{dt} = \frac{E_o}{2}\sum_n d_{mn}C_n(t)\left\{\exp\left[i(\omega_{mn} + \omega)t\right] + \exp\left[i(\omega_{mn} - \omega)t\right]\right\}$$

$$\tag{2.36}$$

$$-i\hbar\frac{\gamma_m}{2}C_m\,,$$

where we have added a relaxation term. In the absence of coupling, the relaxation term will generate a solution

$$|\, C_m(t) \,|^2 = |\, C_m(0) \,|^2 \exp{-\gamma t}\,. \tag{2.37}$$

Obviously, such a decay constant in Schrödinger's equation does not preserve the norm.

2.4 Decay to an Unobserved Level

The effect on the density matrix is as follows:

$$\frac{d\rho_{ij}}{dt} = C'_i \frac{dC'^*_j}{dt} + C^{*\prime}_j \frac{dC'_i}{dt} \tag{2.38}$$

$$= ()_{nodissipation} - \frac{\gamma_i + \gamma_j}{2}\rho_{ij}\,.$$

For population levels, this leads to an exponential decay $\rho_{ii}(t) \propto \exp(-\gamma_i t)$.

As the spontaneous decay process will emit a photon, we find that the number of spontaneously emitted photons will be proportional to $\gamma_i \rho_{ii}$, and the intensity of spontaneously emitted photons will be a measure of the decaying level.

2.5 Decay Between Levels

If we consider the spontaneous emission between the two levels (Fig. 2.1), then the upper state will decay as

$$\frac{d}{dt}\rho_{aa} = -\gamma\rho_{aa} \ . \tag{2.39}$$

On the other hand, this event will increase the population in level b, so

$$\frac{d}{dt}\rho_{bb} = \gamma\rho_{bb} \ . \tag{2.40}$$

The calculation of γ comes from a quantum electrodynamical theory called the Wigner–Weisskopf Theory, that will be covered in a later chapter. It involves the interaction of an atom with an infinite number of harmonic oscillators, at zero temperature, that is in the vacuum state.

On the other hand, the off diagonal term will decay as

$$\frac{d}{dt}\rho_{ab} = -\frac{\gamma}{2}\rho_{ab} \ . \tag{2.41}$$

Equation (2.41) can be proven with a fully quantum mechanical analysis. The Bloch Equations, with losses can be written as

$$\frac{d\rho_{aa}}{dt} = -\frac{i}{2}\nu^* \exp\left[i(\omega_{ba} - \omega)t\right]\rho_{ba} + \frac{i}{2}\nu \exp\left[-i(\omega_{ba} - \omega)t\right]\rho_{ab} \tag{2.42}$$

$$-\gamma\rho_{aa} = -\frac{d\rho_{bb}}{dt} ,$$

$$\frac{d\rho_{ba}}{dt} = \frac{d\rho_{ab}^*}{dt} = \frac{i}{2}\nu \exp\left[-i(\omega_{ba} - \omega)t\right](\rho_{bb} - \rho_{aa}) - \frac{\gamma'}{2}\rho_{ba} ,$$

where $T_1 = \frac{1}{\gamma}$ is the longitudinal relaxation time and $T_2 = \frac{2}{\gamma}$ is the transverse relaxation time, and $\gamma' = \gamma + \gamma_{coll}$, and we have introduced the collision frequency γ_{coll} in a ad-hoc manner.

2.6 Optical Nutation

An interesting case that has exact solution is when $\omega_{ba} = \omega$ and initially $\rho_{aa} = \rho_{ba} = 0$. The solution in this case is

$$\rho_{aa}(t) = \frac{\frac{|\nu|^2}{2}}{\frac{\gamma^2}{2} + |\nu|^2}\left[1 - \left(\cos\lambda t + \frac{3\gamma}{4\lambda}\sin\lambda t\right)\exp-\frac{3\gamma t}{4}\right], \tag{2.43}$$

$$\lambda \equiv \sqrt{|\nu|^2 - \frac{\gamma^2}{16}} \ .$$

The result given in (2.43) is illustrated in the Fig. 2.3.

We notice that the oscillations occur when the Rabi frequency ν is much bigger than the damping γ.

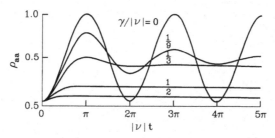

Fig. 2.3. Atomic population of the upper level versus time for various ratios of $\frac{\gamma}{\nu}$

Problem

2.1. Prove that the solution given by (2.43) satisfies (2.35).

References

1. Thyagarajan, K., Ghatak, A.K.: Lasers, Theory and Applications. Plenum Press, NY (1981)
2. Loudon, R.: The Quantum Theory of Light. Clarendon Press, Oxford (1983)
3. Meystre, P., SargentIII, M.: Elements of Quantum Optics. Springer Verlag, Berlin (1990)
4. Stenholm, S.: Lasers in Applied and Fundamental Research. Adam Hilger Ltd, Bristol (1985); See also: Stenholm, S.: Foundations of Laser spectroscopy. J.Wiley, New York (1983)

Further Reading

• Nussenzveig, H.M.: Introduction to Quantum Optics. Gordon and Breach, London (1973)

3. Quantization of the Electromagnetic Field

In this chapter, we quantize the electromagnetic field and find the commutation relations between the various componets of the electric, magnetic fields and the vector potential.

We start with the source-free Maxwell's equations

$$\nabla \cdot \mathbf{B} = 0 \,, \tag{3.1}$$

$$\nabla \times \mathbf{E} = -\frac{\partial \mathbf{B}}{\partial t} \,, \tag{3.2}$$

$$\nabla \cdot \mathbf{E} = 0 \,, \tag{3.3}$$

$$\nabla \times \mathbf{H} = \frac{\partial \mathbf{D}}{\partial t} \,, \tag{3.4}$$

together with

$$\mathbf{B} = \mu_o \mathbf{H} \,, \tag{3.5}$$

$$\mathbf{D} = \varepsilon_o \mathbf{E} \,,$$

where μ_o, ε_o are magnetic permeability and permitivity of free space, obeying the relation $\mu_o \varepsilon_o = c^{-2}$. The (3.1, 3.2) are automatically satisfied when one defines the vector and scalar potentials (A and V)

$$\mathbf{B} = \nabla \times \mathbf{A} \,, \tag{3.6}$$

$$\mathbf{E} = -\frac{\partial \mathbf{A}}{\partial t} - \nabla V \,.$$

As Maxwell's equations are gauge invariant, we choose the Coulomb gauge that is particularly useful when dealing with non-relativistic electrodynamics

$$\nabla \cdot \mathbf{A} = 0 \,, \tag{3.7}$$

$$V = 0 \,.$$

With the above gauge, (3.3) is automatically satisfied and both $\mathbf{B}$ and $\mathbf{E}$ can be expressed in terms of $\mathbf{A}$ only.

If we now substitute (3.6) into (3.4), we get the wave equation for the vector potential

$$\nabla^2 \mathbf{A} = \frac{1}{c^2}\frac{\partial^2 \mathbf{A}}{\partial t^2} \ . \tag{3.8}$$

Now, we perform the standard separation of variables

$$\mathbf{A}(r,t) = \sum_m \sqrt{\frac{\hbar}{2\omega_m \varepsilon_o}} \left[a_m(t)\mathbf{u}_m(\mathbf{r}) + a_m^\dagger(t)\mathbf{u}_m^*(\mathbf{r})\right] \ , \tag{3.9}$$

that after substitution into (3.8) gives

$$\nabla^2 \mathbf{u}_m(\mathbf{r}) + \frac{\omega_m^2}{c^2}\mathbf{u}_m(\mathbf{r}) = 0 \ , \tag{3.10}$$

$$\frac{\partial^2 a_m}{\partial t^2} + \omega_m^2 a_m = 0 \ ,$$

where $\frac{\omega_m^2}{c^2}$ is the separation constant.

Obviously,

$$a_m(t) = a_m \exp(-i\omega_m t) \ , \tag{3.11}$$

$$a_m^\dagger(t) = a_m^\dagger \exp(+i\omega_m t) \ .$$

Both $a_m(t)$ and $a_m^\dagger(t)$ are, for the time being, a pair of complex conjugate numbers. Later on, when we will quantize the field, they will be interpreted as an operator and its harmonic conjugate.

Depending on the boundary conditions, the $\mathbf{u}_m(\mathbf{r})$ functions could be sinusoidal (cavity) or exponentials (traveling waves). For plane waves:

$$\mathbf{u}_m(\mathbf{r}) = \frac{\mathbf{e}_m \exp i\mathbf{k}_m \cdot \mathbf{r}}{\sqrt{v}} \ , \tag{3.12}$$

where $k_m^2 = \frac{\omega_m^2}{c^2}$, v is the volume and the Coulomb gauge condition implies $\mathbf{e}_m \cdot \mathbf{k}_m = 0$, which is the transversality condition for the m-th mode, thus allowing two possible and mutually orthogonal polarizations, contained in a plane perpendicular to $\mathbf{k}_m$.

Therefore, the subscript m signifies the various modes including the two polarization states. The allowed values of k are determined by the boundary conditions. If we take periodic boundary conditions for a cube of volume L^3, then we require that

$$\mathbf{A}(\mathbf{r}+L\hat{\imath}) = \mathbf{A}(\mathbf{r}+L\mathbf{j}) = \mathbf{A}(\mathbf{r}+L\mathbf{k}) = \mathbf{A}(\mathbf{r}) \ , \tag{3.13}$$

which implies

$$\mathbf{k}_m = \frac{2\pi}{L}(m_1\hat{\imath}+m_2\mathbf{j}+m_3\mathbf{k}) \ ,$$

m_1, m_2, m_3, being integer numbers.

The vectors $\mathbf{u}_m(\mathbf{r})$ satisfy the orthogonality condition

$$\int \mathbf{u}_m^*(\mathbf{r})\mathbf{u}_n(\mathbf{r})dv = \delta_{mn} \ . \tag{3.14}$$

The final form for the vector potential $\mathbf{A}$ is

$$\mathbf{A}(r,t) = \sum_m \sqrt{\frac{\hbar}{2\omega_m\varepsilon_o v}}\mathbf{e}_m\{a_m \exp\left[i(\mathbf{k}_m\cdot\mathbf{r}-\omega_m t)\right] + a_m^\dagger \exp\left[-i(\mathbf{k}_m\cdot\mathbf{r}-\omega_m t)\right]\} \ .$$

$$\tag{3.15}$$

From (3.6), we can also write

$$\mathbf{E}(r,t) = i\sum_m \sqrt{\frac{\hbar\omega_m}{2\varepsilon_o v}}\mathbf{e}_m\{a_m \exp\left[i(\mathbf{k}_m\cdot\mathbf{r}-\omega_m t)\right] - a_m^\dagger \exp\left[-i(\mathbf{k}_m\cdot\mathbf{r}-\omega_m t)\right]\} \ ,$$

$$\tag{3.16}$$

$$\mathbf{H}(r,t) = -\frac{i}{c\mu_o}\sum_m \sqrt{\frac{\hbar\omega_m}{2\varepsilon_o v}}\mathbf{e}_m\times\hat{\mathbf{k}}_m\{a_m \exp\left[i(\mathbf{k}_m\cdot\mathbf{r}-\omega_m t)\right]$$

$$-a_m^\dagger \exp-\left[i(\mathbf{k}_m\cdot\mathbf{r}-\omega_m t)\right]\} \ . \tag{3.17}$$

The total energy of the multimode radiation field is given by

$$H = \frac{1}{2}\int(\varepsilon_o\mathbf{E}^2 + \mu_o\mathbf{H}^2)dv \tag{3.18}$$

$$= \frac{1}{2}\int(\varepsilon_o(\frac{\partial\mathbf{A}}{\partial t})^2 + \mu_o^{-1}(\nabla\times\mathbf{A})^2)dv$$

$$= \sum_m \hbar\omega_m(a_m a_m^\dagger + a_m^\dagger a_m) \equiv \sum_m H_m \ .$$

In the last step to obtain (3.18), we used (3.17, 3.14).

We preserved the order of $a_m^\dagger, a_m$, for future purposes. Now they are just numbers.

Now the quantization is trivial. The a_m obey the same differential equation as a harmonic oscillator, so the quantization rule is [1]

$$[a_m, a_n^\dagger] = \delta_{nm} \ , \tag{3.19}$$

$$[a_m, a_n] = 0 \ ,$$

$$[a_m^\dagger, a_n^\dagger] = 0 \ .$$

We remind the reader that the standard connection between the a and $a^\dagger$ operators, with the usual p and q is given by (1 mode)

$$a = \frac{1}{\sqrt{2\hbar\omega}}(\omega q + ip) \ , \tag{3.20}$$

$$a^\dagger = \frac{1}{\sqrt{2\hbar\omega}}(\omega q - ip) \ ,$$

so that:

$$\frac{1}{2}(p^2 + (\omega q)^2) = \hbar\omega\left(a^\dagger a + \frac{1}{2}\right) . \tag{3.21}$$

For many modes, and dropping the zero point energy, the energy and momentum are given by

$$H = \sum_m \hbar\omega_m(a_m^\dagger a_m) , \tag{3.22}$$

$$G = \sum_m \hbar k_m a_m^\dagger a_m . \tag{3.23}$$

We notice that in the Schrödinger picture, the vector potential is [A1]

$$\mathbf{A}(r,0) = \sum_m \sqrt{\frac{\hbar}{2\omega_m\varepsilon_o v}}\mathbf{e}_m \left\{a_m \exp\left[i(\mathbf{k}_m\cdot\mathbf{r})\right] + a_m^\dagger \exp\left[-i(\mathbf{k}_m\cdot\mathbf{r})\right]\right\} . \tag{3.24}$$

Here, we use the notation that a_m is a Schrödinger operator, whereas $a_m(t)$ is its Heisenberg version.

Now, we interpret a_m and $a_m^\dagger$ as the annihilation and creation operators for this particular oscillator, or for this particular mode of the radiation field.

3.1 Fock States

The operator $N_m = a_m^\dagger a_m$ is the photon number of the m-th mode.

We define a basis that can be written as a product of state vectors for each mode, because they are independent, as follows

$$| n_1\rangle | n_2\rangle... | n_\infty\rangle =| n_1, n_2, ..n_\infty\rangle$$

such that both N_m and the Hamiltonian are diagonal in this basis.

For one mode

$$a^\dagger a | n\rangle = n | n\rangle , \tag{3.25}$$

$$H | n\rangle = \hbar\omega\left(a^\dagger a + \frac{1}{2}\right) | n\rangle = \hbar\omega\left(n + \frac{1}{2}\right) | n\rangle = E_n | n\rangle . \tag{3.26}$$

From the commutation rule 3.19, we can simply infer that

$$a^\dagger | n\rangle = \sqrt{n+1} | n+1\rangle , \tag{3.27}$$
$$a | n\rangle = \sqrt{n} | n-1\rangle .$$

Similarly, for a multimode field.

We also notice that the energy of the ground state is

$$\langle 0 \mid H \mid 0 \rangle = \frac{1}{2} \sum_m \hbar \omega_m \ , \tag{3.28}$$

which, of course, diverges, originating a conceptual difficulty with the whole quantization procedure.

In most practical situations, however, one does not measure absolute energies, but rather energy changes, so that the infinite zero-point energy does not generate any divergences.

To generate any Fock state of the k-th mode $\mid n_k \rangle$, from the vacuum, we just have to apply (3.27) several times, getting

$$\mid n_k \rangle = \frac{(a_k^\dagger)^{n_k}}{\sqrt{n_k!}} \mid 0 \rangle \ , \tag{3.29}$$

$$n_k = 0, 1, 2... \ .$$

These Fock or number states are orthogonal

$$\langle n_k \mid m_k \rangle = \delta_{nm} \ , \tag{3.30}$$

and complete

$$\sum_{n_k=0}^{\infty} \mid n_k \rangle \langle n_k \mid = 1 \ . \tag{3.31}$$

3.2 Density of Modes

As we saw,

$$\mathbf{k}_m = \frac{2\pi}{L}(m_1 \hat{\imath} + m_2 \mathbf{j} + m_3 \mathbf{k}) \ , \tag{3.32}$$

so we may ask the following question: How many normal modes are contained in a cavity of volume $v = L^3$?

Each set of integer numbers (m_1, m_2, m_3) correspond to two traveling wave modes, because we have two polarizations. These correspond to a point in the Fig. 3.1:

In an infinitesimal volume element $dm_1 dm_2 dm_3$, the number of modes is

$$dn = 2 dm_1 dm_2 dm_3 \ , \tag{3.33}$$

and according to (3.32), we get

$$dn = 2\left(\frac{L}{2\pi}\right)^3 dk_x dk_y dk_z = 2\left(\frac{L}{2\pi}\right)^3 d\mathbf{k} \ . \tag{3.34}$$

Now, letting $L \to \infty$, the sums become integrals and

$$\frac{1}{L^3} \sum_m [] L \to \infty \to \left(\frac{1}{2\pi}\right)^3 \int \int \int [] d\mathbf{k} \ . \tag{3.35}$$

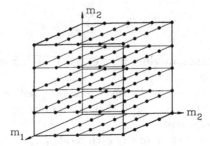

Fig. 3.1. Normal modes in a cavity

If one goes to polar coordinates, we can also write

$$dn = 2 \left(\frac{L}{2\pi} \right)^3 dk_x dk_y dk_z = 2 \left(\frac{L}{2\pi} \right)^3 k^2 dk d\Omega , \qquad (3.36)$$

where $d\Omega$ is an element of solid angle about $\mathbf{k}$.

As $\omega^2 = c^2 \mathbf{k}^2$, one can write

$$dn = 2 \left(\frac{L}{2\pi c} \right)^3 \omega^2 d\omega d\Omega , \qquad (3.37)$$

and the number of normal modes per unit volume , with angular frequencies between ω and $\omega + d\omega$, and per solid angle $d\Omega$ is

$$\frac{dn}{d\Omega L^3} = \frac{2\omega^2}{(2\pi c)^3} d\omega \equiv g(\omega) d\omega , \qquad (3.38)$$

where $g(\omega)$ is the mode density.

3.3 Commutation Relations

Using the commutation relations (3.19) one can write [1]

$$\left[A_i(\mathbf{r}), E_j(\mathbf{r'}) \right] = -\frac{i\hbar}{2v\varepsilon_0} \sum_{l,\sigma} (\mathbf{e}_{l\sigma})_i (\mathbf{e}_{l\sigma})_j \left[\exp(i\mathbf{k}_l.\rho) + cc \right] , \qquad (3.39)$$

$$\rho = \mathbf{r} - \mathbf{r'}.$$

In the above expression, we have replaced $m \rightarrow (l, \sigma)$, in order to separate the propagation vector index (l) from that of the polarization (σ).

Now, the three vectors $\mathbf{e}_{l1}, \mathbf{e}_{l2}, \hat{\mathbf{k}}_l$ are mutually orthogonal. So, in the three-dimensional space, one may write:

$$| \mathbf{e}_{l1} \rangle \langle \mathbf{e}_{l1} | + | \mathbf{e}_{l2} \rangle \langle \mathbf{e}_{l2} | + |\hat{\mathbf{k}}_l \rangle \langle \hat{\mathbf{k}}_l | = 1 , \qquad (3.40)$$

so, taking i and j components of the above relation, one gets

$$\sum_{\sigma=1,2} (\mathbf{e}_{l\sigma})_i (\mathbf{e}_{l\sigma})_j = \delta_{ij} - \frac{(\mathbf{k}_l)_i \, (\mathbf{k}_l)_j}{k_l^2} = \delta_{ij} - (\hat{\mathbf{k}}_l)_i \, (\hat{\mathbf{k}}_l)_j . \qquad (3.41)$$

We substitute the result of (3.41) into the commutation relation (3.39), getting

$$\left[\mathbf{A}_i(\mathbf{r}), \mathbf{E}_j(\mathbf{r}') \right] = -\frac{i\hbar}{v\varepsilon_0} \sum_l \left[\delta_{ij} - (\hat{\mathbf{k}}_l)_i \, (\hat{\mathbf{k}}_l)_j \right] \exp(i\mathbf{k}_l.\rho) , \qquad (3.42)$$

where the sum over l now covers both positive and negative integers, because

$$\hat{\mathbf{k}}_{-l} = -\hat{\mathbf{k}}_l . \qquad (3.43)$$

Letting $L \to \infty$, and making use of (3.35), we can write

$$\left[\mathbf{A}_i(\mathbf{r}), \mathbf{E}_j(\mathbf{r}') \right] = -\frac{i\hbar}{\varepsilon_0} \delta_{ij}^T(\mathbf{r} - \mathbf{r}') , \qquad (3.44)$$

where $\delta_{ij}^T(\mathbf{r} - \mathbf{r}')$ is the transverse δ function defined as

$$\delta_{ij}^T(\mathbf{r} - \mathbf{r}') = \frac{1}{(2\pi)^3} \int \int \int d\mathbf{k} \left[\delta_{ij} - (\hat{\mathbf{k}})_i \, (\hat{\mathbf{k}})_j \right] \exp(i\mathbf{k}.\rho) . \qquad (3.45)$$

Following the same procedure, the reader can show that

$$\left[\mathbf{A}_i(\mathbf{r}, t), \mathbf{A}_j(\mathbf{r}', t) \right] = 0 , \qquad (3.46)$$

$$\left[\mathbf{E}_i(\mathbf{r}, t), \mathbf{E}_j(\mathbf{r}', t) \right] = 0 , \qquad (3.47)$$

$$\left[\mathbf{B}_i(\mathbf{r}, t), \mathbf{B}_j(\mathbf{r}', t) \right] = 0 . \qquad (3.48)$$

Actually, to show the above commutation relation, one can work in the Schrödinger picture, with time-independent operators, and if the above is true, then it is also true in the Heisenberg picture, at equal times.

Problems

3.1. Prove that the $| n \rangle$ state can be expressed in terms of the vacuum state as

$$| n \rangle = \frac{(a^\dagger)^n}{\sqrt{n!}} \, | 0 \rangle .$$

3.2. Show the following commutation relations

$$[a, a^{\dagger n}] = n(a^{\dagger})^{n-1} ,$$

$$[a^n, a^{\dagger}] = n(a)^{n-1} .$$

3.3. The commutation relations of the problem 3.2 can be generalized. Prove that

$$[a, f(a, a^{\dagger})] = \frac{\partial f}{\partial a^{\dagger}} ,$$

$$[a^{\dagger}, f(a, a^{\dagger})] = -\frac{\partial f}{\partial a} ,$$

$$\exp(-\alpha a^{\dagger} a) f(a, a^{\dagger}) \exp(\alpha a^{\dagger} a) = f(ae^{\alpha}, a^{\dagger} e^{-\alpha}).$$

(Appendix A)

3.4. Show that

$$[\mathbf{A}_k(\mathbf{r}), \mathbf{E}_l(\mathbf{r}')] = -i\hbar \frac{\delta_{kl}^T(\mathbf{r} - \mathbf{r}')}{\varepsilon_0} .$$

3.5. Show that

$$[\mathbf{E}_i(\mathbf{r}, t), \mathbf{B}_j(\mathbf{r}', t)] = 0, i = j$$

$$= -i\hbar \frac{\partial}{x_k} \delta(\rho), \ i, j, k = 1, 2, 3$$

$$= i\hbar \frac{\partial}{x_k} \delta(\rho), \ i, j, k = 1, 3, 2 .$$

and $\rho = \mathbf{r} - \mathbf{r}'$.

Reference

1. Dirac, P.A.M.: Proc. Roy. Soc. A, **114**, 243 (1927)

Further Reading

- Cohen-Tannoudji, C., Dupont-Roc, J., Grynberg, G.: Photons and Atoms. Introduction to Quantum Electrodynamics. J. Wiley, NewYork (1989).
- Cohen-Tannoudji, C., Dupont-Roc, J., Grynberg, G.: Atom-Photon Interaction. Wiley, New York (1992)
- Gardiner, C.W.: Quantum Noise. Springer Verlag, Berlin (1991)
- Heitler, W.: The Quantum Theory of Radiation. 2nd ed., Fir Lawn, NJ (1944)
- Itzykson, C., Zuber, J.B.: Quantum Field Theory. McGraw-Hill, NewYork (1980)
- Klauder, J.R., Sudarshan, E.C.G.: Fundamentals of Quantum Optics. W.A.Benjamin, New York (1970).

- Loudon, R.: The Quantum Theory of Light. Clarendon Press, Oxford (1983)
- Louisell, W.H.: Quantum Statistical Properties of Radiation. J.Wiley, New York (1973)
- Mandel, L., Wolf, E.: Optical Coherence and Quantum Optics. Cambridge Univ. Press, Cambridge (1995)
- Milonni, P.W.: The Quantum Vacuum: An Introduction to Quantum Electrodynamics. Academic Press, New York (1994).
- Perina, J.: Quantum Statistics of Linear and Nonlinear Phenomena. Reidel, Dordrecht (1984).
- Power, E.A.: Introductory Quantum Electrodynamics. Longman, London (1964).
- Shore, B.W.: The Theory of Coherent Atomic Excitation. Vols. 1 and 2, J.Wiley, New York (1990)
- Vogel, W., Welsch, D.G.: Lectures on Quantum Optics. Akademie Verlag, Berlin (1994)
- Walls, D.F., Milburn, G.J.: Quantum Optics. Springer Verlag, Berlin (1994)

4. States of the Electromagnetic Field I

In this chapter, we study the coherent states and the thermal radiation.

The coherent states were introduced by **Glauber** [1] and **Sudarshan** [2] defined as the eigenstate of the annihilation operator. For a single mode

$$a \mid \alpha \rangle = \alpha \mid \alpha \rangle, \tag{4.1}$$

where α is a complex number.

Expanding the coherent state in the Fock basis

$$\mid \alpha \rangle = \sum_{n=0}^{\infty} c_n \mid n \rangle, \tag{4.2}$$

we easily get

$$a \mid \alpha \rangle = \sum_{n=1}^{\infty} c_n \sqrt{n} \mid n-1 \rangle \tag{4.3}$$

$$= \alpha \sum_{n=0}^{\infty} c_n \mid n \rangle,$$

from which we get the following recursion relation

$$c_n \sqrt{n} = \alpha c_{n-1}. \tag{4.4}$$

The solution of (4.4) gives

$$c_n = \frac{\alpha^n}{\sqrt{n!}} c_0.$$

The coefficient c_0 is found from normalization

$$\langle \alpha \mid \alpha \rangle = 1 = \mid c_0 \mid^2 \sum_n \frac{\mid \alpha \mid^{2n}}{n!} = \mid c_0 \mid^2 \exp \mid \alpha \mid^2 \tag{4.5}$$

so that we can now write the expansion

$$\mid \alpha \rangle = \exp \left(-\frac{\mid \alpha \mid^2}{2} \right) \sum_n \frac{\alpha^n}{\sqrt{n!}} \mid n \rangle. \tag{4.6}$$

4.1 Further Properties

4.1.1 Coherent States are Minimum Uncertainty States

We write the usual relation between $a, a^\dagger$ and q, p

$$a = \frac{1}{\sqrt{2\hbar\omega}}(\omega x + ip), \tag{4.7}$$

$$a^\dagger = \frac{1}{\sqrt{2\hbar\omega}}(\omega x - ip),$$

with $[x, p] = i\hbar$, or equivalently $(\Delta x)^2(\Delta p)^2 \geq \frac{\hbar^2}{4}$.

By inverting the relations 4.7 and taking the expectation value over a coherent state, one gets

$$\langle x \rangle_\alpha = \sqrt{\frac{\hbar}{2\omega}} \langle \alpha \mid (a + a^\dagger) \mid \alpha \rangle, \tag{4.8}$$

$$= \sqrt{\frac{\hbar}{2\omega}} (\alpha + \alpha^*),$$

and

$$\langle x^2 \rangle_\alpha = \frac{\hbar}{2\omega} \langle \alpha \mid (a + a^\dagger)^2 \mid \alpha \rangle \tag{4.9}$$

$$= \frac{\hbar}{2\omega} (1 + (\alpha + \alpha^*)^2),$$

so

$$(\Delta x)_\alpha^2 = \langle x^2 \rangle_\alpha - \langle x \rangle_\alpha^2 = \frac{\hbar}{2\omega}. \tag{4.10}$$

In a similar way, one finds

$$(\Delta p)_\alpha^2 = \frac{\hbar\omega}{2}, \tag{4.11}$$

so that

$$(\Delta x)_\alpha^2 (\Delta p)_\alpha^2 = \frac{\hbar^2}{4},$$

and the coherent states are Minimum uncertainty states (MUS).

4.1.2 Coherent States are not Orthogonal

$$\langle \alpha \mid \beta \rangle = \exp\left[-\frac{1}{2}(\mid \alpha \mid^2 + \mid \beta \mid^2)\right] \sum_{nm} \langle m \mid \frac{\alpha^n \beta^{*m}}{\sqrt{n!m!}} \mid n \rangle$$

$$= \exp\left[-\frac{1}{2}(\mid \alpha \mid^2 + \mid \beta \mid^2)\right] \exp(\alpha\beta^*),$$

or put it differently

$$\mid \langle \alpha \mid \beta \rangle \mid^2 = \exp\left(- \mid \alpha - \beta \mid^2\right). \tag{4.12}$$

4.1.3 Coherent States are Overcomplete

We now calculate $\int d^2\alpha \mid \alpha\rangle\langle\alpha \mid$ where $d^2\alpha = (dRe\alpha)(dIm\alpha)$:

$$\int d^2\alpha \mid \alpha\rangle\langle\alpha \mid = \sum_{nm} \frac{\mid n\rangle\langle m \mid}{\sqrt{n!m!}} \int d^2\alpha \exp\left(- \mid \alpha \mid^2\right) \alpha^n \alpha^{*m}. \tag{4.13}$$

It is convenient to write $\alpha = r\exp i\phi$, so

$$\int d^2\alpha \mid \alpha\rangle\langle\alpha \mid = \sum_{nm} \frac{\mid n\rangle\langle m \mid}{\sqrt{n!m!}} \int r dr \exp\left(-r^2\right) r^{n+m} \int d\phi \exp i(n - m)\phi, \tag{4.14}$$

but

$$\int d\phi \exp\left[i(n - m)\phi\right] = 2\pi\delta_{nm}$$

so

$$\int d^2\alpha \mid \alpha\rangle\langle\alpha \mid = 2\pi \sum_{n} \frac{\mid n\rangle\langle n \mid}{n!} \int dr \exp\left(-r^2\right) r^{2n+1}, \tag{4.15}$$

and defining $\varepsilon \equiv r^2$, the integral of (4.15) can be written as

$$\frac{1}{2} \int \varepsilon^n e^{-\varepsilon} d\varepsilon = \frac{n!}{2}.$$

The last result combined with (4.15) gives us finally

$$\int d^2\alpha \mid \alpha\rangle\langle\alpha \mid = \pi. \tag{4.16}$$

4.1.4 The Displacement Operator

We define the displacement operator as [1]

$$D(\alpha) = \exp(\alpha a^\dagger - \alpha^* a). \tag{4.17}$$

We make use of the BCH relation (see Appendix A)

$$e^{(A+B)} = e^A e^B e^{-\frac{1}{2}[A,B]}, \tag{4.18}$$

valid if $[A, [A, B]] = [B, [A, B]] = 0$.

We now use the BCH relation to the displacement operator and apply it to the vacuum state

$$D(\alpha) \mid 0\rangle = \exp(\alpha a^\dagger - \alpha^* a) \mid 0\rangle \tag{4.19}$$

$$= \exp(\alpha a^\dagger) \exp(-\alpha^* a) \exp\left(-\frac{\mid \alpha \mid^2}{2}\right) \mid 0\rangle$$

$$= \exp(-\frac{|\alpha|^2}{2}) \exp(\alpha a^\dagger) \exp(-\alpha^* a) \mid 0\rangle$$

$$= \exp(-\frac{|\alpha|^2}{2}) \exp(\alpha a^\dagger) \mid 0\rangle$$

$$= \exp(-\frac{|\alpha|^2}{2}) \sum_{n=0}^{\infty} \frac{\alpha^n}{n!} (a^\dagger)^n \mid 0\rangle,$$

so, finally

$$D(\alpha) \mid 0\rangle = \mid \alpha\rangle. \tag{4.20}$$

In the last step, we used the property

$$(a^\dagger)^n \mid 0\rangle = \sqrt{n!} \mid n\rangle. \tag{4.21}$$

From (4.20), we can see that a coherent state is just the vacuum displaced by $D(\alpha)$.

Other property of the displacement operator can be derived using (Appendix A)

$$\exp(\varepsilon A)B \exp(-\varepsilon A) = B + \varepsilon [A, B] + \frac{\varepsilon^2}{2} [A, [A, B]] + \ldots \tag{4.22}$$

Now, we can easily calculate:

$$D^\dagger(\alpha)aD(\alpha) = a + \alpha,$$
$$D^\dagger(\alpha)a^\dagger D(\alpha) = a^\dagger + \alpha^*.$$

4.1.5 Photon Statistics

By simple inspection of the expansion of the coherent states in terms of Fock states 4.6, one gets

$$P_n \equiv \mid C_n \mid^2 = \mid \langle n \mid \alpha\rangle \mid^2 = \exp(- \mid \alpha \mid^2) \frac{\mid \alpha \mid^{2n}}{n!}. \tag{4.23}$$

Equation (4.23) is saying that the probability of having n photons in a coherent state obeys a Poisson statistics.

We can easily calculate the average photon number and variance

$$\langle n\rangle = \langle \alpha \mid a^\dagger a \mid \alpha\rangle = \mid \alpha \mid^2, \tag{4.24}$$
$$\langle n^2\rangle = \langle \alpha \mid a^\dagger a a^\dagger a \mid \alpha\rangle = \langle \alpha \mid a^\dagger a \mid \alpha\rangle + \langle \alpha \mid a^{\dagger 2} a^2 \mid \alpha\rangle = \mid \alpha \mid^2 + \mid \alpha \mid^4,$$

so that $(\Delta n)^2 = \langle n^2\rangle - \langle n\rangle^2 = \langle n\rangle$, which is expected from the Poisson statistics.

4.1.6 Coordinate Representation

We would like to find the quantity: $\langle q' \mid \alpha \rangle$.

Making use (3.20), we write [3]

$$a \mid \alpha \rangle = \frac{1}{\sqrt{2\hbar\omega}}(\omega q + ip) \mid \alpha \rangle, \tag{4.25}$$

and multiplying (4.25) by $\langle q' \mid$ from the left, we get

$$\langle q' \mid (\omega q + ip) \mid \alpha \rangle = \sqrt{2\hbar\omega}\alpha \langle q' \mid \alpha \rangle \tag{4.26}$$
$$= (\omega q + \hbar\frac{\partial}{\partial q'})\langle q' \mid \alpha \rangle.$$

A more convenient way of writing (4.26) is

$$\frac{d\langle q' \mid \alpha \rangle}{\langle q' \mid \alpha \rangle} = \left[\sqrt{\frac{2\omega}{\hbar}}\alpha - \frac{\omega}{\hbar}q' \right] dq'. \tag{4.27}$$

The solution of (4.27) is

$$\langle q' \mid \alpha \rangle = \left(\frac{\omega}{\pi\hbar} \right)^{\frac{1}{4}} \exp\left(-\frac{\omega}{2\hbar}q'^2 + \sqrt{\frac{2\omega}{\hbar}}\alpha q' - \frac{\mid \alpha \mid^2 + \alpha^2}{2} \right), \tag{4.28}$$

where the result given by 4.28 was obtained using the normalization condition $\int_{-\infty}^{+\infty} dq' \mid \langle q' \mid \alpha \rangle \mid^2 = 1$, and defining $\alpha = r \exp(i\phi)$, with $\phi = 0$.

4.2 Mixed State: Thermal Radiation

A pure state implies a perfect knowledge of the state of our system. If that is not the case, we have the mixed case, where we know our state **only probabilistically.** In general, we write

$$\rho = \sum_R p_R \mid R \rangle \langle R \mid, \tag{4.29}$$

and the expectation value of any operator O can be expressed as

$$\langle O \rangle = \sum_S \langle S \mid \rho O \mid S \rangle = Tr(\rho O) = \tag{4.30}$$
$$\sum_S \langle S \mid \sum_R p_R \mid R \rangle \langle R \mid O \mid S \rangle = \sum_R p_R \langle R \mid O \mid R \rangle,$$

where $\mid S \rangle$ is an arbitrary set of orthogonal and complete states.

A property of ρ is that $Tr\{\rho\} = \sum_S \sum_R P_R \langle S \mid R \rangle \langle R \mid S \rangle = \sum_R P_R = 1$, which just implies that probability conservation should also hold for mixed states.

An example of a mixed state is the thermal radiation. For thermal equilibrium at temperature T, the probability P_n that one mode of the field in excited with n photons is given by the usual Boltzmann factor

$$P_n = \frac{\exp\left(-\frac{E_n}{K_B T}\right)}{\sum_n \exp\left(-\frac{E_n}{K_B T}\right)}. \tag{4.31}$$

The zero point energy cancels when the quantized energy is substituted, and using a shorthand notation

$$Z = \exp\left(-\frac{\hbar\omega}{K_B T}\right), \tag{4.32}$$

the probability can be written as

$$P_n = \frac{Z^n}{\sum_n Z^n}. \tag{4.33}$$

The denominator of the above expression can be easily summed, as a geometrical series

$$\sum_n Z^n = \frac{1}{1-Z},$$

giving

$$P_n = (1-Z)Z^n = \left[1 - \exp\left(-\frac{\hbar\omega}{K_B T}\right)\right]\exp\left(-\frac{n\hbar\omega}{K_B T}\right). \tag{4.34}$$

Therefore, the density operator for a mixed one-mode thermal state is given by

$$\rho_{\text{thermal}} =$$

$$= \left[1 - \exp\left(-\frac{\hbar\omega}{K_B T}\right)\right]\sum_n \exp\left(-\frac{n\hbar\omega}{K_B T}\right) \mid n \rangle \langle n \mid,$$

$$= \left[1 - \exp\left(-\frac{\hbar\omega}{K_B T}\right)\right]\sum_n \exp\left(-\frac{\hbar\omega a^\dagger a}{K_B T}\right) \mid n \rangle \langle n \mid$$

$$= \left[1 - \exp\left(-\frac{\hbar\omega}{K_B T}\right)\right]\exp\left(-\frac{\hbar\omega a^\dagger a}{K_B T}\right)$$

$$= \sum_n \frac{\langle n \rangle_{\text{th}}^n}{(1 + \langle n \rangle_{th})^{n+1}} \mid n \rangle \langle n \mid. \tag{4.35}$$

In the last line, we used the relation

$$P_n = \frac{\langle n \rangle_{th}^n}{(1 + \langle n \rangle_{th})^{n+1}}, \tag{4.36}$$

which can be easily proven as follows

$$\langle n \rangle_{th} = \sum_n n P_n = (1 - Z) \sum_n n Z^n \tag{4.37}$$

$$= (1 - Z) Z \frac{\partial}{\partial Z} \sum_n Z^n$$

$$= \frac{Z}{1 - Z},$$

thus, the average photon number is

$$\langle n \rangle_{th} = \frac{1}{\exp\left(\frac{\hbar\omega}{K_B T}\right) - 1}. \tag{4.38}$$

From (4.38), the reader can easily verify (4.36), for the photon statistics of a one-mode thermal state.

The obvious extension of (4.35) is, for the multimode case,

$$\rho_{thermal} = \sum_{\{n_k\}} \Pi_k \frac{(\langle n_k \rangle_{th})^{n_k}}{(1 + \langle n_k \rangle_{th})^{n_k+1}} \mid \{n_k\} \rangle \langle \{n_k\} \mid . \tag{4.39}$$

Problems

4.1. Show that the eigenstate of the creation operator does not exist.

4.2. Show that

$$a^\dagger \mid \alpha \rangle \langle \alpha \mid = (\alpha^* + \frac{\partial}{\partial \alpha}) \mid \alpha \rangle \langle \alpha \mid,$$

and

$$\mid \alpha \rangle \langle \alpha \mid a = (\alpha + \frac{\partial}{\partial \alpha^*}) \mid \alpha \rangle \langle \alpha \mid .$$

4.3. Show that if a state is initially coherent

$$\mid \psi, 0 \rangle = \mid \alpha \rangle,$$

then at $t = t$ it will still be coherent

$$\mid \psi, t \rangle = \mid \alpha \exp(-i\omega t) \rangle.$$

4.4. For a collection of oscillators in thermal equilibrium at temperature T, one can write

$$P(q, T) = \sum_{n=0}^{\infty} P_n \mid \psi_n(q) \mid^2 = \frac{\sum_n \exp(-\beta E_n) \mid \psi_n(q) \mid^2}{\sum_n \exp(-\beta E_n)},$$

$$P(p,T) = \sum_{n=0}^{\infty} P_n \mid \phi_n(p) \mid^2 = \frac{\sum_n \exp(-\beta E_n) \mid \phi_n(p) \mid^2}{\sum_n \exp(-\beta E_n)},$$

with $\beta = \frac{1}{K_B T}$.

Show that the result is

$$P(q,T) = \frac{\exp\left(-\frac{q^2}{2\sigma_q^2}\right)}{\sqrt{2\pi\sigma_q^2}},$$

$$P(p,T) = \frac{\exp\left(-\frac{p^2}{2\sigma_p^2}\right)}{\sqrt{2\pi\sigma_p^2}},$$

with

$$\sigma_q^2 = \frac{\hbar}{2m\omega} \coth\left(\frac{\hbar\omega}{2K_B T}\right),$$

$$\sigma_p^2 = \frac{\hbar m\omega}{2} \coth(\frac{\hbar\omega}{2K_B T}).$$

Also, verify that

$$\sigma_q \sigma_p = \frac{K_B T}{\omega},$$

for $K_B T \gg \hbar\omega$ and

$$\sigma_q \sigma_p = \frac{\hbar}{4\pi}$$

for $K_B T \ll \hbar\omega$.

4.5. Show that for a pure state, the condition

$$\rho^2 = \rho$$

is a necessary and sufficient one.

Also, verify that for a mixed state

$$Tr\{\rho^2\} < 1.$$

4.6. Coherent state with an unknown phase.

Let $\alpha = \mid \alpha \mid \exp(i\varphi)$, with φ unknown and uniformly distributed. Then show that

$$\rho = \frac{1}{2\pi} \int_0^{2\pi} \mid\mid \alpha \mid \exp(i\varphi)\rangle\langle\mid \alpha \mid \exp(i\varphi) \mid d\varphi$$

$$= \sum_{n=0}^{\infty} \exp(-\mid \alpha \mid^2) \frac{\mid \alpha \mid^{2n}}{n!} \mid n\rangle\langle n \mid.$$

We can see that the phase ignorance washes out the off diagonal elements.

4.7. Define the characteristic function or "momentum generating function" (see also Chap. 7) as:

$$C_A(\xi) = \sum_{n=0}^{\infty} \frac{(i\xi)^n}{n!} \langle A^n \rangle.$$

Show that

$$\langle A^n \rangle = (\frac{\partial}{\partial(i\xi)})^n C_A(\xi) \mid_{\xi=0},$$

$$C_A(\xi) = Tr(\rho \exp(i\xi A)),$$

$$C_A(\xi) = \int P(A'/\rho) \exp(i\xi A') dA',$$

where A' is an eigenvalue of A and $P(A'/\rho)$ the corresponding probability density.

Hint: To prove the last property, use the second one for a continuous spectrum.

4.8. Let the operator A in the problem 4.7 be

$$A = \gamma a + \gamma^* a^\dagger.$$

For a harmonic oscillator in a pure state α, show that

$$C_A(\xi) = \exp\left[-\frac{1}{2}\xi^2 \gamma\gamma^* + i(\alpha^*\gamma^* + \alpha\gamma)\right],$$

$$\langle A \rangle = \alpha^*\gamma^* + \alpha\gamma,$$

$$\sigma_A^2 = \langle A^2 \rangle - \langle A \rangle^2 = \mid \gamma \mid^2,$$

$$P(A'/\rho) = \frac{1}{\sqrt{2\pi\sigma_A^2}} \exp\left(-\frac{(A' - \langle A \rangle)^2}{2\sigma_A^2}\right).$$

The last property shows that the distribution is Gaussian.

References

1. Glauber, R.J.: Phys. Rev, **130**, 2529 (1963); Glauber, R.J.: Phys. Rev., **131**, 2766 (1963), Glauber, R.J.: Phys. Rev. Lett., **10**, 84 (1963).
2. Sudarshan, E.C.G.: Phys. Rev. Lett. **10**, 277 (1963)
3. Louisell, W.H.: Quantum Statistical Properties of Radiation. John Wiley, New York (1973)

Further Reading

• Glauber R.J.: In Quantum Optics and Electronics, Les Houches, De-Wit, C., Blandin, A., Cohen-Tannoudji, C. (eds) Gordon and Breach, New York (1965)

5. States of the Electromagnetic Field II

In this chapter, we deal with the general properties of squeezed states. We also describe two methods of detection of these states.

5.1 Squeezed States: General Properties and Detection

We define the quadrature operators X and Y [1]

$$X = \frac{a + a^\dagger}{2} = \sqrt{\frac{\omega}{2\hbar}} q \, , \tag{5.1}$$

$$Y = \frac{a - a^\dagger}{2i} = \sqrt{\frac{1}{2\hbar\omega}} p \, .$$

The name "quadrature" appears naturally if one replaces the expressions (5.1) in the quantized electric field

$$E(1 \text{ mode}) = i\sqrt{\frac{\hbar\omega}{2\varepsilon_o v}} \mathbf{e}_m \left[a \exp(-i\omega t + ik \cdot \mathbf{r}) - a^\dagger \exp(i\omega t - ik \cdot \mathbf{r}) \right] \, , \tag{5.2}$$

$$= 2\sqrt{\frac{\hbar\omega}{2\varepsilon_o v}} \mathbf{e}_m \left[X \sin(\omega t - k \cdot \mathbf{r}) - Y \cos(\omega t - k \cdot \mathbf{r}) \right] \, ,$$

thus appearing as factor-operators in front of the sin and cos functions.

The X and Y are Hermitian operators obeying the commutation relation

$$[X, Y] = \frac{i}{2} \, , \tag{5.3}$$

or

$$\langle (\Delta X)^2 \rangle \langle (\Delta Y)^2 \rangle \geq \frac{1}{16} \, . \tag{5.4}$$

In the case of the coherent states, according to (4.10, 4.11), one can write

$$\langle (\Delta X)^2 \rangle = \frac{1}{4}, \langle (\Delta Y)^2 \rangle = \frac{1}{4} \, . \tag{5.5}$$

From the quadrature perspective, a coherent state is a minimum uncertainty state with **equal fluctuations in both quadratures**.

Furthermore, since

$$\langle (X + iY) \rangle_\alpha = \langle a \rangle_\alpha = \alpha , \tag{5.6}$$

so

$$\langle X \rangle_\alpha = \mathrm{Re}\,\{\alpha\} , \tag{5.7}$$
$$\langle Y \rangle_\alpha = \mathrm{Im}\,\{\alpha\} .$$

Pictorially, a coherent state can be represented, in the complex plane, as an error circle of diameter $\frac{1}{2}$ and its center displaced by α (Fig. 5.1).

If there is a state for which either X or Y has a dispersion less than $\frac{1}{4}$, at the expense of the other quadrature, then its representation in the complex plane takes the form of an ellipse and we call this state a **squeezed state** [2, 3] .Of course, we may generalize our treatment, not only to squeeze along the X or Y axis, but along any pair of axes

$$X_1 = \frac{ae^{-i\phi} + a^\dagger e^{i\phi}}{2} , \tag{5.8}$$

$$Y_1 = \frac{ae^{-i\phi} - a^\dagger e^{i\phi}}{2i} .$$

One can check easily that X_1, Y_1 obey the same commutation relation as X and Y

$$[X_1, Y_1] = \frac{i}{2} , \tag{5.9}$$

$$\langle (\Delta X_1)^2 \rangle \langle (\Delta Y_1)^2 \rangle \geq \frac{1}{16} .$$

From (5.8), we show that

$$\Delta X_1 = \Delta X \cos\phi + \Delta Y \sin\phi , \tag{5.10}$$
$$\Delta Y_1 = -\Delta X \sin\phi + \Delta Y \cos\phi ,$$

where $\Delta O \equiv O - \langle O \rangle$.

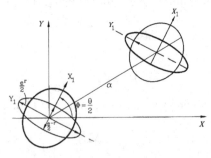

Fig. 5.1. Phase space representation of a coherent (*displaced circle*) and a squeezed state (*displaced ellipse*)

Then, we define a squeezed state by $\langle (\Delta X_1)^2 \rangle$ or $\langle (\Delta Y_1)^2 \rangle$ less than $\frac{1}{4}$, for some ϕ.

We denote by $\langle : X_1{}^2 : \rangle$ the normal ordered average, then

$$\langle : X_1{}^2 : \rangle = \frac{1}{4} \left[\langle a^{\dagger 2} \rangle \exp(2i\phi) + \langle a^2 \rangle \exp(-2i\phi) + 2\langle a^{\dagger} a \rangle \right] , \qquad (5.11)$$

and

$$\langle X_1{}^2 \rangle = \frac{1}{4} \left[\langle a^{\dagger 2} \rangle \exp(2i\phi) + \langle a^2 \rangle \exp(-2i\phi) + 2\langle a^{\dagger} a \rangle + 1 \right] , \qquad (5.12)$$

so that $\langle : \Delta X_1{}^2 : \rangle = \langle \Delta X_1{}^2 \rangle - \frac{1}{4}$, and for a squeezed state $\langle : \Delta X_1{}^2 : \rangle < 0$.

If we represent the density operator ρ as (we will see the details in the Chap. 7)

$$\rho = \int P(\alpha) \mid \alpha \rangle \langle \alpha \mid d^2\alpha ,$$

then

$$\langle : \Delta X_1{}^2 : \rangle = \int P(\alpha)(\Delta X_1{}^2)_\alpha d^2\alpha , \qquad (5.13)$$

where $(\Delta X_1{}^2)_\alpha$ is just $\Delta X_1{}^2$ with $a \longrightarrow \alpha, a^{\dagger} \longrightarrow \alpha^*$.

Thus

$$(\Delta X_1{}^2)_\alpha = \frac{1}{4} \left[\Delta\alpha^* \exp(i\phi) + \Delta\alpha \exp(-i\phi) \right]^2 , \qquad (5.14)$$

and the squeezing condition can be written as

$$\langle : \Delta X_1{}^2 : \rangle = \frac{1}{4} \int P(\alpha) \left[\Delta\alpha^* \exp(i\phi) + \Delta\alpha \exp(-i\phi) \right]^2 d^2\alpha < 0 , \qquad (5.15)$$

and since $\left[\Delta\alpha^* \exp(i\phi) + \Delta\alpha \exp(-i\phi) \right]^2$ is real and positive, squeezing can only take place if $P(\alpha)$ is not a positive definite probability density.

5.1.1 The Squeeze Operator and the Squeezed State

We define a squeeze operator as

$$S(\xi) = \exp \left[\frac{1}{2} (\xi^* a^2 - \xi a^{\dagger 2}) \right] , \qquad (5.16)$$

which is a unitary transformation , and $\xi \equiv r \exp(i\theta)$ is called the **squeeze parameter.**

With the help of (4.22), we can define a **generalized annihilation operator** as

$$A = S(\xi) a S^{\dagger}(\xi) \qquad (5.17)$$
$$= a \cosh r + a^{\dagger} \exp i\theta \sinh r$$
$$\equiv \mu a + \nu a^{\dagger} .$$

As we can see $\mu \equiv \cosh r, \nu \equiv \exp(i\theta) \sinh r$ and $\left[A, A^{\dagger} \right] = 1$.

The **coherent squeezed state** is defined as (one mode) [4]:

$$\mid \alpha, \xi \rangle = D(\alpha) S(\xi) \mid 0 \rangle . \qquad (5.18)$$

Inverting the 5.17, one can also write

$$a = \mu A - \nu A^\dagger \,, \tag{5.19}$$
$$a^\dagger = \mu A^\dagger - \nu^* A \,.$$

5.1.2 The Squeezed State is an Eigenstate of A

$$A \,|\, \alpha, \xi \rangle = A D(\alpha) S(\xi) \,|\, 0 \rangle \tag{5.20}$$
$$= S(\xi) a S^\dagger(\xi) D(\alpha) S(\xi) \,|\, 0 \rangle \,.$$

One can prove that

$$D(\alpha) S(\xi) = S(\xi) D(\beta) \,, \tag{5.21}$$

with

$$\beta \equiv \alpha \cosh r + \alpha^* \exp(i\theta) \sinh r \,. \tag{5.22}$$

Replacing (5.21) in (5.20), we get

$$A \,|\, \alpha, \xi \rangle \tag{5.23}$$
$$= S(\xi) a D(\beta) \,|\, 0 \rangle = S(\xi) a \,|\, \beta \rangle$$
$$= \beta S(\xi) \,|\, \beta \rangle = \beta S(\xi) D(\beta) \,|\, 0 \rangle = \beta D(\alpha) S(\xi) \,|\, 0 \rangle$$
$$= \beta \,|\, \alpha, \xi \rangle \,.$$

Pictorially, a squeezed state in phase space is the vacuum (represented by a circle in the origin), which is first squeezed into an ellipse, tilted by $\frac{\theta}{2}$ and then displaced by α (See Fig. 5.1).

5.1.3 Calculation of Moments with Squeezed States

With the help of (5.19, 5.21), we find

$$\langle \alpha, \xi \,|\, a \,|\, \alpha, \xi \rangle = \langle \alpha, \xi \,|\, \mu A - \nu A^\dagger \,|\, \alpha, \xi \rangle \tag{5.24}$$
$$= \mu \beta - \nu \beta^*$$
$$= \alpha.$$

In a similar way, one can find

$$\langle \alpha, \xi \,|\, a^\dagger a \,|\, \alpha, \xi \rangle \equiv \langle n \rangle_{sq} \tag{5.25}$$
$$= \langle 0 \,|\, S^\dagger(\xi) D^\dagger(\alpha) a^\dagger a D(\alpha) S(\xi) \,|\, 0 \rangle$$
$$= \langle 0 \,|\, S^\dagger(\xi) D^\dagger(\alpha) a^\dagger D(\alpha) D^\dagger(\alpha) a D(\alpha) S(\xi) \,|\, 0 \rangle$$
$$= \langle 0 \,|\, S^\dagger(\xi) (a^\dagger + \alpha^*)(a + \alpha) S(\xi) \,|\, 0 \rangle \,,$$

and since $\langle 0 \mid S^\dagger(\xi)a^\dagger S(\xi) \mid 0\rangle, \langle 0 \mid S^\dagger(\xi)aS(\xi) \mid 0\rangle$ are linear combinations of a and $a\dagger$ averaged over the vacuum state, will give no contribution, and

$$
\begin{aligned}
\langle n\rangle_{sq} &= \langle 0 \mid S^\dagger(\xi)(a^\dagger a)S(\xi) \mid 0\rangle + \mid \alpha \mid^2 \qquad\qquad (5.26)\\
&= \langle 0 \mid S^\dagger(\xi)a^\dagger S(\xi)S^\dagger(\xi)aS(\xi) \mid 0\rangle + \mid \alpha \mid^2 \\
&= \langle 0 \mid [a^\dagger \cosh r - a \exp(-i\theta)\sinh r]\,[a \cosh r - a^\dagger \exp(i\theta)\sinh r] \mid 0\rangle \\
&\quad + \mid \alpha \mid^2 \\
&= \sinh^2 r + \mid \alpha \mid^2 \ .
\end{aligned}
$$

Finally, using a similar procedure, one can verify that

$$
\langle a^2\rangle_{sq} = \alpha^2 - \cosh r \sinh r \exp(i\theta) \ . \qquad\qquad (5.27)
$$

5.1.4 Quadrature Fluctuations

We calculate

$$
\begin{aligned}
\langle \Delta X_1^2\rangle_{sq} &= \langle 0 \mid S^\dagger(\xi)D^\dagger(\alpha)\Delta X_1 D(\alpha)D^\dagger(\alpha)\Delta X_1 D(\alpha)S(\xi) \mid 0\rangle \qquad (5.28)\\
&= \langle 0 \mid S^\dagger(\xi)\frac{1}{4}\,[\exp(-i\phi)(a + \alpha - \langle a\rangle) \\
&\quad + \exp(i\phi)(a^\dagger + \alpha^* - \langle a^\dagger\rangle)]^2\,S(\xi) \mid 0\rangle \\
&= \frac{1}{4}\langle 0 \mid S^\dagger(\xi)\,[\exp(-i\phi)a + \exp(i\phi)a^\dagger]^2\,S(\xi) \mid 0\rangle \\[2mm]
&= \frac{1}{4}\,[\exp(-2i\phi)\langle\xi,0 \mid a^2 \mid \xi,0\rangle + \exp(2i\phi)\langle\xi,0 \mid a^{\dagger 2} \mid \xi,0\rangle \\
&\quad + 1 + 2\langle\xi,0 \mid a^\dagger a \mid \xi,0\rangle] \ .
\end{aligned}
$$

We take $\theta = 2\phi$, thus getting

$$
\begin{aligned}
\langle \Delta X_1^2\rangle_{sq} &= \frac{1}{4}\,(1 + 2\sinh^2 r - 2\cosh r \sinh r) \qquad\qquad (5.29)\\
&= \frac{e^{-2r}}{4} \ .
\end{aligned}
$$

Similarly

$$
\langle \Delta Y_1^2\rangle_{sq} = \frac{e^{2r}}{4} \ . \qquad\qquad (5.30)
$$

If we now go back to the original quadratures, we readily find

$$
\langle (\Delta X)^2\rangle_{sq} = \frac{1}{4}\left[\exp(-2r)\cos^2\frac{\theta}{2} + \exp(2r)\sin^2\frac{\theta}{2}\right] , \qquad (5.31)
$$

$$
\langle (\Delta Y)^2\rangle_{sq} = \frac{1}{4}\left[\exp(-2r)\sin^2\frac{\theta}{2} + \exp(2r)\cos^2\frac{\theta}{2}\right] .
$$

We notice, from the above expressions, that the variances are independent of the coherent amplitude α.

The squeezing condition for the quadrature operator X

$$\langle (\Delta X)^2 \rangle_{sq} < \frac{1}{4} ,$$

is satisfied if

$$\cos \theta > \tanh r ,$$

and its minimum value is

$$\langle (\Delta X)^2 \rangle_{sq} = \frac{\exp(-2r)}{4} ,$$

for $\theta = 0$.

The squeezing condition for Y is

$$\cos \theta < - \tanh r.$$

Finally, it is easy to verify that:

$$\langle (\Delta X)^2 \rangle_{sq} \langle (\Delta Y)^2 \rangle_{sq} = \frac{1}{16} \left[\cosh^2(2r) \sin^2 \theta + \cos^2 \theta \right] .$$

The above formula takes the minimum uncertainty value (MUS), for $\theta = 0$ or $\theta = \pi$,

$$\langle (\Delta X)^2 \rangle_{sq} \langle (\Delta Y)^2 \rangle_{sq} = \frac{1}{4} .$$

5.1.5 Photon Statistics

The photon distribution of a coherent squeezed state can be written as

$$P_n = | \langle n \mid \alpha, \xi \rangle |^2 ,$$

where

$$\langle n \mid \alpha, \xi \rangle = \frac{1}{\sqrt{n! \cosh r}} \left[\frac{1}{2} \exp(i\theta) \tanh r \right]^{\frac{n}{2}}$$
$$\times \exp \left[-\frac{1}{2} (| \alpha |^2 + \alpha^{*2} \exp i\theta \tanh r) \right]$$
$$\times H_n \left[\frac{\alpha + \alpha^* \exp i\theta \tanh r}{(2 \exp i\theta \tanh r)^{\frac{1}{2}}} \right] ,$$

where H_n is the Hermite polynomial of degree n [5, 6].

5.2 Multimode Squeezed States

In general, the squeezed states experimentally generated are not single mode, but rather, they cover a certain frequency band.

We consider, here, for example, a two-mode squeezed state, defined as

$$| \alpha_+, \alpha_-, \xi \rangle = D_+(\alpha_+)D_-(\alpha_-)S_{+-}(\xi) | 0 \rangle , \qquad (5.32)$$

where

$$D_\pm(\alpha_\pm) = \exp \left[\alpha_\pm a_\pm^\dagger - \alpha_\pm^* a_\pm \right] , \qquad (5.33)$$

are the coherent displacement operators for the two modes described by the destruction operators a_+ and a_-, and

$$S_{+-}(\xi) = \exp \left(\xi^* a_+ a_- - \xi a_+^\dagger a_-^\dagger \right) , \qquad (5.34)$$

which is the two-mode squeezing operator and $| 0 \rangle$ the two-mode vacuum state.

Similarly to (5.17), we have

$$S_{+-}(\xi) a_\pm S_{+-}(\xi)^\dagger = a_\pm \cosh r + a_\mp^\dagger \exp(i\theta) \sinh r . \qquad (5.35)$$

Using the above properties enable us to calculate various expectation values of combinations of creation and destruction operators

$$\langle a_\pm \rangle = \alpha_\pm , \qquad (5.36)$$
$$\langle a_+ a_- \rangle = \alpha_+ \alpha_- - \exp(i\theta) \sinh r \cosh r = \langle a_- a_+ \rangle,$$
$$\langle a_+ a_+ \rangle = \alpha_+^2 ,$$
$$\langle a_- a_- \rangle = \alpha_-^2 ,$$
$$\langle a_+^\dagger a_+ \rangle = | \alpha_+ |^2 + \sinh^2 r ,$$
$$\langle a_-^\dagger a_- \rangle = | \alpha_- |^2 + \sinh^2 r .$$

As we can see from the above results, the squeezing affects only the diagonal photon number for each mode and the off diagonal two-mode expectation values.

5.3 Detection of Squeezed States

As we saw in the previous section, the quadrature fluctuations of squeezed light have a dependence on the phase θ of the squeezing parameter.

In principle, we can detect the squeezed signal by four different methods:

(a) Direct photodetection,
(b) Ordinary homodyne detection,

(c) Balanced homodyne detection,
(d) Heterodyne detection.

The first method of direct photodetection is not the most convenient one, although simple, because the advantage of the phase-dependent squeezed light is lost, and one can associate antibunching or sub-Poissonian photon counting statistics to an incoming squeezed signal. However, both effects can also be measured with non-squeezed light, so we really need a phase-sensitive method to display the characteristics of the squeezed input.

Therefore, our discussion will be centered on the two homodyne detection methods, and we will also mention some aspects of the heterodyne detection.

5.3.1 Ordinary Homodyne Detection

In the Fig. 5.2, we show the schematic arrangement of the homodyne detection. In the case of ordinary homodyne detection [7, 8, 9], only three of the four ports will be used, and in the case of balanced homodyne detection, all four ports are used.

A lossless symmetric beam splitter mixes the squeezed signal a_{sig} with a local oscillator a_{LO}, with t and r being the transmission and reflection coefficients respectively, so one can write

$$\begin{pmatrix} d_1 \\ d_2 \end{pmatrix} = \begin{pmatrix} r & t \\ t & r \end{pmatrix} \begin{pmatrix} a_{LO} \\ a_{sig} \end{pmatrix}, \tag{5.37}$$

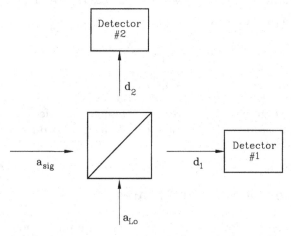

Fig. 5.2. Schematic arrangement of a homodyne detector

and the unitarity of the $\begin{pmatrix} r & t \\ t & r \end{pmatrix}$ matrix imposes the conditions

$$|r|^2 + |t|^2 = 1, \qquad (5.38)$$
$$rt^* + r^*t = 0.$$

If we write $r = |r| \exp(i\theta_r)$, $t = |t| \exp(i\theta_t)$, the second condition from (5.38) becomes

$$|r||t| \{\exp[i(\theta_r - \theta_t)] + \exp[-i(\theta_r - \theta_t)]\} = 0,$$

or, since $|r|, |t| \neq 0$

$$\theta_r - \theta_t = \frac{\pi}{2}. \qquad (5.39)$$

In the ordinary homodyne detection $|r| << |t|$ while in the balanced detection $|r| = |t| = \frac{1}{\sqrt{2}}$.

Now, we calculate the number of photons $\langle d_1 d_1 \rangle$ measured in the detector #1, assumed 100% efficient, and also assuming that the local oscillator is in a coherent state $|\alpha_{LO} = |\alpha_{LO}| \exp(i\phi_{LO})\rangle$

$$\langle d_1^\dagger d_1 \rangle = \langle n_1 \rangle \qquad (5.40)$$
$$= \langle (t^* a_{sig}^\dagger + r^* a_{LO}^\dagger)(t a_{sig} + r a_{LO}) \rangle$$
$$= |r|^2 |\alpha_{LO}|^2 + 2|r||t||\alpha_{LO}| \langle X_1(\phi) \rangle + |t|^2 \langle a_{sig}^\dagger a_{sig} \rangle,$$

with

$$X_1(\phi) = \frac{1}{2} \left[a_{sig} \exp(-i\phi) + a_{sig}^\dagger \exp(i\phi) \right], \qquad (5.41)$$
$$\phi = \phi_{LO} + \theta_r - \theta_t = \frac{\pi}{2} + \phi_{LO}.$$

Normally, the homodyne detectors use strong local oscillators, that is

$$|r|^2 |\alpha_{LO}|^2 >> |t|^2 \langle a_{sig}^\dagger a_{sig} \rangle, \qquad (5.42)$$

and one can approximately write

$$\langle d_1^\dagger d_1 \rangle = \langle n_1 \rangle = |r|^2 |\alpha_{LO}|^2 + 2|r||t||\alpha_{LO}| \langle X_1(\phi) \rangle. \qquad (5.43)$$

One can also calculate the variance of n_1. The result is

$$\langle \Delta n_1^2 \rangle = |r|^2 |\alpha_{LO}|^2 \left[|r|^2 + 4|t|^2 \langle \Delta X_1(\phi)^2 \rangle \right]. \qquad (5.44)$$

As we can see, with $r << t$, the photon number fluctuations at the detector #1 is determined by the quadrature fluctuations of the phase-dependent squeezed signal.

5.3.2 Balanced Homodyne Detection

An alternative detection scheme that eliminates the large local oscillator term of the fluctuations in the direct homodyne case is to take the photocurrent difference between the two exit ports [10, 11, 12] that is (also in this balanced case $r = t = \frac{1}{\sqrt{2}}$)

$$n_{12} = d_1^\dagger d_1 - d_2^\dagger d_2 \tag{5.45}$$
$$= i(a_{sig}^\dagger a_{LO} - a_{sig} a_{LO}^\dagger) .$$

We leave the reader to show that

$$\langle n_{12} \rangle = 2 \mid \alpha_{LO} \mid \langle X_1(\phi) \rangle , \tag{5.46}$$
$$\langle \Delta n_{12}^2 \rangle = 4 \mid \alpha_{LO} \mid^2 \langle \Delta X_1^2(\phi) \rangle ,$$

where again we have assumed that the local oscillator is coherent and much stronger than the input signal.

We notice that for a squeezed signal, $\langle \Delta X_1(\phi)^2 \rangle < \frac{1}{4}$, and therefore, the fluctuations of the photon number difference $\langle \Delta n_{12}^2 \rangle < \mid \alpha_{LO} \mid^2$ is sub-Poissonian.

This result would seem no different from the direct or ordinary homodyne detection scheme; however, here we have the phase dependence to check that the origin of the sub-Poissonian nature of the photocurrent is due to the squeezed field, on one hand, and we have been able to eliminate the large coherent amplitude of the local oscillator, on the other hand.

If we consider a more realistic detector, with a quantum efficiency η, the above result is modified [1], and calling m_{12} the photoncount

$$\langle m_{12} \rangle = 2\eta \mid \alpha_{LO} \mid \langle X_1(\phi) \rangle , \tag{5.47}$$
$$\langle \Delta m_{12}^2 \rangle = \eta \mid \alpha_{LO} \mid^2 \left\{ 1 + \eta \left[4\langle \Delta X_1^2(\phi) \rangle - 1 \right] \right\} .$$

These results coincide with (5.46) for $\eta = 1$.

5.3.3 Heterodyne Detection

Heterodyne detection is appropriate when dealing with a two-mode squeezed state. For more details, the interested reader is referred to [1].

Problems

5.1. Prove that for an ideal squeezed state

$$(\Delta n)^2 = \mid \alpha \mid^2 \left[\exp(-2r) \cos^2(\phi - \frac{\theta}{2}) + \exp(2r) \sin^2(\phi - \frac{\theta}{2}) \right]$$
$$+ 2 \sinh^2 r \cosh^2 r ,$$

where $\alpha = \mid \alpha \mid \exp i\phi$.

5.2. If we define

$$g^{(2)}(0) = \frac{\langle a^\dagger a^\dagger a a \rangle}{\langle a^\dagger a \rangle^2} \,,$$

and

$$(Q)_{\text{Mandel}} = \frac{\langle (\Delta n)^2 \rangle - \langle n \rangle}{\langle n \rangle} \,,$$

prove that for a squeezed vacuum

$$g^{(2)}(0)_{sq} = 3 + \frac{1}{\langle n \rangle} \,,$$

$$(Q)_{\text{Mandel,sq}} = 2\langle n \rangle + 1 \,,$$

so that the squeezed vacuum shows photon bunching and Super-Poissonian statistics (Chap. 6).

5.3. Show that in balanced homodyne detection, with a quantum efficiency η, the average and fluctuations of the photocouns are

$$\langle m_{12} \rangle = 2\eta \mid \alpha_{LO} \mid \langle X_1(\phi) \rangle \,,$$

$$\langle \Delta m_{12}^2 \rangle = \eta \mid \alpha_{LO} \mid^2 \left\{ 1 + \eta \left[4\langle \Delta X_1^2(\phi) \rangle - 1 \right] \right\} \,.$$

5.4. If $\mid X \rangle$ are eigenstates of X, show that, for $\theta = 0$

$$\mid \langle X \mid \alpha, r \rangle \mid^2 = \left(\frac{2 \exp 2r}{\pi} \right)^{\frac{1}{2}} \exp \left[-2(X - \text{Re}\,\{\alpha\})^2 \exp(2r) \right] \,,$$

thus showing that squeezed states are Gaussian wavepackets.

5.5. Show that for a squeezed state, the maximum and minimum values of the Q_{Mandel} factor are given by [1]

$$Q_{\max} = \exp(2r) - 1, \phi = \frac{\theta + \pi}{2} \,,$$

$$Q_{\min} = \exp(-2r) - 1, \phi = \frac{\theta}{2} \,.$$

References

1. A good review of squeezed states if found in: Loudon R., and Knight, P., J. Mod. Opt., **34**, 709 (1989). Also in Walls, D.F.: Nature, **306**, 141 (1983)
2. Stoler, D.: Phys. Rev. D, **1**, 3217 (1970); Stoler, D.: Phys. Rev. D, **4**, 1925 (1971)
3. Yuen, H.P.: Phys. Lett. A, **51**, 1 (1975); Yuen, H.P.: Phys. Rev. A, **13**, 2226 (1976)
4. This definition, slightly different from Yuen and Stoler, is due to C. Caves, see for example: Caves, C.M.: Phys. Rev. Lett., **45**, 75 (1980); Caves, C.M.: Phys. Rev. D, **23**, 1693 (1982)

5. Hirota, O.: Squeezed Light. Elsevier, Amsterdam (1992).
6. Yuen, H.P.: Phys. Rev. A, **13**, 2226 (1976)
7. Yuen, H.P., Shapiro, J.H.: IEEE, Tran. Inf. Theory, **26**, 78 (1980)
8. Yuen, H.P., Shapiro, J.H.: IEEE, Tran. Inf. Theory, **24**, 657 (1978)
9. Yuen, H.P., Shapiro, J.H., Machado, J.A.: Mata. IEEE, Tran. Inf. Theory, **25**, 179 (1979)
10. Yuen, H.P., Chan. V.W.: Opt. Lett., **8**, 177 (1983)
11. Shapiro J.H.: IEEE, J. Quantum. Elect, **QE21**, 237 (1985)
12. Schumaker B.L.: Opt. Lett., **9**, 189 (1984)

6. Quantum Theory of Coherence

In this chapter, we study Glauber's quantum theory of coherence and photon counting.

This theory was formulated originally by **Glauber** [1], where he considers the process of photon detection, which plays a central role.

The basic process involved in the detection is the absorption of a photon and the corresponding generation of a photoelectron, measured through an electric current. This type of detector is insensitive to phases and spontaneous emission. This simple Glauber's model is an ideal detector, sensitive to what we define as the positive frequency component of the field (proportional to the annihilation operator of the field)

$$\mathbf{E}^+(\mathbf{r}, t) = i \sum_k \sqrt{\frac{\hbar \omega_\mathbf{k}}{2\epsilon_o}} a_\mathbf{k} \mathbf{u}_{k,\lambda}(\mathbf{r}) \exp(-i\omega_\mathbf{k} t) , \tag{6.1}$$

$$\mathbf{u}_{k,\lambda}(\mathbf{r}) = \frac{\mathbf{e}^\lambda \exp(i\mathbf{k} \cdot \mathbf{r})}{\sqrt{v}} , \tag{6.2}$$

$$\mathbf{E} = \mathbf{E}^+ + \mathbf{E}^- , \tag{6.3}$$

$$\mathbf{E}^- = (\mathbf{E}^+)^\dagger . \tag{6.4}$$

with $\mathbf{e}^\lambda$ being the polarization vector. Also, in this model, the detector atoms are in the ground state, so that only absorption takes place. As it is only the annihilation part $\mathbf{E}^+(\mathbf{r}, t)$ of the field that takes place in the photodetection process, there is a real asymmetry between $\mathbf{E}^+(\mathbf{r}, t)$ and $\mathbf{E}^-(\mathbf{r}, t)$ in a way that the actual detection is more closely related to $\mathbf{E}^+(\mathbf{r}, t)$ than the total field $\mathbf{E}$.

An ideal photodetector would also have an infinite band, responding to a field at time t and a negligible spatial extension.

The transition probability from an initial state $\mid \psi_i \rangle$ and a final state $\mid \psi_f \rangle$ is proportional to

$$W_{if} = \mid \langle \psi_f \mid \mathbf{E}^+ \mid \psi_i \rangle \mid^2 . \tag{6.5}$$

As we will see, this is a first-order approximation.

In general, the final state of the field is not known, so we have to sum over all the possible final states

$$I_i(\mathbf{r}, t) = \sum_f |\langle \psi_f | \mathbf{E}^+ | \psi_i \rangle|^2 \tag{6.6}$$

$$= \sum_f \langle \psi_i | \mathbf{E}^- | \psi_f \rangle \langle \psi_f | \mathbf{E}^+ | \psi_i \rangle$$

$$= \langle \psi_i | \mathbf{E}^- \mathbf{E}^+ | \psi_i \rangle \ ,$$

giving us an average field intensity. In the last step, we made use of the completeness of the final states.

If the initial state is a mixed one, then we have to use the density matrix, and write

$$\langle I_i(\mathbf{r}, t) \rangle = Tr \left\{ \rho \mathbf{E}^-(\mathbf{r}, t) \mathbf{E}^+(\mathbf{r}, t) \right\} \ . \tag{6.7}$$

We now define the first-order coherence function:

$$G^{(1)}(x, x') = Tr \left\{ \rho \mathbf{E}^-(x) \mathbf{E}^+(x') \right\} \ , \tag{6.8}$$

where x and x' are $x = (\mathbf{r}, t)$ and $x' = (\mathbf{r}', t')$.

The first-order coherence function appears typically in the interference experiments. To describe more sophisticated experiments, like the coincidence experiments of Handbury Brown and Twiss, it is useful to define an n-th order coherence function

$$G^{(n)}(x_1, x_2..x_n; x_{n+1}, ...x_{2n}) = Tr(\rho \mathbf{E}^-(x_1) ... \mathbf{E}^-(x_n) \mathbf{E}^+(x_{n+1}) ... \mathbf{E}^+(x_{2n})) \ . \tag{6.9}$$

We will later discuss the analytical properties of these functions.

One, in principle, could have a more general definition of coherence functions with unequal number of creation and annihilation operators. However, these functions are not particularly useful in the typical photoncounting measurement.

6.1 One-Atom Detector

We now consider the detailed photodetection process . Imagine, for simplicity, as a first approach to the detection problem, that we have an one-atom photodetector, which can undergo photoabsorption transitions such as the photoelectric effect. We will calculate the probability for this type of transition to occur, within a given time interval. The Hamiltonian of the system is

$$H = H_{0,at} + H_{o,f} + H_I \ , \tag{6.10}$$

where $H_{0,at}$ is the free atom, $H_{0,f}$ the free field and H_I the interaction between both. H_I is in the Schrödinger picture and therefore time independent. However, when one goes into the interaction picture, it becomes time-dependent:

$$H_1 = \exp\left[\frac{i(H_{0,at} + H_{0,f})t}{\hbar}\right] H_I \exp\left[-\frac{i(H_{0,at} + H_{0,f})t}{\hbar}\right] \quad (6.11)$$

$$= -e \sum_\gamma \mathbf{q}_\gamma(t) \cdot \mathbf{E}(\mathbf{r}, t) \,,$$

where $\mathbf{r}$ represents the position of the atomic nucleus and $\mathbf{q}_\gamma$ the position of the γ electron, relative to the nucleus.

The Schrödinger equation for the system is

$$i\hbar\frac{\partial}{\partial t} \mid \psi(t)\rangle = H_1 \mid \psi(t)\rangle \,, \quad (6.12)$$

and its solution, to first order is

$$\mid \psi(t)\rangle = U(t, t_o) \mid \psi(t_o)\rangle \quad (6.13)$$

$$\simeq \left\{1 + \frac{1}{i\hbar}\int_{t_o}^t dt' H_1(t')\right\} \mid \psi(t_o)\rangle \,.$$

Now, suppose that initially the system is in the state $\mid b\rangle \mid i\rangle$, where $\mid i\rangle$ is the initial state of the field and $\mid b\rangle$ the ground state of the atom. We ask for the probability for the system to be in the excited state $\mid a\rangle$ and with a final state of the field $\mid f\rangle$. Using Schrödinger Equation with the interaction Hamiltonian H_1, we can write

$$\langle a, f \mid U(t, t_o) \mid b, i\rangle \quad (6.14)$$

$$= \frac{1}{i\hbar}\int_{t_o}^t dt'\langle a, f \mid H_1(t') \mid b, i\rangle$$

$$= \frac{ie}{\hbar}\sum_\gamma \int_{t_o}^t dt'\langle a \mid \mathbf{q}_\gamma(t') \mid b\rangle\langle f \mid \mathbf{E}(\mathbf{r}', t') \mid i\rangle \,.$$

The atomic matrix element can be evaluated starting from

$$\mathbf{q}_\gamma(t') = \exp\left[\frac{iH_{o,at}t'}{\hbar}\right] \mathbf{q}_\gamma(0) \exp\left[-\frac{iH_{o,at}t'}{\hbar}\right] \,, \quad (6.15)$$

and therefore

$$\langle a \mid \sum_\gamma \mathbf{q}_\gamma(t') \mid b\rangle = \mathbf{d}_{ab}\exp(i\omega_{ab}t') \,, \quad (6.16)$$

$$\mathbf{d}_{ab} = \langle a \mid \sum_\gamma \mathbf{q}_\gamma(0) \mid b\rangle \,.$$

Thus, using (6.14 and 6.16)

$$\langle a, f \mid U(t, t_o) \mid b, i\rangle = \frac{ie}{\hbar}\sum_\gamma \int_{t_o}^t dt'\exp(i\omega_{ab}t')\mathbf{d}_{ab} \cdot \langle f \mid \mathbf{E}(\mathbf{r}', t') \mid i\rangle \,. \quad (6.17)$$

We can now replace the electric field in the above expression, which consists in the sum of two operators. The emission operator $\mathbf{E}^-(\mathbf{r}, t)$ contains the negative frequencies of the form $\exp(i\omega t)$ for $\omega > 0$, which will produce a rapidly oscillating term in (6.17), and can be safely neglected, when compared with the annihilation operator $\mathbf{E}^+(\mathbf{r}, t)$.

We now calculate the square of the absolute value of (6.17) and sum over the final states, getting

$$\sum_f |\langle a, f \mid U(t, t_o) \mid b, i\rangle|^2 \tag{6.18}$$

$$= \left(\frac{e}{\hbar}\right)^2 \int_{t_o}^t \int_{t_o}^t dt' dt'' \exp[i\omega_{ab}(t'' - t')]$$

$$\sum_{\mu,v} \mathbf{d}_{ab,\mu}^* \mathbf{d}_{ab,v} \langle i \mid \mathbf{E}_\mu^-(\mathbf{r}, t') \mathbf{E}_v^+(\mathbf{r}, t'') \mid i\rangle ,$$

where, to derive the last result, we used the completeness of the final states and the relation $\langle f \mid \mathbf{E}^+ \mid i\rangle = \langle i \mid \mathbf{E}^- \mid f\rangle^*$.

As the initial state $\mid i\rangle$ is rarely known, we must add over all initial states, we finally get the transition probability

$$p_{b \to a} = \left[\sum_f |\langle a, f \mid U(t, t_o) \mid b, i\rangle|^2 \right]_{av(i)} \tag{6.19}$$

$$= \left(\frac{e}{\hbar}\right)^2 \sum_{\mu,v} \int_{t_o}^t \int_{t_o}^t dt' dt'' \exp\left[i\omega_{ab}(t'' - t')\right] \mathbf{d}_{ab,\mu}^* \mathbf{d}_{ab,v}$$

$$Tr\left[\rho \mathbf{E}_\mu^-(\mathbf{r}, t') \mathbf{E}_v^+(\mathbf{r}, t'')\right]$$

$$= \left(\frac{e}{\hbar}\right)^2 \sum_{\mu,v} \int_{t_o}^t \int_{t_o}^t dt' dt'' \exp\left[i\omega_{ab}(t'' - t')\right] \mathbf{d}_{ab,\mu}^* \mathbf{d}_{ab,v}$$

$$G_{\mu,v}^{(1)}(\mathbf{r}t', \mathbf{r}t'') .$$

So far, we have discussed the case of discrete final electronic states. Perhaps a more realistic model is to consider a continuum of states, characterized by a density of states $g(\omega_{ab})$, so (6.19) should be replaced by

$$p(t) = \int g(\omega_{ab}) p_{b \to a}(t) d\omega_{ab} . \tag{6.20}$$

For a broad-band detector, g is practically a constant and the integral over frequencies gives us

$$\int_{-\infty}^{+\infty} d\omega_{ab} \exp[i\omega_{ab}(t'' - t')] = 2\pi\delta(t'' - t') , \tag{6.21}$$

and

$$p(t) = \sum_{\mu,\nu} S_{\mu\nu} \int_{t_0}^{t} G_{\mu\nu}^{(1)}(\mathbf{r}, t'; \mathbf{r}, t') dt' , \tag{6.22}$$

with

$$S_{\mu,\nu} \equiv 2\pi \left(\frac{e}{\hbar}\right)^2 \sum_a R(a) d_{ab,\mu}^* d_{ab,\nu} \delta(\omega - \omega_{ab}) .$$

We notice that in the last expression, we have averaged over all final states, using $R(a)$ as a weight.

Differentiating (6.22), we get, for the rate of transition probability , or counting rate

$$w^{(1)} = \frac{dp(t)}{dt} = \sum_{\mu,\nu} S_{\mu\nu} G_{\mu\nu}^{(1)}(\mathbf{r}, t; \mathbf{r}, t) . \tag{6.23}$$

If we finally put a polarization filter in front of the counter, then

$$w^{(1)} = s G^{(1)}(\mathbf{r}, t; \mathbf{r}, t) \tag{6.24}$$

with $s = S_{ii}$, i being the direction of the polarization.

The ideal photon counter is, thus, proportional to the first-order correlation function, evaluated at a single point and at a single time. AA real detector, of course, has many atoms.

6.2 The n-Atom Detector

The photon counter we have discussed so far consisted in a single atom. We will see that a many-atom detector will be useful to study higher-order correlation functions of the field.

The type of experiment we are thinking in is the coincidence type, where n-atom detectors are placed in positions $\mathbf{r}_1, \mathbf{r}_2 ... \mathbf{r}_n$. There will be a shutter opening at $t = 0$ and closing at $t = t$. We ask for the probability $p_n(t)$ for each atom to absorb a photon, after a time interval t. It is clear that for this purpose, we must apply n-th order perturbation theory.

The time evolution operator

$$U(t, 0) = \sum_{n=0}^{\infty} \frac{(-i)^n}{n!} \int_0^t dt_1 ... \int_0^t dt_n T\left[H_1(t_1)...H_1(t_n)\right] \tag{6.25}$$

$$\equiv \sum_{n=0}^{\infty} U^{(n)}(t, 0) , \tag{6.26}$$

where T is Dyson's time ordering operator.

Now we assume that there is no direct interaction among the atoms. Then the interaction Hamiltonian, assuming the field linearly polarized in the x-direction, is

$$H_1(t) = -e \sum_j x_j(t)E(r_j, t) \equiv \sum_j H_{1,j}(t) \,. \tag{6.27}$$

Replacing the Hamiltonian in Dyson's expansion, we get different type of terms. The ones with repeated $H_{1,j}(t)$ correspond to atoms that absorbed more than one photon and do not contribute to $p_n(t)$. There are, on the other hand, $n!$ terms in which $H_{1,j}(t)$ appears only once, and they are all equal after time ordering. Thus, save for this factor

$$U^{(n)}(t,0) \propto (-i)^n \int_0^t dt_1 ... \int_0^t dt_n T\left[H_{1,1}(t_1)...H_{1,n}(t_n)\right] \tag{6.28}$$

$$= (ie)^n \int_0^t dt_1 ... \int_0^t dt_n T\left[\prod_j^n x_j(t_j)\mathbf{E}^{(+)}(\mathbf{r}_j, t_j)\right] \,,$$

where only the positive frequency part was considered, in an approximation similar to the one-atom detector case.

Taking the square modulus of the matrix elements of (6.28) between an initial and a final state, summing over those final states and averaging over the initial ones, we readily get for the probability $p_n(t)$:

$$p_n(t) = s^n \int_0^t dt_1' ... \int_0^t dt_n' G^{(n)}(r_1, t_1'...r_n, t_n'; r_n, t_n'...r_1, t_1') \,, \tag{6.29}$$

where all detectors are broad band with sensitivity s. Also

$$G^{(n)}(x_1...x_n; x_{n+1}...x_{2n}) \equiv \tag{6.30}$$

$$Tr\left[\rho E^{(-)}(x_1)...E^{(-)}(x_n)E^{(+)}(x_{n+1})...E^{(+)}(x_{2n})\right] \,,$$

where, again $x_j = (\mathbf{r}_j, t_j)$.

We thus far considered that the n atoms undergoing absorption as a part of a single detector, which in a way is similar to having n detectors, each consisting of a single atom.

Now, if instead, the shutter in each atom is closed at different times, that is the j-th atoms shutter closes at t_j, then instead of (6.29) we get

$$p_n(t_1...t_n) = s^n \int_0^{t_1} dt_1' ... \int_0^{t_n} dt_n' G^{(n)}(\mathbf{r}_1, t_1'...r_n, t_n'; \mathbf{r}_n, t_n'...r_1, t_1') \,, \tag{6.31}$$

and the n-th fold coincidence rate is given by

$$w^{(n)}(t_1...t_n) = \frac{\partial^n}{\partial t_1...\partial t_n} p_n(t_1...t_n) \tag{6.32}$$

$$= s^n G^{(n)}(\mathbf{r}_1, t_1...r_n, t_n; \mathbf{r}_n, t_n...r_1, t_1) \,.$$

Equation (6.32) is telling us that a coincidence experiment with ideal detectors give us a measure of the higher-order correlation functions.

We remark here that the present theory is only approximate to the lowest order. Higher-order corrections are, however, typically, extremely small.

6.3 General Properties of the Correlation Functions

The n-th order correlation function was defined as the expectation value

$$G^{(n)}(x_1...x_n; x_{n+1}...x_{2n}) \equiv Tr[\rho E^{(-)}(x_1)...E^{(-)}(x_n)E^{(+)}(x_{n+1})...E^{(+)}(x_{2n})] \ .$$

As a first property, we notice that if there is an upper bound on the number of photons present in the field, then $G^{(n)}(x_1...x_n; x_{n+1}...x_{2n})$ must vanish identically for n larger than the upper bound M.

To be more specific, if the field density operator is written as

$$\rho = \sum_{n,m} C_{n,m} \mid n \rangle\langle m \mid , \tag{6.33}$$

and if $C_{n,m} = 0$ for n or m larger than M, then

$$E^{(+)}(x_1)...E^{(+)}(x_p)\rho = 0 , \tag{6.34}$$

for $p > M$, simply because the number of times the annihilation operator is applied to the density matrix is larger than the number of photons available in the field. Thus, $G^{(p)} = 0$ for $p > M$.

Another property can be derived from the identity

$$Tr(A^\dagger) = Tr(A)^* \tag{6.35}$$

valid for any linear operator A.

Applying this identity to $G^{(n)}(x_1...x_n; x_{n+1}...x_{2n})$, we get

$$\left[G^{(n)}(x_1...x_n; x_{n+1}...x_{2n}) \right]^* = Tr \left[E^{(-)}(x_{2n})...E^{(-)}(x_{n+1})E^{(+)}(x_n)...E^{(+)}(x_1)\rho \right]$$

$$= Tr \left[\rho E^{(-)}(x_{2n})...E^{(-)}(x_{n+1})E^{(+)}(x_n)...E^{(+)}(x_1) \right] \tag{6.36}$$

$$= \left[G^{(n)}(x_{2n}...x_{n+1}; x_n...x_1) \right] ,$$

where we made use of the Hermitian character of ρ and the invariance of the trace under cyclic permutation.

As a consequence of the commutation properties of $E^{(-)}$ and $E^{(+)}$, we can freely permute the arguments $(x_1, x_2, ...x_n)$ and $(x_{n+1}, x_{n+2}, ...x_{2n})$ without changing $G^{(n)}$, but we cannot interchange any of the first n arguments with any of the remaining n, because the corresponding operators do not commute.

Another set of properties can be derived from the positive definite character of the operator $A^\dagger A$, so that

$$Tr(A^\dagger A) \geq 0 , \tag{6.37}$$

for any linear operator A.

To show the above inequality, we write

$$Tr(\rho A^\dagger A) = \sum_k p_k \langle k \mid A^\dagger A \mid k \rangle \tag{6.38}$$

$$= \sum_{k,m} p_k \langle k \mid A^\dagger \mid m \rangle \langle m \mid A \mid k \rangle$$

$$= \sum_{k,m} p_k \mid \langle m \mid A \mid k \rangle \mid^2 \geq 0 ,$$

because p_k and $\mid \langle m \mid A \mid k \rangle \mid^2 \geq 0$.

There are several interesting cases:

a) $A = E^{(+)}(x_1)$, then applying the inequality (6.37), we get

$$G^{(1)}(x_1, x_1) \geq 0 . \tag{6.39}$$

b) $A = E^{(+)}(x_1)...E^{(+)}(x_n)$,

we get directly

$$G^{(n)}(x_1...x_n; x_n...x_1) \geq 0 . \tag{6.40}$$

c) $A = \sum_{j=1}^n \lambda_j E^{(+)}(x_j)$,

where λ_j is a set of arbitrary complex numbers. In this case, we get

$$\sum_{i,j} \lambda_i^* \lambda_j G^{(1)}(x_i, x_j) \geq 0 , \tag{6.41}$$

thus the set of correlation functions $G^{(1)}(x_i, x_j)$ forms a matrix coefficient for the quadratic form of the λs. Such a matrix has a positive determinant, thus

For $n = 1$ we get (6.39).

For $n = 2$ we get

$$G^{(1)}(x_1, x_1) G^{(1)}(x_2, x_2) \geq \mid G^{(1)}(x_1, x_2) \mid^2 , \tag{6.42}$$

which is a simple generalization of Schwartz's identity.

6.4 Young's Interference and First-Order Correlation

Consider Young's experiment in the Fig. 6.1.

We consider the positive frequency component of the field

$$\mathbf{E}^+(\mathbf{r}, t) = \mathbf{E}_1^+(\mathbf{r}, t) + \mathbf{E}_2^+(\mathbf{r}, t) , \tag{6.43}$$

where $\mathbf{E}_i^+(\mathbf{r}, t)$ is the spherical wave field produced at the pinhole i, observed at the screen 2

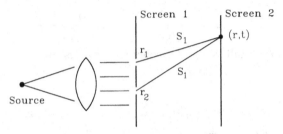

Fig. 6.1. Young's double slit experiment

$$\mathbf{E}_i^+(\mathbf{r},t) = \mathbf{E}_i^+\left(\mathbf{r}_i, t - \frac{s_i}{c}\right)\left(\frac{1}{s_i}\right)\exp i(ks_i - \omega t), \qquad (6.44)$$

and $\mathbf{E}_i^+\left(\mathbf{r_i}, t - \frac{s_i}{c}\right)$ is the field at the pinholes. Denoting

$$x_i = \left(r_i, t - \frac{s_i}{c}\right), \qquad (6.45)$$
$$i = 1, 2$$

and $s_1 \approx s_2 = R$, then

$$\mathbf{E}^+(\mathbf{r},t) = \frac{1}{R}\left[\mathbf{E}_1^+(\mathbf{x}_1) + \mathbf{E}_2^+(\mathbf{x}_2)\right], \qquad (6.46)$$

and one can write the intensity as

$$I = \eta Tr\left[\rho\mathbf{E}^-(\mathbf{r},t)\mathbf{E}^+(\mathbf{r},t)\right] \qquad (6.47)$$
$$= \eta\left[G^{(1)}(x_1,x_1) + G^{(1)}(x_2,x_2) + 2\,\mathrm{Re}\exp\left[ik(s_1 - s_2)\right]G^{(1)}(x_1,x_2)\right],$$

where η scales as $\frac{1}{R^2}$.

The first two terms are the intensities from each pinhole, with the other one blocked and the third term is the interference.

Writing

$$G^{(1)}(x_1,x_2) = |\,G^{(1)}(x_1,x_2)\,|\exp(i\Psi), \qquad (6.48)$$

then

$$I = \eta\left[G^{(1)}(x_1,x_1) + G^{(1)}(x_2,x_2) + 2\cos(\phi)\,|\,G^{(1)}(x_1,x_2)\,|\right] \qquad (6.49)$$

with $\phi = \Psi + k(s_1 - s_2)$.

To have a maximum interference term or maximum visibility, we have to maximize $|\,G^{(1)}(x_1,x_2)\,|$. However, this quantity is limited by the inequality

$$|\,G^{(1)}(x_1,x_2)\,| \leq \sqrt{G^{(1)}(x_1,x_1)G^{(1)}(x_2,x_2)}, \qquad (6.50)$$

which leads us to the definition of the first-order normalized correlation function

$$g^{(1)}(x_1, x_2) = \frac{G^{(1)}(x_1, x_2)}{\sqrt{G^{(1)}(x_1, x_1)G^{(1)}(x_2, x_2)}} \,. \tag{6.51}$$

The condition of full first-order coherence is satisfied if

$$g^{(1)}(x_1, x_2) = \exp(i\Psi) \,, \tag{6.52}$$

or

$$\mid g^{(1)}(x_1, x_2) \mid = 1 \,. \tag{6.53}$$

One usually defines a quantity called the visibility, as

$$v = \frac{I_{\max} - I_{\min}}{I_{\max} + I_{\min}} \tag{6.54}$$

$$= \frac{2 \mid G^{(1)}(x_1, x_2) \mid}{G^{(1)}(x_1, x_1) + G^{(1)}(x_2, x_2)} \,,$$

and for equal intensities in the two pinholes

$$v = \mid g^{(1)}(x_1, x_2) \mid \,. \tag{6.55}$$

For full first-order coherence, $v = 1$ and it corresponds to the maximum visibility.

A more general definition of coherence is related to the factorization of the correlation functions.

For a first-order coherence, the first-order correlation function factorizes

$$G^{(1)}(x_1, x_2) = \varepsilon(x_1)\varepsilon(x_2) \,. \tag{6.56}$$

Obviously, for a state that is an eigenstate of $\mathbf{E}^+(\mathbf{x})$, that is an eigenstate of the annihilation operator, this factorization holds. This is precisely the case of the coherent states.

In a similar way, n-th order optical coherence implies

$$G^{(n)}(x_1, x_2..x_n, ..x_{2n}) = \varepsilon(x_1)\varepsilon(x_2)...\varepsilon(x_{2n}) \,, \tag{6.57}$$

which again is satisfied by coherent states.

To finish this section, we point out that although the first-order correlation function can be evaluated quantum mechanically, the difference between the classical and the quantum predictions in first-order coherence may be difficult to detect. In both cases, $0 \leq \mid g^{(1)}(x_1, x_2) \mid \leq 1$.

In the second-order coherence effects, the differences are more striking.

6.5 Second-Order Correlations: Photon Bunching and Antibunching

The second-order normalized correlation function is defined as [6]

$$g^{(2)}(r_1 t_1, r_2 t_2; r_2 t_2, r_1 t_1) = \frac{\langle \mathbf{E}^-(\mathbf{r}_1, t_1)\mathbf{E}^-(\mathbf{r}_2, t_2)\mathbf{E}^+(\mathbf{r}_2, t_2)\mathbf{E}^+(\mathbf{r}_1, t_1)\rangle}{\langle \mathbf{E}^-(\mathbf{r}_1, t_1)\mathbf{E}^+(\mathbf{r}_1, t_1)\rangle\langle \mathbf{E}^-(\mathbf{r}_2, t_2)\mathbf{E}^+(\mathbf{r}_2, t_2)\rangle} \,. \tag{6.58}$$

In this section, we will consider only parallel light beams (z-direction), so that the space time coordinates $(z_1, t), (z_2, t_2)$ enter in $g^{(2)}$ only as a phase difference:

$$\tau = t_2 - t_1 + \frac{z_1 - z_2}{c} . \tag{6.59}$$

We start the subject with a brief review of some classical aspects.

6.5.1 Classical Second-Order Coherence

We consider a beam of light described by a classical intensity $I_1(t)$, which is time-dependent and averaged over each cycle.

In general, the intensity will show random fluctuations, if one is dealing, for example, with a source of chaotic light.

We will assume that the light sources under study are stationary and ergodic, in such a way that ensemble averages are equal to time averages.

Classically, the second-order correlation function may be defined as

$$g_{11}^{(2)}(t) = \frac{\langle I_1(t) I_1(0) \rangle}{\overline{I_1}^2} , \tag{6.60}$$

where the average is over a long series of pairs of intensity measurements separated by a fixed time t and $\overline{I_1} = \langle I_1 \rangle$ is a time-independent average, due to the stationary assumption.

In a different type of experiment, we may measure intensities at different positions $\mathbf{r}_1, \mathbf{r}_2$, then the relevant second-order correlation function is

$$g_{12}^{(2)}(t) = \frac{\langle I_1(t) I_2(0) \rangle}{\overline{I_1} \overline{I_2}} . \tag{6.61}$$

The classical correlation functions satisfy a series of inequalities

(a) As the intensity is positive

$$g_{12}^{(2)}(t) \geq 0 . \tag{6.62}$$

(b) As $\langle I_1^2 \rangle \geq \overline{I_1}^2$, then

$$g_{11}^{(2)}(0) \geq 1 . \tag{6.63}$$

(c) In a more general case, and according to the Cauchy's inequality

$$\langle I_1^2 \rangle \langle I_2^2 \rangle \geq \langle I_1 I_2 \rangle^2 . \tag{6.64}$$

Fixing the time t, between the two measurements on the two beams, then

$$g_{11}^{(2)}(0) g_{22}^{(2)}(0) \geq \left[g_{12}^{(2)}(t) \right]^2 , \tag{6.65}$$

and for a single beam, $g_{11}^{(2)}(0) = g_{22}^{(2)}(0)$ and

$$g_{11}^{(2)}(0) \geq g_{12}^{(2)}(t) \,. \tag{6.66}$$

In many cases, the fluctuations of the cycle-averaged intensities are too rapid for direct observation, and the measurement reflects some average of the fluctuations over some typical response time of the detector. However, we do have nowadays fast detectors, and let us assume that its response time is much faster than the coherence time of the light, so that effectively, we have instantaneous measurements of the intensity.

If, furthermore, the ergodic hypothesis is satisfied, then the time average may be replaced by statistical averages, denoted by angle brackets.

We take a model of chaotic light emitted by a collision broadened light source.

In this model, the elastic collisions break up the wave radiated by single atoms, in discrete sections, where each section has a constant phase that abruptly ends with a collision.

Suppose that we have light of intensity I_0 from n radiating atoms, the phase of the field emitted from the i-th atom being a random variable ϕ_i. Then, one can write

$$E(t) = E_1(t) + E_2(t) + \ldots + E_n(t)$$
$$= E_0 \{ \exp[i\phi_1(t)] + \exp i [\phi_2(t)] + \ldots + \exp i [\phi_n(t)] \} \,,$$

where each atom has been associated with the same amplitude and field frequency but with phases that are completely independent.

The instantaneous average value of the square of the intensity is

$$\overline{I_1^2(0)} = I_0^2 \overline{| \exp(i\phi_1) + \exp(i\phi_2) + \ldots + \exp(i\phi_n) |^4} \,. \tag{6.67}$$

The only non-zero contributions come from the terms in which each factor is multiplied by its complex conjugate. These are

$$\overline{I_1^2(0)} = I_0^2 \left[\sum_{i=1}^{n} | \exp(i\phi_i) |^4 + \sum_{i \neq j} | 2 \exp i(\phi_i - \phi_j) |^2 \right] \tag{6.68}$$
$$= I_0^2 \left[2n^2 - n \right] \,.$$

If we further average, considering a Poissonian distribution of incoming atoms, with a mean $\overline{n}$, and $\overline{n^2} = \overline{n}^2 + \overline{n}$, then

$$\langle \overline{I_1^2(0)} \rangle_{\text{Poiss}} = I_0^2 (2\overline{n}^2 + \overline{n}) \,. \tag{6.69}$$

Also, as $\langle \overline{I_1(0)} \rangle_{\text{Poiss}} = \overline{n} I_0$, we have

$$g_{11}^{(2)}(0) = 2 + \frac{1}{\overline{n}} \,. \tag{6.70}$$

The standard theory of chaotic light considers a very large number of atoms radiating, that is, the limit $\bar{n} \to \infty$, $g_{11}^{(2)}(0) = 2$.

More generally, one can consider a large number of radiating atoms, and the summation over the phases is treated as a random walk. As a result of such a theory, one gets the probability distribution for the instantaneous intensity I_1

$$P(I) = \frac{1}{\bar{I_1}} \exp -\frac{I_1}{\bar{I_1}} , \tag{6.71}$$

giving $g_{11}^{(2)}(0) = 2$, in agreement with our previous result.

Normally, when one deals with a single beam of light, we skip the lower indices, and for chaotic light, we will just write $g^{(2)}(0) = 2$.

6.5.2 Quantum Theory of Second-Order Coherence

The quantum mechanical normalized second-order correlation function $g^2(\tau)$ is positive, so that the inequality

$$\infty \geq g^{(2)}(\tau) \geq 0 , \tag{6.72}$$

is identical to the classical range.

However, the classical inequalities given in (6.63 and 6.66) are, in general, no longer true.

Even for zero time delay, in the quantum mechanical case, the only true inequality is

$$\infty \geq g^{(2)}(0) \geq 0 . \tag{6.73}$$

and as classically $g_2(0)|_{class} \geq 1$, there is an interesting range:

$$1 > g^{(2)}(0) \geq 0 , \tag{6.74}$$

that is a purely quantum mechanical range.

For a single mode field, the normalized correlation functions become simpler, and one can write

$$g^{(2)}(\tau) = \frac{\langle a^\dagger a^\dagger a a \rangle}{\langle a^\dagger a \rangle^2} , \tag{6.75}$$

which can also be written in terms of the photon-number operator

$$g^{(2)}(\tau) = \frac{\langle n(n-1) \rangle}{\langle n \rangle^2} \tag{6.76}$$

$$= 1 + \frac{\langle (\Delta n)^2 \rangle - \langle n \rangle}{\langle n \rangle^2} .$$

We observe that for a single mode field, there is no time dependence (the τ-dependent phase factor cancels) in $g^{(2)}(\tau)$.

A few simple examples of $g^{(2)}(\tau)$ are

(a) For an $| n \rangle$ state

$$g^{(2)}(\tau) = \frac{(n-1)}{n}, n \geq 2 \tag{6.77}$$

and $g^{(2)}(\tau) = 0$ for $n = 0, 1$.

(b) For a coherent state $| \alpha \rangle$, $\langle (\Delta n)^2 \rangle = \langle n \rangle$ and $g^{(2)}(\tau) = 1$.

It is convenient to define a momentum generating function Q(s) as

$$Q(s) = \sum_{n=0}^{\infty} (1-s)^n P(n) , \tag{6.78}$$

where $P(n)$ is the probability of having n photons in the field. One immediately sees that

$$\langle n \rangle = -\frac{d}{ds}Q(s) \mid_{s=0}, \tag{6.79}$$

$$\langle (\Delta n)^2 \rangle = \langle n^2 \rangle - \langle n \rangle^2$$

$$= (\frac{d}{ds})^2 Q(s) \mid_{s=0} - \langle n \rangle (\langle n \rangle - 1) ,$$

and also

$$g^{(2)} = \frac{1}{\langle n \rangle^2} (\frac{d}{ds})^2 Q(s) \mid_{s=0} . \tag{6.80}$$

In general, light with $g^{(2)} = 1$ is second-order coherent or Poissonian (as in the case of the coherent state), $g^{(2)} > 1$ super-Poissonian and $g^{(2)} < 1$ sub-Poissonian.

(c) Squeezed States.

From Chap. 5, we saw that

$$(\Delta n)^2 = | \alpha |^2 \left[\exp(-2r) \cos^2 \left(\phi - \frac{\theta}{2} \right) + \exp(2r) \sin^2 \left(\phi - \frac{\theta}{2} \right) \right] \tag{6.81}$$

$$+2 \sinh^2 r \cosh^2 r ,$$

with

$$\alpha = | \alpha | \exp i\phi . \tag{6.82}$$

With the above expression and (6.76), one can evaluate $g^{(2)}(0)$.

In the Fig. 6.2, we show the second-order correlation function of the squeezing parameter r, with constant average photon number. We can see that for $\phi = \frac{\theta}{2}$ one can minimize $\langle \Delta n \rangle_{sq}^2$ and we can get sub-Poissonian light. On the other hand, for very many other combinations of the parameters, the light is super-Poissonian.

If $g^{(2)}(\tau) < g^{(2)}(0)$, there is a tendency for photons to arrive in pairs, a situation referred to as *photon bunching*.

The reverse situation $g^{(2)}(\tau) > g^{(2)}(0)$ is called *photon antibunching*, occurring typically , when an atom emits a photon and right after that, there

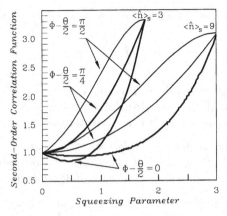

Fig. 6.2. The second-order correlation function of the coherent squeezed state (After [5])

is an anticorrelation for a second photon to be emitted, considering that the atom requires a finite time to go back to its excited state to be ready to emit a second photon.

For very long times, there is no longer any correlation and $g^{(2)}(\tau)\mid_{\tau\to\infty}\to 1$.

Thus, a field with $g^{(2)}(0) < 1$ will always be antibunched over some time scale, which is the quantum mechanical case with no classical analog.

Photon antibunching and sub-Poissonian statistics get sometimes mixed up in the literature, getting the wrong impression that they correspond to the same thing. Although they are related, they are not the same.

Mandel [2] derived a formula, for stationary fields

$$V(n) - \langle n \rangle = \frac{\langle n \rangle^2}{T^2} \int_{-T}^{+T} d\tau (T - |\tau|) \left[g^{(2)}(\tau) - 1 \right] , \qquad (6.83)$$

with $V(n) = \langle n^2 \rangle - \langle n \rangle^2$. When a field has $g^{(2)}(\tau) < 1$ for all τ, then $V(n) - \langle n \rangle < 0$ and exhibits a sub-Poissonian statistics. However, we may have the case $g^{(2)}(\tau) > g^{(2)}(0)$ (antibunching), which exhibits super-Poissonian statistics $(g^{(2)}(\tau), g^{(2)}(0) > 1)$, for some time interval τ.

6.6 Photon Counting

The probability distribution $p(n, t, T)$ of registering n photoelectrons in the interval $t, t + T$ is given by the relation

$$p(n, T) = \int_0^\infty \frac{\left[\alpha \bar{I}(t) T \right]^n}{n!} \exp\left[-\alpha \bar{I}(t) T \right] P\left[\bar{I}(t) \right] d\bar{I}(t) , \qquad (6.84)$$

where α is a quantum efficiency of the detector and $\bar{I}(t)$ is the average intensity

$$\bar{I}(t) = \frac{1}{T} \int_t^{t+T} dt' I(t') . \tag{6.85}$$

$P(\bar{I}(t))$ is the probability density of $\bar{I}(t)$, considered as a random variable. This formula was derived in **Mandel** [3], using classical arguments.

The two basic assumptions in this derivation are

(a) The probability of registering a photoelectric count in a short time interval Δt is linear with Δt and with the instantaneous intensity I(t).

(b) Different photon counts are statistically independent.

However, as the photoelectric effect is a quantum mechanical phenomena, the above assumptions were not completely satisfactory and Mandel et al. (6.84), using first-order perturbation theory, with a Semiclassical model, where the field is classical and the detector quantum mechanical[4].

He considered a model of photodetector that consisted in a group of independent atoms interacting with the radiation field. The result showed that the probability of photoemission is proportional to I(t)

$$P(t)\Delta t = \alpha I(t)\Delta t , \tag{6.86}$$

α, again being the quantum efficiency that depends on the detector parameters.

We also assume that the light falling on the detector is quasi monochromatic and that Δt is much smaller than the coherence time of the light t_c. This coherence time gives the time scale over which intensity changes take place.

From the assumption that different photoelectric emissions are statistically independent events, it follows that the probability to have n photoelectric emissions, in a finite time interval $t, t + T$ is a Poisson distribution

$$p_r(n, t, T) = \frac{[\alpha\bar{I}(t, T)T]^n}{n!} \exp\left[-\alpha\bar{I}(t, T)T\right] . \tag{6.87}$$

To see how (6.87) follows from (6.86) [5], we divide the interval $t, t+T$ in a large number N of subintervals, with $\Delta t = \frac{T}{N}$.

Let z_k be a random variable taking the values 0 or 1, depending respectively, whether or not there has been a photoemission in the interval $t + (k-1)\Delta t, t + (k)\Delta t$, for $k = 1, 2 \ldots N$. The total number of photoemissions is then

$$n = \sum_{k=1}^{N} z_k . \tag{6.88}$$

Now, we define a generating function $G_r(\lambda, t, T)$ as

$$G_r(\lambda, t, T) = \sum_n (1 - \lambda)^n p_r(n, t, T) \, . \tag{6.89}$$

If we assume that all the z_k are independent, we find that

$$\begin{aligned} G_r(\lambda, t, T) &= \langle (1 - \lambda)^n \rangle \\ &= \langle \sqcap_{k=1}^N (1 - \lambda)^{z_k} \rangle \\ &= \sqcap_{k=1}^N \langle (1 - \lambda)^{z_k} \rangle \, . \end{aligned} \tag{6.90}$$

As z_k is either 0 or 1, $(1 - \lambda)^{z_k} = 1 - \lambda z_k$, and therefore

$$\begin{aligned} G_r(\lambda, t, T) &= \sqcap_{k=1}^N [1 - \lambda p(z_k = 1)] \\ &= \sqcap_{k=1}^N [1 - \lambda \alpha \bar{I}(t + k\Delta t)\Delta t] \end{aligned} \tag{6.91}$$

$$\xrightarrow[N \to \infty]{} \exp \left[-\alpha\lambda \int_t^{t+T} I(t')dt' \right] = \exp \left[-\alpha\lambda T \bar{I}(t) \right] \, ,$$

and as

$$p_r(n, t, T) = \frac{(-1)^n}{n!} \frac{\partial^n}{\partial\lambda^n} G_r(\lambda, t, T) |_{\lambda=1} \, , \tag{6.92}$$

we get (6.87).

The probability $p_r(n, t, T)$ represents the distribution of readings of the photon count obtained in a series of experiments, all of them with the same initial time t. Normally in experiments, the situation is rather different. Measurements are not in parallel but in series, that is, one conducts only one counting measurement at a time, followed by successive counting periods, consecutively rather than simultaneously, and the outcome of such a sequence of measurements is an average of $p_r(n, t, T)$ over the starting times t. Thus, we write

$$p(n, T) = \langle \frac{[\alpha\bar{I}(t, T)T]^n}{n!} \exp\left[-\alpha\bar{I}(t, T)T)] \rangle_t \tag{6.93}$$

where $\langle\rangle_t$ means an average over the initial times t.

The photon count distribution can be further simplified, in the case of counting times $T \ll t_c$, in which case the intensity is basically a constant during the counting time, and we can write

$$\bar{I}(t, T) = \bar{I}(t) \, . \tag{6.94}$$

With the usual ergodic hypothesis that time averages can be replaced by ensemble averages, and considering the intensity distribution $P(\bar{I}(t))$, one can write

$$p(n, T) = \int d\bar{I}(t) P(\bar{I}(t)) \frac{[\alpha\bar{I}(t)T]^n}{n!} \exp[-\alpha\bar{I}(t)T] \, ,$$

which is precisely (6.84).

Our discussion shows that the fluctuations in the photoelectric emission may be regarded as due to two causes

(1) Intrinsic fluctuations in the detection process. This is due to the random ejection of photoelectrons, regardless of the intensity fluctuations of the light falling on the detector, resulting in a Poisson distribution.
(2) Fluctuations in the intensity.

As a result, usually *the photoelectron distribution is not a Poissonian.*

Making use of the expression for $p(n, T)$, we can calculate the average number of photon counts, as well as various moments.

For the average

$$\langle n \rangle = \sum_n n p(n, T) \tag{6.95}$$

$$= \int_0^\infty \sum_{n=0}^\infty d\bar{I}(t) P(\bar{I}(t)) n \frac{[\alpha \bar{I}(t) T]^n}{n!} \exp\left[-\alpha \bar{I}(t) T\right]$$

$$= \int_0^\infty d\bar{I}(t) P(\bar{I}(t)) \alpha T \bar{I}(t) \sum_{n=1}^\infty \frac{[\alpha \bar{I}(t) T]^{n-1}}{(n-1)!} \exp\left[-\alpha \bar{I}(t) T\right]$$

$$= \alpha T \langle \bar{I}(t) \rangle .$$

Similarly, one finds

$$\langle n^2 \rangle = \alpha^2 T^2 \langle \bar{I}(t)^2 \rangle + \alpha T \langle \bar{I}(t) \rangle, \tag{6.96}$$

$$\langle (\Delta n)^2 \rangle = \langle n^2 \rangle - \langle n \rangle^2 = \alpha^2 T^2 \left[\langle \bar{I}(t)^2 \rangle - \langle \bar{I}(t) \rangle^2 \right] + \alpha T \langle \bar{I}(t) \rangle .$$

6.6.1 Some Simple Examples

1) Constant intensity
   $I(t) = I$
   In this case, the averaging procedure is unnecessary, and one gets the Poisson distribution

$$p(n, T) = \frac{\langle n \rangle^n}{n!} \exp(-\langle n \rangle) . \tag{6.97}$$

with

$$\langle n \rangle = \alpha T \bar{I} . \tag{6.98}$$

2) $P(\bar{I}(t)) = \dfrac{\exp\left(-\frac{\bar{I}(t)}{\bar{I}}\right)}{\bar{I}}$.

This corresponds to a chaotic light source. The photon count probability is

$$p(n, T) = \frac{1}{\bar{I}} \int d\bar{I}(t) \exp\left(-\frac{\bar{I}(t)}{\bar{I}}\right) \left(\left\{ \frac{[\alpha \bar{I}(t) T]^n}{n!} \right\} \right) \exp\left[-\alpha \bar{I}(t) T\right] \tag{6.99}$$

$$= \frac{\langle n \rangle^n}{(1 + \langle n \rangle)^{n+1}} .$$

Thus, for short counting times, the photon count distribution for chaotic light is similar to the photon distribution of a single-mode thermal light. However, the difference is that this photon count distribution applies to any chaotic light, in general with many modes.

6.6.2 Quantum Mechanical Photon Count Distribution

A fully quantum mechanical description of photon-counting was first derived by Kelly and Kleiner [6, 7]. The result is similar to the classical expression

$$p(n, T) = \langle : \frac{\left[\alpha \bar{I}(T)T \right]^n}{n!} \exp\left[-\alpha \bar{I}(T)T \right] : \rangle . \tag{6.100}$$

The only difference with the classical expression is the :: symbol, indicating normal ordering. In (6.100), $\bar{I}(T)$ is defined as

$$\bar{I}(T) = \frac{1}{T} \int_0^T I(t) dt \tag{6.101}$$

$$= \frac{1}{T} \int_0^T E^{(-)}(\mathbf{r}, t) E^{(+)}(\mathbf{r}, t) dt .$$

For the case of a single radiation mode, the above formula simplifies to

$$p(n, T) = Tr \left[\rho : \frac{(\xi a^\dagger a)^n}{n!} \exp(-\xi a^\dagger a) : \right] , \tag{6.102}$$

with

$$\xi \equiv \alpha \frac{\hbar \omega T}{2 \varepsilon_0 \upsilon} . \tag{6.103}$$

The parameter ξ is usually called the quantum efficiency of the detector. We also notice that in the case of a single mode, $p(n, T)$ is time independent. Expanding the exponential in (6.102), one can write

$$p(m, T) = \sum_n P_n \left(\frac{\xi^m}{m!} \right) \sum_{l=0}^\infty (-1)^l \frac{\xi^l}{l!} \langle n \mid (a^\dagger)^{m+l} (a)^{m+l} \mid n \rangle , \tag{6.104}$$

where $P_n = \langle n \mid \rho \mid n \rangle$.

By making use of (3.27), the above result can be simplified to

$$p(m, T) = \sum_n P_n \left(\frac{\xi^m}{m!} \right) \sum_{l=0}^{n-m} (-1)^l \frac{\xi^l}{l!} \frac{n!}{(n - m - l)!} , \tag{6.105}$$

$$= \sum_{n=m}^\infty P_n \binom{n}{m} \xi^m (1 - \xi)^{n-m} .$$

This form of the photocount distribution is the *Bernoulli Distribution* and has a physical interpretation as follows: the probability of a photon being counted during the period T is the quantum efficiency ξ, thus the probability of counting m out of n photons is proportional to the probability of counting m photons ξ^m times the probability of not counting n-m photons $(1 - \xi)^{n-m}$. Now, the only fixed number in this analysis is m, the number of detected photons after the given interval, so the total probability of counting m-photons has to involve a sum from $n = m$ to ∞, weighted by two factors. The first one is the probability of having n photons to start with P_n, and the second factor is related to the indistinguishability of the photons.

6.6.3 Particular Examples

It is simple to calculate the following cases:
Coherent state,

$$P_n = \frac{(\overline{n})^n}{n!} \exp(-\overline{n}), \tag{6.106}$$

$$p(m, T) = \frac{(\xi \overline{n})^m}{m!} \exp(-\xi \overline{n}) .$$

Filtered single-mode chaotic light,

$$P_n = \frac{(\overline{n})^n}{(1 + \overline{n})^{1+n}} , \tag{6.107}$$

$$p(m, T) = \frac{(\xi \overline{n})^m}{(1 + \xi \overline{n})^{m+1}} .$$

Problems

6.1. Prove (6.81) and (6.82).
6.2. Prove (6.96).

References

1. Glauber, R.J.: Phys. Rev., **130**, 2529 (1963); Glauber, R.J.: Phys. Rev. **131**, 2766 (1963); Glauber, R.J.: Phys. Rev. Lett. **10**, 84, (1963).
2. Mandel, L.: Phys. Rev. Lett., **49**, 136 (1982)
3. Mandel, L.: Proc. Phys. Soc. London, **72**, 1037 (1958)
4. Mandel, L., Sudarshan, E.C.G., Wolf, E.: Proc. Phys. Soc. London, **84**, 435 (1964)
5. Mehta, C.L.: In: Progress in Optics. Vol. 8, 373, Wolf, E. (ed.) North Holland, Amsterdam (1970)
6. Loudon, R.: The Quantum Theory of Light. Clarendon Press, Oxford (1983)
7. Kelley, P.L., Kleiner, W.H.: Phys. Rev. A, **136**, 316 (1964)

Further Reading

• Born, M., Wolf, E.: Principles of Optics. 2rd ed. Pergamon, London (1965)

7. Phase Space Description

In this chapter, we consider an alternative way of studying quantum phenomena, using c-number functions in the phase space.

In general, a full description of the state of the electromagnetic field is given by the density operator. However, there are various ways in which the field can be described by complex functions of α. These methods are of considerable practical interest, because one is dealing with functions rather than operators.

The most common representations in phase space are the P Glauber's distribution, the Q Wigner representations.

The Wigner distribution is the one that resembles more closely to the classical probability. On the other hand, the P distribution originates from representing the density operator as an ensemble of coherent states and the Q representation is a description of the density matrix through its diagonal elements, again, in a coherent state basis.

7.1 Q-Representation: Antinormal Ordering

The definition of the Q distribution function is

$$Q(\alpha, \alpha^*) \equiv \frac{1}{\pi} \langle \alpha \mid \rho \mid \alpha \rangle . \tag{7.1}$$

There are a number of properties associated to the Q-function.

7.1.1 Normalization

$$Tr\{\rho\} = 1 = Tr\left(\frac{1}{\pi} \int d^2\alpha \mid \alpha\rangle\langle \alpha \mid \rho\right)$$
$$= \frac{1}{\pi} \int d^2\alpha \langle \alpha \mid \rho \mid \alpha\rangle ,$$

which implies

$$\int d^2\alpha Q(\alpha, \alpha^*) = 1 . \tag{7.2}$$

7.1.2 Average of Antinormally Ordered Products

We calculate the averages of the type

$$\langle a^r (a^\dagger)^s \rangle = Tr \left[a^r (a^\dagger)^s \rho \right]$$
$$= Tr \left[\frac{1}{\pi} \int d^2\alpha \mid \alpha \rangle \langle \alpha \mid (a^\dagger)^s \rho a^r \right]$$
$$= \frac{1}{\pi} \int d^2\alpha Tr \left[\mid \alpha \rangle \langle \alpha \mid (a^\dagger)^s \rho a^r \right]$$
$$= \frac{1}{\pi} \int d^2\alpha \langle \alpha \mid (a^\dagger)^s \rho a^r \mid \alpha \rangle ,$$

so, we finally write

$$\langle a^r (a^\dagger)^s \rangle = \int d^2\alpha (\alpha^*)^s \alpha^r Q(\alpha, \alpha^*) . \qquad (7.3)$$

7.1.3 Some Examples

Coherent state

$$\rho = \mid \alpha_0 \rangle \langle \alpha_0 \mid ,$$

which gives

$$Q(\alpha, \alpha^*) = \frac{1}{\pi} \mid \langle \alpha_0 \mid \alpha \rangle \mid^2 = \frac{1}{\pi} \exp \left(- \mid \alpha - \alpha_0 \mid^2 \right) . \qquad (7.4)$$

Number state

$$\rho = \mid n \rangle \langle n \mid ,$$

$$Q(\alpha, \alpha^*) = \frac{1}{\pi} \mid \langle \alpha \mid n \rangle \mid^2 = \frac{1}{\pi} \exp(- \mid \alpha \mid^2) \left[\frac{(\mid \alpha \mid^2)^n}{n!} \right] . \qquad (7.5)$$

We notice that the Q function is independent of the phase of α, and the maximum is located at $\mid \alpha \mid^2 = n$.

Thermal state

$$\rho = \left[1 - \exp \left(-\frac{\hbar\omega}{kT} \right) \right] \sum_n \mid n \rangle \langle n \mid \exp \left(-\frac{n\hbar\omega}{kT} \right) , \qquad (7.6)$$

$$Q(\alpha, \alpha^*) = \frac{1 - \exp \left(-\frac{\hbar\omega}{kT} \right)}{\pi} \sum_n \exp \left(-\frac{n\hbar\omega}{kT} \right) \left[\exp \left(- \mid \alpha \mid^2 \right) \right] \frac{\mid \alpha \mid^{2n}}{n!} ,$$

so that

$$Q(\alpha, \alpha^*) = \frac{1 - \exp \left(-\frac{\hbar\omega}{kT} \right)}{\pi} \exp \left\{ - \mid \alpha \mid^2 \left[1 - \exp \left(-\frac{\hbar\omega}{kT} \right) \right] \right\} . \qquad (7.7)$$

It is simple to show that

$$\langle \alpha^p \rangle = \langle \alpha^{*p} \rangle = 0 \,, \tag{7.8}$$

by observing that

$$\langle \alpha^p \rangle = \langle a^p \rangle \propto \langle n \mid a^p \mid n \rangle = 0 \,.$$

Also,

$$\langle aa^\dagger \rangle = \langle \mid \alpha \mid^2 \rangle = \left[1 - \exp\left(-\frac{\hbar\omega}{kT} \right) \right]^{-1} \,. \tag{7.9}$$

In the particular limit $kT \to 0$, we get $\langle aa^\dagger \rangle = \langle \mid \alpha \mid^2 \rangle = 1$, which is the correct answer, because at $T = 0$, $\langle n \rangle = 0$.

7.1.4 The Density Operator in Terms of the Function *Q*

We pose the following question: Is it possible to construct the density matrix, once the *Q* function is known?

Starting from a coherent state

$$\mid \alpha \rangle = \exp\left(\frac{- \mid \alpha \mid^2}{2} \right) \sum_n \frac{\alpha^n}{\sqrt{n!}} \mid n \rangle \,,$$

and using the definition of the *Q* function (7.1), we write

$$Q(\alpha, \alpha^*) \equiv \frac{1}{\pi} \langle \alpha \mid \rho \mid \alpha \rangle \tag{7.10}$$

$$= \frac{\exp(- \mid \alpha \mid^2)}{\pi} \sum_{n,m} \frac{\langle n \mid \rho \mid m \rangle}{\sqrt{n!m!}} \alpha^m \alpha^{*n}$$

$$\equiv \sum_{n,m} Q_{n,m} \alpha^m \alpha^{*n} \,.$$

From (7.10), we get

$$Q(\alpha, \alpha^*) \exp \mid \alpha \mid^2$$

$$= \sum_{n',m',r} \frac{Q_{n',m'} \alpha^{m'+r} \alpha^{*n'+r}}{r!}$$

$$= \frac{1}{\pi} \sum_{n,m} \frac{\langle n \mid \rho \mid m \rangle \alpha^m \alpha^{*n}}{\sqrt{n!m!}} \,.$$

Comparing equal powers in α and α^*, we get

$$\sum_r \frac{Q_{n-r,m-r}}{r!} \pi \sqrt{n!m!} = \langle n \mid \rho \mid m \rangle \,, \tag{7.11}$$

thus answering the question posed at the beginning of this section.

7.2 Characteristic Function

There are three characteristic functions, defined in a normal, antinormal and symmetric (Wigner) way

$$X_N(\eta) = Tr\left[\rho \exp(\eta a^\dagger) \exp(-\eta^* a)\right], \tag{7.12}$$
$$X_A(\eta) = Tr\left[\rho \exp(-\eta^* a) \exp(\eta a^\dagger)\right],$$
$$X_W(\eta) = Tr\left[\rho \exp(\eta a^\dagger - \eta^* a)\right].$$

The antinormal characteristic function is related to the Q distribution:

$$X_A(\eta) = Tr\left[\rho \exp(-\eta^* a) \exp(\eta a^\dagger)\right] \tag{7.13}$$
$$= \frac{1}{\pi} \int d^2\alpha \langle \alpha \mid \exp(\eta a^\dagger)\rho \exp(-\eta^* a) \mid \alpha \rangle$$
$$= \int d^2\alpha \exp(-\eta^*\alpha + \eta\alpha^*)Q(\alpha),$$

thus, the two functions are Fourier transforms of each other, in a two dimensional space.

7.3 P Representation: Normal Ordering

The Glauber-Sudarshan P function is defined as

$$\rho = \int d^2\alpha P(\alpha, \alpha^*) \mid \alpha \rangle\langle \alpha \mid . \tag{7.14}$$

If one allows $P(\alpha, \alpha^*)$ to be singular, this representation always exists for any density operator. However, for certain quantum states, $P(\alpha, \alpha^*)$ may become negative, thus in general, this function cannot be interpreted as a probability density.

7.3.1 Normalization

Starting from

$$Tr(\rho) = 1$$
$$= \int d^2\alpha P(\alpha, \alpha^*) \sum_n \langle n \mid \alpha \rangle\langle \alpha \mid n \rangle$$
$$= \int d^2\alpha P(\alpha, \alpha^*)\langle \alpha \mid \alpha \rangle,$$

we conclude that

$$\int d^2\alpha P(\alpha, \alpha^*) = 1. \tag{7.15}$$

7.3.2 Averages of Normally Ordered Products

Here, we compute the normally ordered averages

$$\langle a^{\dagger r} a^s \rangle = Tr \left(a^{\dagger r} a^s \rho \right) = Tr \left(a^s \rho a^{\dagger r} \right)$$
$$= \int d^2\alpha P(\alpha, \alpha^*) Tr \left(a^s \mid \alpha \rangle \langle \alpha \mid a^{\dagger r} \right) ,$$

so, we get

$$\langle a^{\dagger r} a^s \rangle = \int d^2\alpha P(\alpha, \alpha^*) \alpha^s \alpha^{*r} . \tag{7.16}$$

As we can see, the average of a normally ordered product can be written as a *c*-number integral in the two-dimensional complex plane.

7.3.3 Some Interesting Properties

The *Q* Function is a Gaussian Convolution of *P*

As per the definition of the *Q*-function

$$Q(\alpha) = \frac{1}{\pi} \langle \alpha \mid \rho \mid \alpha \rangle$$
$$= \frac{1}{\pi} \int \langle \alpha \mid \beta \rangle \langle \beta \mid P(\beta) d^2\beta \mid \alpha \rangle ,$$

or

$$Q(\alpha) = \frac{1}{\pi} \int \exp(- \mid \alpha - \beta \mid^2) P(\beta) d^2\beta . \tag{7.17}$$

P is the Fourier Transform of X_N

$$X_N(\eta) = Tr \left[\rho \exp(\eta a^{\dagger}) \exp(-\eta^* a) \right]$$
$$= Tr \left[\int d^2\alpha P(\alpha, \alpha^*) \mid \alpha \rangle \langle \alpha \mid \exp(\eta a^{\dagger}) \exp(-\eta^* a) \right] ,$$

or finally

$$X_N(\eta) = \left[\int d^2\alpha P(\alpha, \alpha^*) \exp(\eta \alpha^* - \eta^* \alpha) \right] . \tag{7.18}$$

7.3.4 Some Examples

Coherent State

$$\rho = \mid \alpha_0 \rangle \langle \alpha_0 \mid , \tag{7.19}$$
$$P(\alpha, \alpha^*) = \delta(\alpha - \alpha_0) .$$

Thermal State

To obtain $P(\alpha, \alpha^*)$, we first calculate X_A and X_N. Because

$$\rho = \left[1 - \exp - \left(\frac{\hbar\omega}{kT}\right)\right] \sum_n |n\rangle\langle n| \exp\left(-\frac{n\hbar\omega}{kT}\right) \, ,$$

one can define

$$s = \left[1 - \exp\left(-\frac{\hbar\omega}{kT}\right)\right] \, , \tag{7.20}$$

$$\eta \equiv x + iy \, ,$$

$$\alpha = r + ik \, ,$$

then

$$X_A(\eta) = \frac{s}{\pi} \int d^2\alpha \exp(\eta\alpha^* - \eta^*\alpha) \exp(-s \mid \alpha \mid^2)$$

$$= \frac{s}{\pi} \int dr \int dk \exp\left[-s\left(r - \frac{iy}{s}\right)^2 - s\left(k + \frac{ix}{s}\right)^2\right] \exp\left(-\frac{x^2 + y^2}{s}\right)$$

or

$$X_A(\eta) = \exp\left(-\frac{\mid \eta \mid^2}{s}\right) \, . \tag{7.21}$$

Now, we proceed to calculate the normally ordered characteristic function. By definition

$$X_N(\eta) = Tr\left[\rho \exp(\eta a^\dagger) \exp(-\eta^* a)\right]$$

$$= Tr\left[\rho \exp(-\eta^* a) \exp(\eta a^\dagger)\right] \exp \mid \eta \mid^2 \, ,$$

so that

$$X_N(\eta) = X_A(\eta) \exp \mid \eta \mid^2 \, . \tag{7.22}$$

so, for the thermal state, we get

$$X_N(\eta) = \exp\left(-\frac{\mid \eta \mid^2 (1 - s)}{s}.\right) \tag{7.23}$$

Finally, we calculate $P(\alpha)$ as the Fourier transform of the normally ordered characteristic function

$$P(\alpha) = \frac{1}{\pi^2} \int d^2\eta \exp(\alpha\eta^* - \alpha^*\eta) \exp\left[-\frac{\mid \eta \mid^2 (1 - s)}{s}\right]$$

$$= \frac{1}{\pi^2} \int dx \int dy \exp\left[2i(kx - ry) - \frac{(1 - s)}{s}(x^2 + y^2)\right]$$

$$= \frac{s}{\pi(1 - s)} \exp\left(-\frac{\mid \alpha \mid^2 s}{1 - s}\right) \, ,$$

or written in a different way

$$P(\alpha) = \frac{1}{\pi} \left[\exp\left(\frac{\hbar\omega}{kT}\right) - 1 \right] \exp\left\{ -\mid\alpha\mid^2 \left[\exp\left(\frac{\hbar\omega}{kT}\right) - 1 \right] \right\} . \qquad (7.24)$$

For this system, P is a well-behaved function, a Gaussian [1].

When T$\rightarrow$ 0, $P(\alpha)$ becomes a very sharp function of α. In the other extreme, that is when T$\rightarrow \infty$

$$P(\alpha) \rightarrow \frac{\hbar\omega}{kT} \exp\left(-\mid\alpha\mid^2 \frac{\hbar\omega}{kT} \right) ,$$

which is the same as the high temperature limit of the Q function and basically corresponds to a classical Boltzmann distribution.

Number state

This is a case where one cannot find a functional form solution for the P function, but only in terms of derivatives of the delta function. The result is (for an $\mid n \rangle$ state)

$$P(\alpha) = \frac{\exp\mid\alpha\mid^2}{n!} \left(\frac{\partial^2}{\partial\alpha\partial\alpha^*} \right)^n \delta^2(\alpha) . \qquad (7.25)$$

For a squeezed state, $P(\alpha)$ is negative.

This was already shown in (5.15).

7.4 The Wigner Distribution: Symmetric Ordering

The first quasi probability distribution was introduced by **Wigner** [2], to study quantum corrections to classical statistical mechanics. We will designate the Wigner distribution by W.

The original idea was to reformulate Schrödinger's equation, and it found many applications in quantum chemistry, statistical mechanics and quantum optics.

In our context, we define the Wigner function $W(\alpha, \alpha^*)$ as the Fourier transform of the symmetric characteristic function X_W

$$W(\alpha, \alpha^*) = \frac{1}{\pi^2} \int d^2\eta \exp(-\eta\alpha^* + \eta^*\alpha) X_W(\eta, \eta^*) . \qquad (7.26)$$

7.4.1 Moments

The moments of $W(\alpha, \alpha^*)$ are equal to the averages of symmetrically ordered products of creation and annihilation operators. These products denoted by $\{a^r a^{\dagger s}\}_{\text{sym}}$ are defined as the expansion coefficient of $\eta^s(-\eta^{*r})$ in $(\eta a^\dagger - \eta^* a)^{r+s}$.

We give here some examples

$$\{a^2 a^{\dagger 2}\}_{\text{sym}} = \frac{1}{6}\left(a^{\dagger 2}a^2 + a^\dagger aa^\dagger a + a^\dagger a^2 a^\dagger + aa^{\dagger 2}a + aa^\dagger aa^\dagger + a^2 a^{\dagger 2} \right),$$

$$\{aa^{\dagger 2}\}_{\text{sym}} = \frac{1}{3}\left(a^{\dagger 2}a + a^\dagger aa^\dagger + aa^{\dagger 2} \right)$$

etc.

With the above definition, one can write

$$\exp(\eta a^\dagger - \eta^* a) = \sum_{r,s} \frac{\eta^s(-\eta^*)^r}{r!s!} \{a^r a^{\dagger s}\}_{\text{sym}} . \tag{7.27}$$

Now, by partial integration of (7.26), we get

$$\int d^2\alpha\, \alpha^r \alpha^{*s} W(\alpha, \alpha^*) = \left(\frac{\partial}{\partial\eta}\right)^s \left(-\frac{\partial}{\partial\eta^*}\right)^r X_W(\eta, \eta^*)\,|_{\eta=0} , \tag{7.28}$$

and making use of (7.12), we readily get

$$\langle\{a^r a^{\dagger s}\}_{\text{sym}}\rangle = \int d^2\alpha\, \alpha^r \alpha^{*s} W(\alpha, \alpha^*) .$$

Problems

7.1 Normal ordering.
   Let
   $$A = \sum_{n,m} A_{nm} a^{\dagger n} a^m$$

   be normally ordered.
   Show that
   $$\langle A \rangle = \frac{1}{\pi} \int P(\alpha) \sum_{n,m} A_{nm} \alpha^{*n} \alpha^m d^2\alpha.$$

7.2. Show that if
   $$A = (a^\dagger a)^2,$$

   then its normal and antinormal ordered version are

   $$\left[(a^\dagger a)^2\right]_{\text{normal}} = a^{\dagger 2}a^2 + (a^\dagger a),$$

   $$\left[(a^\dagger a)^2\right]_{\text{antinormal}} = a^2 a^{\dagger 2} - 3(aa^\dagger) + 1.$$

7.3. Let us assume that an operator A is normally ordered. Then, we substitute

$$a \to z,$$
$$a^\dagger \to z^*$$

and define the result as

$$A^{(n)}(z, z^*).$$

Similarly, if A is antinormally ordered, we define a $A^{(a)}(z, z^*)$. For example, from the previous problem, we can write

$$A^{(n)}(z, z^*) = |z|^4 + |z|^2,$$
$$A^{(a)}(z, z^*) = |z|^4 - 3|z|^2 + 1.$$

Now, we define the $\mathcal{N}$ and $\mathcal{A}$ operators whose effect in a given operator is to rearrange it as to transform it to normal and antinormal form respectively. For example

$$\mathcal{N}[a^{\dagger 2}a^2 + (a^\dagger a)] = a^{\dagger 2}a^2 + (a^\dagger a),$$

$$\mathcal{A}[a^{\dagger 2}a^2 + (a^\dagger a)] = a^2 a^{\dagger 2} + (aa^\dagger).$$

Show that

$$aa^{\dagger n}a^m = \mathcal{N}\left[a + \frac{\partial}{\partial a^\dagger}\right]a^{\dagger n}a^m.$$

Hint: use the commutation relation

$$[a, a^{\dagger n}] = na^{\dagger n-1}.$$

7.4. If $F^{(n)}(a^\dagger, a)$ is a normally ordered operator, show, using the notation of the Prob(7.3), that

$$aF^{(n)}(a^\dagger, a) = \mathcal{N}\left(a + \frac{\partial}{\partial a^\dagger}\right)F^{(n)}(a^\dagger, a),$$

$$F^{(n)}(a^\dagger, a)a^\dagger = \mathcal{N}\left(a^\dagger + \frac{\partial}{\partial a}\right)F^{(n)}(a^\dagger, a).$$

7.5 Show, using the same notation as in the Prob (7.3) and (7.4), that for a product of two normally ordered operators (the product is of course not normally ordered)

$$F^{(n)}(a^\dagger, a)G^{(n)}(a^\dagger, a) = \mathcal{N}F^{(n)}\left(a^\dagger, a + \frac{\partial}{\partial a^\dagger}\right)F^{(n)}(a^\dagger, a).$$

References

1. Hillery, M.O., O'Connell, R.F., Scully, M.O., Wigner, E.P.: Phys. Rep., **106**, 121 (1984)
2. Wigner, E.P.: Phys. Rev., **40**, 749 (1932)

Further Reading

- Loudon, R.: The Quantum Theory of Light. Clarendon Press, Oxford (1983)
- Louisell, W.H.: Quantum Statistical Properties of Radiation. J.Wiley, New York (1973)
- Walls, D.F., Milburn, G.J.: Quantum Optics. Springer Verlag, Berlin (1994)

8. Atom–Field Interaction

In this Chapter, we study the atom–field interaction in the usual and dressed picture. We also address the problems of Rabi oscillations and the collapse and revivals.

8.1 Atom–Field Hamiltonian and the Dipole Approximation

The Hamiltonian for an atom interacting with an electromagnetic field, may be written as

$$H = \frac{1}{2m} \left[\mathbf{p} - e\mathbf{A}(\mathbf{r}, t) \right]^2 + eV(\mathbf{r}) + H_r , \tag{8.1}$$

where $\mathbf{p}$ is the momentum of the electron, $V(\mathbf{r})$ the Coulomb potential, $\mathbf{A}$ the vector potential of the field and H_r the free radiation field.

We now make use of a unitary transformation [1]:

$$| \psi(t) \rangle = \exp \left[\frac{ie\mathbf{r}}{\hbar} \cdot \mathbf{A}(\mathbf{r}, t) \right] | \chi(t) \rangle \equiv U | \psi(t) \rangle , \tag{8.2}$$

so Schrödinger equation can be written as

$$i\hbar \frac{\partial | \psi(t) \rangle}{\partial t} = H | \psi(t) \rangle \tag{8.3}$$

$$i\hbar U \frac{\partial | \chi(t) \rangle}{\partial t} + i\hbar \frac{\partial U}{\partial t} | \chi(t) \rangle = HU | \chi(t) \rangle .$$

Multiplying (8.3) by $U^\dagger = U^{-1}$ from the left, we get

$$i\hbar \frac{\partial | \chi(t) \rangle}{\partial t} = H' | \chi(t) \rangle, \tag{8.4}$$

$$H' \equiv U^\dagger H U - i\hbar U^\dagger \frac{\partial U}{\partial t} . \tag{8.5}$$

The second term in (8.5) can be written as

$$-i\hbar U^\dagger \frac{\partial U}{\partial t} = e\mathbf{r} \cdot \frac{\partial \mathbf{A}}{\partial t} = -e\mathbf{r} \cdot \mathbf{E}(\mathbf{r}, t) , \tag{8.6}$$

where in the last equation we used (3.6, 3.7), and H′ becomes

$$H' = U^\dagger H U - e\mathbf{r} \cdot \mathbf{E}(\mathbf{r}, t) \,. \tag{8.7}$$

We have to calculate $U^\dagger H U$.

$$U^\dagger H U = \frac{1}{2m} U^\dagger \mathbf{p}^2 U - \frac{e}{m}\mathbf{A} \cdot U^\dagger \mathbf{p} U + \frac{e^2}{2m}\mathbf{A}^2 + eV(\mathbf{r}) + H_r \,, \tag{8.8}$$

where we used the fact that only the p-dependent terms are affected by the transformation and that $\sum_{l=1}^{3} [p_l, A_l] = -i\hbar \sum_l \frac{\partial A_l}{\partial x_l} = 0$ in the Coulomb gauge, thus $\mathbf{p} \cdot \mathbf{A} = \mathbf{A} \cdot \mathbf{p}$.

As

$$[p_i, U] = -i\hbar \frac{\partial U}{\partial x_i} = eU \frac{\partial(\mathbf{r} \cdot \mathbf{A})}{\partial x_i} \,, \tag{8.9}$$

therefore

$$-\frac{e}{m}\mathbf{A}\cdot(U^\dagger \mathbf{p} U) = -\frac{e}{m}\mathbf{A} \cdot U^\dagger [U\mathbf{p} + eU\nabla(\mathbf{r} \cdot \mathbf{A})] \tag{8.10}$$

$$= -\frac{e}{m}\mathbf{A} \cdot \mathbf{p} - \frac{e^2}{m}A \cdot \nabla(\mathbf{r} \cdot \mathbf{A}) \,.$$

Next, we calculate

$$\sum_{i=1}^{3} [p_i^2, U] = \sum_i \left(p_i [p_i, U] + [p_i, U] p_i \right) \tag{8.11}$$

$$= \sum_i \left([p_i, [p_i, U]] + 2[p_i, U] p_i \right) \,.$$

Now, from (8.9), one can write

$$[p_i, [p_i, U]] = e\left[p_i, U\frac{\partial(\mathbf{r} \cdot \mathbf{A})}{\partial x_i} \right] = -i e\hbar \frac{\partial}{\partial x_i}\left[U\frac{\partial(\mathbf{r} \cdot \mathbf{A})}{\partial x_i} \right] \tag{8.12}$$

$$= U\left\{ -i e\hbar \frac{\partial^2(\mathbf{r} \cdot \mathbf{A})}{\partial^2 x_i} + e^2 \left[\frac{\partial(\mathbf{r} \cdot \mathbf{A})}{\partial x_i} \right]^2 \right\} \,,$$

so

$$[\mathbf{p}^2, U] = U\left\{ -i e\hbar \nabla^2(\mathbf{r} \cdot \mathbf{A}) + e^2 [\nabla(\mathbf{r} \cdot \mathbf{A})]^2 + 2e\nabla(\mathbf{r} \cdot \mathbf{A}) \cdot \mathbf{p} \right\} \,. \tag{8.13}$$

With (8.9 and 8.13) we get for $U^\dagger H U$.

$$U^\dagger H U. = \frac{e^2}{2m}\mathbf{A}^2 + eV(\mathbf{r}). + H_r + \frac{\mathbf{p}^2}{2m} + \frac{e\hbar}{2im}\nabla^2(\mathbf{r} \cdot \mathbf{A}) + \frac{e^2}{2m}[\nabla(\mathbf{r} \cdot \mathbf{A})]^2 \tag{8.14}$$

$$+ \frac{e}{m}\nabla(\mathbf{r} \cdot \mathbf{A}) \cdot \mathbf{p} - \frac{e}{m}\mathbf{A} \cdot \mathbf{p} - \frac{e^2}{m}\mathbf{A} \cdot [\nabla(\mathbf{r} \cdot \mathbf{A})] \,.$$

Now

$$\frac{\partial}{\partial x_i}(\mathbf{r} \cdot \mathbf{A}) = \mathbf{A}_i + \mathbf{r} \cdot \frac{\partial \mathbf{A}}{\partial x_i}, \tag{8.15}$$

$$\sum_i \left[\frac{\partial}{\partial x_i}(\mathbf{r} \cdot \mathbf{A}) \right]^2 = \mathbf{A}^2 + \sum_{i,j} 2A_i x_j \frac{\partial A_j}{\partial x_i} + \sum_{i,j,k} x_j x_k \frac{\partial A_j}{\partial x_i} \frac{\partial A_k}{\partial x_i},$$

$$\nabla^2(\mathbf{r} \cdot \mathbf{A}) = \sum_{i,j} x_i \frac{\partial^2 \mathbf{A}}{\partial x_j^2} \,.$$

In the last term, we used $\nabla \cdot \mathbf{A} = 0$.
Finally, one can write

$$U^\dagger H U = eV(\mathbf{r}). + \frac{\mathbf{p}^2}{2m} + H_r + \frac{e}{m} \sum x_i \frac{\partial A_i}{\partial x_j} p_j \tag{8.16}$$

$$+ \frac{e\hbar}{2mi} \sum_{i,j} x_i \frac{\partial^2 A_i}{\partial x_j^2} + \frac{e^2}{2m} \sum_{i,j,k} x_i x_j \frac{\partial A_i}{\partial x_k} \frac{\partial A_j}{\partial x_k}$$

$$= H' + e\mathbf{r} \cdot \mathbf{E}.$$

The standard dipole approximation assumes that, in the case of plane waves,

$$\mathbf{A} = \mathbf{A}_o \exp(-i\omega t + i\mathbf{k} \cdot \mathbf{r}) \tag{8.17}$$
$$\approx \mathbf{A}_o \exp(-i\omega t + i\mathbf{k} \cdot \mathbf{r}_0) = \mathbf{A}(\mathbf{r}_0) \,,$$

where $\mathbf{r}_0$ is the position of the atomic nucleus, provided the radiation wavelength is several orders of magnitude larger than the atomic size. In that case, all the derivatives of the vector potential are neglected, and one gets

$$H' = eV(\mathbf{r}). + \frac{\mathbf{p}^2}{2m} + H_r - e\mathbf{r} \cdot \mathbf{E}(\mathbf{r}, t) \tag{8.18}$$

or

$$H' = \hbar \sum_i \omega_i \mid i \rangle \langle i \mid + H_r - e\mathbf{r} \cdot \mathbf{E}(\mathbf{r}, t) \,,$$

where $\mid i \rangle$ are the unperturbed atomic states with eigenenergies $\hbar \omega_i$.

8.2 A Two-Level Atom Interacting with a Single Field Mode

In the case of a two-level atom interacting with one mode of the field, the interaction Hamiltonian in the dipole approximation can be written as:

$$H_1 = -e\mathbf{r} \cdot \mathbf{E}(\mathbf{r}) \,. \tag{8.19}$$

We can assume one mode for instance when we have a high quality electromagnetic cavity. Thus, for this particular case, is more convenient to have stationary rather than traveling waves. The one mode electric field can be written as

$$E(z,t) = \varepsilon(a + a^\dagger) \sin kz , \qquad (8.20)$$

where $\varepsilon \equiv \sqrt{\frac{\hbar\omega}{\epsilon_o v}}$ is the **field per photon**, and the Hamiltonian can be written as

$$H_1 = \hbar g(\sigma_+ + \sigma_-)(a + a^\dagger) , \qquad (8.21)$$

with $g \equiv -\frac{\varepsilon d}{\hbar} \sin kz$, $d = er_{ab} \cdot e$ and the σ are the usual Pauli spin matrices defined as

$$\sigma_+ = \begin{bmatrix} 0 & 1 \\ 0 & 0 \end{bmatrix} , \sigma_- = \begin{bmatrix} 0 & 0 \\ 1 & 0 \end{bmatrix} , \qquad (8.22)$$

$$\sigma_x = \begin{bmatrix} 0 & 1 \\ 1 & 0 \end{bmatrix} , \sigma_y = \begin{bmatrix} 0 & -i \\ i & 0 \end{bmatrix} ,$$

$$\sigma_z = \begin{bmatrix} 1 & 0 \\ 0 & -1 \end{bmatrix} .$$

Same as in the Semiclassical case, we have assumed that the dipole induced by the field has only non-diagonal matrix elements and that the two states considered here have opposite parity so that $r_{aa} = r_{bb} = 0$, thus the **r** operator is proportional to $(\sigma_+ + \sigma_-)$. The complete Hamiltonian is

$$H = \frac{\hbar\omega_{ab}}{2}\sigma_z + \hbar\omega a^\dagger a + \hbar g(\sigma_+ + \sigma_-)(a + a^\dagger) . \qquad (8.23)$$

We notice, in the above expression, that the zero energy level was taken halfway between the two atomic levels, so that the unperturbed atomic energies are $\pm\frac{\hbar\omega_{ab}}{2}$.

The four terms appearing in the interaction part of the Hamiltonian have the following simple interpretation:

$a\sigma_+$: one photon is absorbed and the atom is excited from state b→state a.

$a^\dagger\sigma_-$: emission of a photon and de-excitation of the atom.

These two processes are energy conserving. We will show that for a very weak coupling constant, they vary slowly in time. On the other hand, the terms $a^\dagger\sigma_+$ and $a\sigma_-$ do not conserve the energy. They represent:

$a^\dagger\sigma_+$: one photon is emitted and the atom is excited.

$a\sigma_-$: one photon is absorbed and the atom gets de-excited.

These processes are shown in the Fig. 8.1.

The straight arrows represent atomic levels and the wavy ones photons. Lines towards (away) the interaction point correspond to destruction (creation) of states.

To see the time dependence of all these processes, we go to the interaction picture

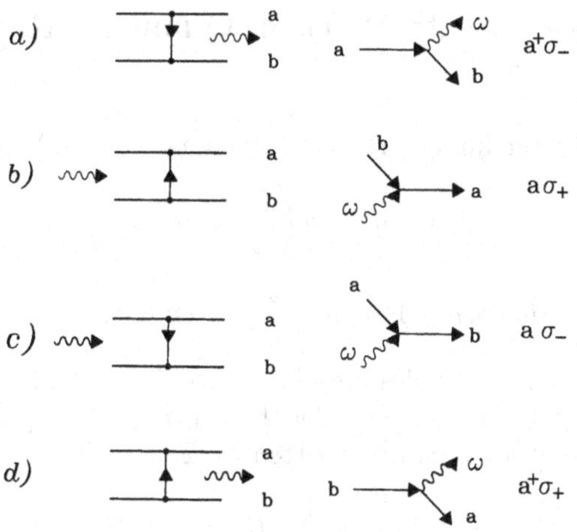

Fig. 8.1. Four processes in the atom–field interaction. The processes **a** and **b** conserve the energy

$$H_1^{(I)} = \hbar g \exp(i\omega t a^\dagger a)(a + a^\dagger)\exp(-i\omega t a^\dagger a) \qquad (8.24)$$

$$\exp\left[it\begin{pmatrix} \frac{\omega_{ab}}{2} & 0 \\ 0 & -\frac{\omega_{ab}}{2} \end{pmatrix}\right]\begin{bmatrix} 0 & 1 \\ 1 & 0 \end{bmatrix}\exp\left[-it\begin{pmatrix} \frac{\omega_{ab}}{2} & 0 \\ 0 & -\frac{\omega_{ab}}{2} \end{pmatrix}\right] .$$

Making use of the properties:

$$\exp(i\omega t a^\dagger a)(a + a^\dagger)\exp(-i\omega t a^\dagger a) = a\exp(-i\omega t) + a^\dagger \exp(i\omega t) ,$$

$$\exp\left[it\begin{pmatrix} \frac{\omega_{ab}}{2} & 0 \\ 0 & -\frac{\omega_{ab}}{2} \end{pmatrix}\right]\begin{bmatrix} 0 & 1 \\ 1 & 0 \end{bmatrix}\exp\left[-it\begin{pmatrix} \frac{\omega_{ab}}{2} & 0 \\ 0 & -\frac{\omega_{ab}}{2} \end{pmatrix}\right]$$

$$= \sigma_+ \exp(i\omega_{ab}t) + \sigma_- \exp(-i\omega_{ab}t) .$$

Replacing the above results into $H_1^{(I)}$, we get

$$H_1^{(I)}(t) = \hbar g\{\sigma_+ a\exp\left[-i(\omega - \omega_{ab})t\right] + \sigma_- a^\dagger \exp\left[i(\omega - \omega_{ab})t\right] \quad (8.25)$$
$$+ \sigma_- a\exp\left[-i(\omega + \omega_{ab})t\right] + a^\dagger \sigma_+ \exp\left[i(\omega + \omega_{ab})t\right]\} .$$

The rotating wave approximation, as in the Semiclassical theory, consists of neglecting the rapidly oscillating terms $\sigma_- a\exp\left[-i(\omega + \omega_{ab})t\right] + a^\dagger \sigma_+ \exp\left[i(\omega + \omega_{ab})t\right]$. Going back to the Schrödinger picture, the Hamiltonian in the dipole and rotating wave approximations is

$$H = H_o + H_1 = \frac{\hbar\omega_{ab}}{2}\sigma_z + \hbar\omega a^\dagger a + \hbar g(a\sigma_+ + \sigma_- a^\dagger) . \qquad (8.26)$$

This is the Jaynes–Cummings Hamiltonian, and it will be very useful in describing various physical effects.

8.3 The Dressed State Picture: Quantum Rabi Oscillations

We begin by taking the unperturbed states $|\,a, n\rangle, |\,b, n+1\rangle$, eigenstates of H_0

$$H_0 \,|\, a, n\rangle = \hbar \left(\frac{\omega_{ab}}{2} + n\omega\right) , \tag{8.27}$$

$$H_0 \,|\, b, n+1\rangle = \hbar \left[-\frac{\omega_{ab}}{2} + (n+1)\omega\right] . \tag{8.28}$$

Now, the interaction couples only $|\,a, n\rangle$ to $|\,b, n+1\rangle$, for each n, and no other states; therefore, we can consider the subspace $\epsilon_n = \{|\,a, n\rangle, |\,b, n+1\rangle\}$ and the total Hamiltonian can be written as [2, 3]

$$H = \sum_n H_n , \tag{8.29}$$

where H_n acts only in ϵ_n and can be written as

$$H_n = \hbar\omega \left(n + \frac{1}{2}\right) \begin{bmatrix} 1 & 0 \\ 0 & 1 \end{bmatrix} + \frac{\hbar}{2} \begin{bmatrix} \delta & 2g\sqrt{n+1} \\ 2g\sqrt{n+1} & -\delta \end{bmatrix} , \tag{8.30}$$

and one can easily diagonalize the above Hamiltonian, getting the following eigenvalues

$$E_{1n} = \hbar\omega \left(n + \frac{1}{2}\right) + \frac{\hbar}{2} R_n , \tag{8.31}$$

$$E_{2n} = \hbar\omega \left(n + \frac{1}{2}\right) - \frac{\hbar}{2} R_n , \tag{8.32}$$

with

$$\delta = \omega_{ab} - \omega, \tag{8.33}$$
$$R_n = \sqrt{\delta^2 + 4g^2(n+1)} .$$

The corresponding eigenstates are

$$|\,1n\rangle = \cos\theta_n \,|\, a, n\rangle + \sin\theta_n \,|\, b, n+1\rangle , \tag{8.34}$$
$$|\,2n\rangle = -\sin\theta_n \,|\, a, n\rangle + \cos\theta_n \,|\, b, n+1\rangle ,$$

with

$$\cos\theta_n = \frac{2g\sqrt{n+1}}{\sqrt{(R_n - \delta)^2 + 4g^2(n+1)}} . \tag{8.35}$$

Also, it is simple to prove that

$$\sin 2\theta_n = \frac{2g\sqrt{n+1}}{R_n} , \tag{8.36}$$

$$\cos 2\theta_n = \frac{\delta}{R_n} ,$$

$$\tan 2\theta_n = \frac{2g\sqrt{n+1}}{\delta} .$$

We can write

$$\begin{bmatrix} |\, 1n\rangle \\ |\, 2n\rangle \end{bmatrix} = \begin{bmatrix} \cos\theta_n & \sin\theta_n \\ -\sin\theta_n & \cos\theta_n \end{bmatrix} \begin{bmatrix} |\, a, n\rangle \\ |\, b, n+1\rangle \end{bmatrix} \tag{8.37}$$

$$\equiv R(\theta_n) \begin{bmatrix} |\, a, n\rangle \\ |\, b, n+1\rangle \end{bmatrix} .$$

These are the dressed states, as opposed to the 'bare' states $|\, a, n\rangle$ and $|\, b, n+1\rangle$.

Diagrammatically, the dressed state eigenvalues are pictured in the Fig. 8.2, where we plot energy versus ω_{ab} for different n-values.

In a particularly simple case $\delta = 0$ and $\sin\theta_n = \cos\theta_n = \frac{1}{\sqrt{2}}$, the states and eigenvalues are

$$|\, 1n\rangle = [|\, a, n\rangle + |\, b, n+1\rangle] \frac{1}{\sqrt{2}} , \tag{8.38}$$

$$|\, 2n\rangle = (-|\, a, n\rangle + |\, b, n+1\rangle) \frac{1}{\sqrt{2}} , \tag{8.39}$$

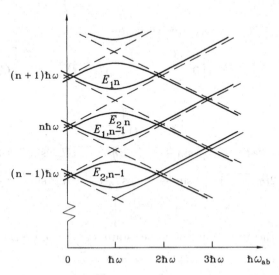

Fig. 8.2. Energy levels of the dressed and bare states versus ω_{ab}, for different n subspaces

and

$$E_{1n} = \hbar\omega\left(n + \frac{1}{2}\right) + \hbar g\sqrt{n+1}\,, \tag{8.40}$$

$$E_{2n} = \hbar\omega\left(n + \frac{1}{2}\right) - \hbar g\sqrt{n+1}\,. \tag{8.41}$$

Finally, we will look, in this section, into the problem of the quantum Rabi oscillations.

If one writes

$$|\,\psi(t)\rangle = \exp\left(-\frac{iHt}{\hbar}\right)|\,\psi(0)\rangle\,, \tag{8.42}$$

and introducing a unit operator in terms of the dressed states, we readily get

$$|\,\psi(t)\rangle = \sum_{n=0}^{\infty}\sum_{j=1}^{2}\exp\left(-\frac{iE_jt}{\hbar}\right)|\,j,n\rangle\langle j,n\,|\,\psi(0)\rangle\,, \tag{8.43}$$

where E_j are the eigenvalues corresponding to the dressed states.

Now, we can write the state vector in terms of both dressed and bare basis, as follows:

$$|\,\psi(t)\rangle = \sum_{n}[C_{an}\,|\,a,n\rangle + C_{b,n+1}\,|\,b,n+1\rangle] \tag{8.44}$$

$$= \sum_{n}[C_{1n}\,|\,1n\rangle + C_{2,n}\,|\,2n\rangle]\,.$$

In the rotating frame at a frequency $\left(n + \frac{1}{2}\right)\omega$, we can write (8.43) as

$$\begin{pmatrix} C_{1n}(t) \\ C_{2n}(t) \end{pmatrix} = \begin{pmatrix} \exp\left(-i\frac{R_n}{2}t\right) & 0 \\ 0 & \exp\left(i\frac{R_n}{2}t\right) \end{pmatrix}\begin{pmatrix} C_{1n}(0) \\ C_{2n}(0) \end{pmatrix} \tag{8.45}$$

and also making use of (8.37), we can write

$$\begin{bmatrix} C_{an}(t) \\ C_{bn+1}(t) \end{bmatrix} = R^{-1}(\theta_n)\begin{bmatrix} \exp\left(-i\frac{R_n}{2}t\right) & 0 \\ 0 & \exp\left(i\frac{R_n}{2}t\right) \end{bmatrix}R(\theta_n)\begin{bmatrix} C_{an}(0) \\ C_{bn+1}(0) \end{bmatrix} \tag{8.46}$$

or

$$\begin{bmatrix} C_{an}(t) \\ C_{bn+1}(t) \end{bmatrix} = \tag{8.47}$$

$$\begin{bmatrix} \cos\frac{R_n}{2}t - i\delta R_n^{-1}\sin\frac{R_n}{2}t & -2ig\sqrt{n+1}R_n^{-1}\sin\frac{R_n}{2}t \\ -2ig\sqrt{n+1}R_n^{-1}\sin\frac{R_n}{2}t & \cos\frac{R_n}{2}t + i\delta R_n^{-1}\sin\frac{R_n}{2}t \end{bmatrix}$$

$$\begin{bmatrix} C_{an}(0) \\ C_{bn+1}(0) \end{bmatrix}$$

If, initially, the atom is in the upper state and $\delta = 0$, we get

$$|\,C_{an}(t)\,|^2 = \cos^2 g\sqrt{n+1}t\,, \tag{8.48}$$

$$|\,C_{bn+1}(t)\,|^2 = \sin^2 g\sqrt{n+1}t\,. \tag{8.49}$$

This is the quantum Rabi oscillation.

8.4 Collapse and Revivals

With the Jaynes–Cummings Hamiltonian, one has, in principle, the time evolution of the system. Of course, there are also the initial conditions [4].

The simplest case is when one knows precisely the energy level of the atom, which is suddenly brought into a cavity with a definite photon number.

Usually in experiments, one can only specify, for example, the statistics of the cavity field, that is the probability of having a given number of photons.

We will deal with the general case. Consider the state vector $\psi(n, t)$ that evolves from an initial state with exactly n photons. When this field has an unknown photon number, specified only by a probability P_m for having m photons, then at time t, the probability of being in a given k state ($k = a, b$) is

$$P_k(t) = \sum_{m=0}^{\infty} P_m \mid \langle m, \psi_k \mid \psi(m, t) \rangle \mid^2 = \sum_{m=0}^{\infty} P_m \mid C_{k,m}(t) \mid^2 , \qquad (8.50)$$

and the photon distribution

$$p_n(t) = \sum m = 0 \qquad (8.51)$$

$$k = a, b^{\infty} P_m \mid \langle n, \psi_k \mid \psi(m, t) \rangle \mid^2 \qquad (8.52)$$

$$= P_n \mid C_{a,n}(t) \mid^2 + P_{n-1} \mid C_{b,n}(t) \mid^2 .$$

One would expect that the superposition of periodic solutions might produce destructive interference, thus a collapse. This indeed occurs [5].

An interesting example is the case of a two-level atom that encounters a cavity at temperature T, whose photon number distribution is the one mode Bose–Einstein distribution, with a probability for m photons, given by

$$P_m(T) = \frac{1}{1 + \overline{n}} \left(\frac{\overline{n}}{1 + \overline{n}} \right)^m , \qquad (8.53)$$

$$\overline{n} = \left[\exp \left(\frac{\hbar \omega}{kT} \right) - 1 \right]^{-1} ,$$

$$\langle \Delta n^2 \rangle = \langle n \rangle^2 + \langle n \rangle .$$

In the Fig. 8.3, we show the population inversion $w(t) = P_a(t) - P_b(t)$ for an initially excited two-level atom interacting with a single-mode thermal field. The time axis has been scaled to an adimensional time $\tau = \frac{gt\sqrt{\overline{n}}}{2}$. The first curve shows the Rabi's oscillations when the atom enters an empty cavity. The subsequent curves show an increasing average photon number. As we can observe, there is a wide range of Rabi frequencies, for which there is no trace of population inversion. The collapse time, for large $\overline{n}$ is of the order of [6]

$$t_c^{-1} \cong g\sqrt{\overline{n}} . \qquad (8.54)$$

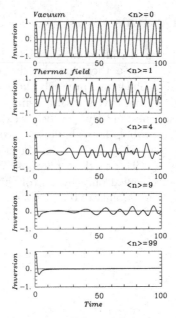

Fig. 8.3. The population inversion for a two-level atom initially in the upper state interacting with a thermal field and mean photon numbers $\langle n \rangle_{\text{th}} = 0, 1, 4, 9, 99$. (after [4])

The time scale of the Fig. 8.3 has been chosen in such a way that in each case, the collapse occurs at $\tau = 2$.

A particular and interesting example where the collapse can be studied in detail is the case when the initial field is a coherent state, where one can study also the **revivals** [7].

In this case, the probability for an excited atom is

$$P_a(t) = \sum_n P_n \mid C_{a,n}(t) \mid^2 = \exp - \mid \alpha \mid^2 \sum_n \frac{\mid \alpha \mid^{2n}}{n!} \cos^2(g\sqrt{n+1}t) , \quad (8.55)$$

and for short times $(gt <<\mid \alpha \mid)$ can be approximated to

$$P_a(t) = \frac{1}{2} + \frac{1}{2}\cos(2gt \mid \alpha \mid) \exp\left[-(gt)^2\right] . \quad (8.56)$$

This case is described in the Fig. 8.4.

As we can see, as the field becomes more intense, the Rabi oscillations persist for longer intervals, until the destructive interference between the oscillations takes over. Basically, the relevant range of Rabi frequencies is $g\sqrt{n+\Delta n} \rightarrow g\sqrt{n-\Delta n}$,

the inverse of which is the collapse time, so

$$t_c^{-1} \approx g , \quad (8.57)$$

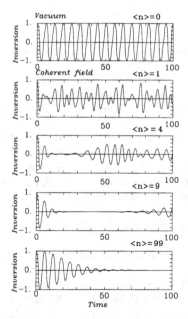

Fig. 8.4. Population inversion of a two-level atom interacting with a single-mode coherent field, for $\langle n \rangle_{\text{th}} = 0, 1, 4, 9, 99$. We observe a collapse and then a revival. (after [4])

which is independent of the average photon number.

Also, the effect of revival is quite remarkable, that is, after a certain time, the Rabi's oscillations reappear. This was studied first by Eberly et al. (1980) and others [8, 9, 10]. The revival time is $t_r = t_c 2\pi \sqrt{\bar{n}}$.

As we can see, the behaviour of a two-level atom interacting with a single electromagnetic mode (coherent) is surprisingly rich. We have Rabi's oscillations that collapse and remain quiescent, revive, then collapse again. The Jaynes–Cummings model, in its linear and non-linear version, has been used extensively in connection with trapped ions with quantized vibrational motion [11, 12, 13, 14, 15].

Problems

8.1. Starting from the Hamiltonian H_n, find the eigenvalues $E_{i,n}$ for $i = 1, 2$, and the dressed states $| 1n \rangle, | 2n \rangle$. That is, verify (8.31–8.35).

8.2. Prove (8.47).

8.3. Show that the summation for the probability of the excited atom (8.55) reduces to the (8.56), when $gt \ll | \alpha |$.

References

1. Power, E.A.: Introductory Quantum Electrodynamics. Longman, London (1964)
2. Cohen-Tannoudji, C., Dupont-Roc, J., Grynberg, G.: Photons and Atoms. Introduction to Quantum Electrodynamics. J.Wiley, New York (1989)
3. Meystre, P., SargentIII, M.: Elements of Quantum Optics. Springer Verlag, Berlin (1991)
4. Shore, B.W., Knight P.L.: J. Mod. Optics, **40**, 1195 (1993)
5. Cummings, F.W.: Phys. Rev., **140**, 1051 (1965)
6. Barnett, S.M., Filipowicz, P., Javanainen, J., Knight, P.L., In: Frontiers in Quantum Optics. Pike, E.R., Sarkar, S., (eds) Adam Hilger, London (1986)
7. Eberly, J.H., Narozny, N.B., Sanchez-Mondragon, J.J.: Phys. Rev. Lett., **44**, 1323 (1980)
8. Eberly, J.H., Narozhny, N.B., Sanchez-Mondragon, J.J.: Phys. Rev. Lett., **44**, 1323 (1980)
9. Narozhny, N.B., Sanchez-Mondragon, J.J., Eberly, J.H.: Phys. Rev. A, **23**, 236 (1981)
10. Yoo, H.I., Eberly, J.H.: Phys. Rep., **118**, 239 (1985)
11. Vogel, W., MatosFilho, R.L.: Phys. Rev. A, **52**, 4214 (1995)
12. Meekhof, D.M., Monroe, C., King, B.E., Itano, W.M., Wineland, D.J.: Phys. Rev. Lett., **76**, 1796 (1996)
13. Gou, S.C., Knight, P.L.: Phys. Rev. A, **54**, 1682 (1996)
14. Wu, Y., Yang, X.: Phys. Rev. Lett., **78**, 30861 (1997)
15. Wallentowitz, S., Vogel, W.: Phys. Rev. A, **58**, 679 (1998)

Further Reading

• Allen, L., Eberly, J.H.: Optical Resonance and Two-Level Atoms. J.Wiley, New York (1975)
• Cohen-Tannoudji, C., Dupont-Roc, J., Grynberg, G.: *Atom-Photon Interaction.* Wiley, New York (1992)
• Gardiner, C.W.: Quantum Noise. Springer Verlag, Berlin (1991)
• Heitler, W.: The Quantum Theory of Radiation. 2nd ed. Fir Lawn, NJ (1944).
• Itzykson, C., Zuber, J.B.: Quantum Field Theory. McGraw-Hill, New York (1980)
• Jaynes, E.T., Cummings, F.W.: Proc, IEEE, **51**, 89 (1963)
• Klauder, J.R., Sudarshan, E.C.G.: Fundamentals of Quantum Optics. W.A.Benjamin, New York (1970)
• Knight, P.L., Milonni, P.W.: Phys. Rep., **66**, 21 (1980)
• Louisell, W.H., Quantum Statistical Properties of Radiation. J.Wiley, New York (1973)
• Loudon, R.: The Quantum Theory of Light. Clarendon Press, Oxford (1983)
• Mandel, L., Wolf, E.: Optical Coherence and Quantum Optics. Cambridge Univ. Press, Cambridge (1995)
• Milonni, P.W.: The Quantum Vacuum: An Introduction to Quantum Electrodynamics. Academic Press, New York (1994)

- Nussenzveig, H.M.: Introduction to Quantum Optics. Gordon and Breach, London (1973)
- Perina, J.: Quantum Statistics of Linear and Nonlinear Phenomena. Reidel, Dordrecht (1984)
- SargentIII, M., Scully, M.O., Lamb, W.E.: Laser Physics. AddisonWesley, Mass. USA (1974)
- Scully, M.O., Zubairy, M.S.: Quantum Optics. Cambridge University Press, Cambridge (1997)

- Shore, B.W.: The Theory of Coherent Atomic Excitation. Vols. 1 and 2, J.Wiley, New York (1990)
- Stenholm, S.: Foundations of Laser Spectroscopy. J.Wiley, New York (1984)
- Stenholm, S.: Phys. Rep. **6**, 1 (1973)
- Vogel, W.: Welsch, D.G.: Lectures on Quantum Optics. Akademie Verlag, Berlin (1994)
- Walls, D.F., Milburn, G.J.: Quantum Optics. Springer Verlag, Berlin (1994)

9. System–Reservoir Interactions

In this chapter, we learn how to introduce losses in our optical systems.

Losses play an important role in physics, and, in general, they cannot be avoided. For example, the decay of an atom can be described as a small or relevant system (the atom) interacting with a large reservoir consisting in an infinite number of harmonic oscillators or electromagnetic modes. It appears quite surprising that starting from a time reversible dynamics, one ends up in an irreversible situation, such as the natural decay of an excited atom. Of course, as we shall see, this is closely related to the type of approximation (Markov) used.

Here we present a quantum theory of damping where the system consists in a single harmonic oscillator coupled to a reservoir of a large number of oscillators. Using the density matrix approach, we eliminate the reservoir variables, obtaining a differential equation for the reduced density matrix in the Schrödinger picture. We also study the problem in the Heisenberg picture, thus introducing the concepts of Langevin equations and noise operators [1, 2].

9.1 Quantum Theory of Damping

The main purpose, in this section, is to study the fluctuations and relaxations of quantum systems. We will find that the expectation values of some relevant operators do relax as their classical counterparts. Also, and very importantly, relaxation phenomena is always accompanied by statistical fluctuations. In other words, a pure state does not relax and a system connected to a bath or reservoir, even if starting from a pure state, will always become a mixed state.

The method that we are going to use is the following one. Starting from Liouville's equation, under a certain approximation scheme, and tracing over the reservoir variables, we end up with a differential equation for a reduced density matrix, the so-called master equation, which is still an operator equation. Then, we use the phase space techniques learned in the Chap. 7, to get a c number differential equation, for example, for the Glauber P distribution. This is the Fokker–Planck Equation.

Consider a harmonic oscillator

$$H_{0,A} = \hbar \omega a^\dagger a , \tag{9.1}$$

$$[a, a^\dagger] = 1 ,$$

that we define as the **system**, and a set of harmonic oscillators:

$$H_{0,B} = \sum_j \hbar \omega_j b_j^\dagger b_j ,$$

$$\left[b_j, b_k^\dagger \right] = \delta_{jk} ,$$

which is the **bath or reservoir**.

We further assume a very general type of system–reservoir coupling of the type XX_j, which contains terms such as $ab_j^\dagger, a^\dagger b_j, a^\dagger b_j^\dagger, ab_j$, and using the rotating-wave approximation arguments, already discussed previously, we only keep the counter-rotating terms. Therefore, the complete Hamiltonian reads

$$H = H_0 + H_1 = \hbar \omega a^\dagger a + \sum_j \hbar \omega_j b_j^\dagger b_j \tag{9.2}$$

$$+ \sum_j g_j(a^\dagger b_j + ab_j^\dagger) ,$$

where the g_j are taken as reals.

Now, we define ρ_{AB} as the density matrix of the complete system, whereas $\rho_A = Tr_B(\rho_{AB}) = \sum_B \langle B \mid \rho_{AB} \mid B \rangle$ and $\rho_B = Tr_A(\rho_{AB}) = \sum_A \langle A \mid \rho_{AB} \mid A \rangle$ are the reduced density matrices for the A and B system, respectively, and obtained simply by tracing over the other variable.

The Liouville equation for the complete system is

$$i\hbar \frac{d\rho_{AB}}{dt} = [H, \rho_{AB}] . \tag{9.3}$$

It is convenient to work in the interaction picture. The density matrix is

$$\widetilde{\rho_{AB}} = \exp\left(\frac{i}{\hbar} H_0 t\right) \rho_{AB} \exp\left(-\frac{i}{\hbar} H_0 t\right) ,$$

and differentiating with respect to time, we get

$$\frac{d\widetilde{\rho_{AB}}}{dt} = \frac{i}{\hbar}[H_0, \widetilde{\rho_{AB}}] + \exp\left(\frac{i}{\hbar} H_0 t\right) \frac{\partial \rho_{AB}}{\partial t} \exp\left(-\frac{i}{\hbar} H_0 t\right) \tag{9.4}$$

$$= \frac{i}{\hbar}\left\{ \left[H_0 - \tilde{H}, \widetilde{\rho_{AB}} \right] \right\} = -\frac{i}{\hbar}\left[\tilde{H}_1, \widetilde{\rho_{AB}} \right] ,$$

where

$$\tilde{H}_1(t) = \exp\left(\frac{i}{\hbar} H_0 t\right) H_1 \exp\left(-\frac{i}{\hbar} H_0 t\right)$$

$$= \exp\left(i\omega t a^\dagger a + i\sum_j \omega_j t b_j^\dagger b_j\right)$$

$$\times \sum_j \hbar g_j (a^\dagger b_j + a b_j^\dagger) \exp\left(-i\omega t a^\dagger a - i\sum_j \omega_j t b_j^\dagger b_j\right)$$

$$= \sum_j \hbar g_j \left\{a^\dagger b_j \exp\left[i(\omega - \omega_j)t\right] + a b_j^\dagger \exp\left[-i(\omega - \omega_j)t\right]\right\}$$

$$= \hbar(G(t)a^\dagger + G^\dagger(t)a) ,$$

where

$$G(t) \equiv \sum_j g_j b_j \exp\left[i(\omega - \omega_j)t\right] . \tag{9.5}$$

Next, we formally integrate the Liouville equation

$$\widetilde{\rho_{AB}}(t) = \widetilde{\rho_{AB}}(0) + \frac{1}{i\hbar}\int_0^t \left[\widetilde{H}_1(t'), \widetilde{\rho_{AB}}(t')\right] dt' \tag{9.6}$$

and substituting back in the (16.26), we get

$$\frac{d\widetilde{\rho_{AB}}}{dt} = \frac{1}{i\hbar}[H_1, \widetilde{\rho_{AB}}(0)] - \frac{1}{\hbar^2}\int_0^t \left[\widetilde{H}_1(t), \left[\widetilde{H}_1(t'), \widetilde{\rho_{AB}}(t')\right]\right] dt' . \tag{9.7}$$

We trace over the reservoir variables to get

$$\frac{d\widetilde{\rho_A}}{dt} = -\frac{1}{\hbar^2}\int_0^t Tr_B \left[\widetilde{H}_1(t), \left[\widetilde{H}_1(t'), \widetilde{\rho_{AB}}(t')\right]\right] dt' . \tag{9.8}$$

In the last step, we assumed that

$$Tr_B \left[\widetilde{H}_1(t), \widetilde{\rho_{AB}}(0)\right] = 0 \tag{9.9}$$

To justify (9.9), assume that at $t = 0$ there is no correlation between the system and the bath, or $\widetilde{\rho_{AB}}(0) = \widetilde{\rho_A}(0) \otimes \widetilde{\rho_B}(0)$, where

$$\widetilde{\rho_B}(0) = \frac{\prod_j \exp\left(-\frac{\hbar\omega_j b_j^\dagger b_j}{K_B T}\right)}{Tr_B \prod_j \exp\left(-\frac{\hbar\omega_j b_j^\dagger b_j}{K_B T}\right)} . \tag{9.10}$$

Now, $\widetilde{H}_1(t)$ contains linear terms in b_j and $b_j^\dagger$ and $Tr_{B_j} b_j$ $\exp\left(-\frac{\hbar\omega_j b_j^\dagger b_j}{K_B T}\right) = 0$.

The final step to find the master equation for the damped harmonic oscillator is to evaluate the double commutator appearing in (9.8). Also, a fundamental step is to assume that $\widetilde{\rho_{AB}}(t) = \widetilde{\rho_A}(t) \otimes \widetilde{\rho_B}(0)$, which is the Markovian assumption .

After a straightforward calculation, one finds

$$\frac{d\widetilde{\rho_A}}{dt} = -i\Delta\omega \left[a^\dagger a, \widetilde{\rho_A}(t)\right] + A\left[a, \widetilde{\rho_A}(t)a^\dagger\right] + A\left[a\widetilde{\rho_A}(t), a^\dagger\right] \quad (9.11)$$
$$+ B\left[a^\dagger, \widetilde{\rho_A}(t)a\right] + B\left[a^\dagger\widetilde{\rho_A}(t), a\right] .$$

where to derive (9.11), we used the following properties:

$$Tr_B(b_j^\dagger b_k \widetilde{\rho_B}(0)) = \delta_{jk}\langle n_j\rangle, \quad (9.12)$$
$$Tr_B(b_j b_j \widetilde{\rho_B}(0)) = 0,$$

$$\Delta\omega = P\int_0^\infty \frac{g(\omega_j)^2 D(\omega_j)}{\omega - \omega_j} d\omega_j ,$$

and $D(\omega)$ is the density function that converts $\sum_j \to \int D(\omega_j)d\omega_j$, and

$$\int_0^t dt' \exp \pm i(\omega - \omega_j)t \approx \pi\delta(\omega - \omega_j) \pm P\left(\frac{1}{\omega - \omega_j}\right) , \quad (9.13)$$

$$A \equiv \pi g(\omega)^2 D(\omega)(1 + \langle n(\omega)\rangle) , \quad (9.14)$$
$$B = \pi g(\omega)^2 D(\omega)\langle n(\omega)\rangle .$$

A more standard form of the master equation (9.11) is

$$\frac{d\widetilde{\rho_A}}{dt} = -i\Delta\omega \left[a^\dagger a, \widetilde{\rho_A}(t)\right] \quad (9.15)$$
$$-\frac{\gamma}{2}(1 + \langle n(\omega)\rangle)(\widetilde{\rho_A}(t)a^\dagger a + a^\dagger a\widetilde{\rho_A}(t) - 2a\widetilde{\rho_A}(t)a^\dagger)$$
$$-\frac{\gamma}{2}\langle n(\omega)\rangle(\widetilde{\rho_A}(t)aa^\dagger + aa^\dagger\widetilde{\rho_A}(t) - 2a^\dagger\widetilde{\rho_A}(t)a) .$$

For the particular case $T = 0$ or $\langle n(\omega)\rangle = 0$ and $\Delta\omega \approx 0$, we get the simpler version of (9.15):

$$\frac{d\widetilde{\rho_A}}{dt} = -\frac{\gamma}{2}(\widetilde{\rho_A}(t)a^\dagger a + a^\dagger a\widetilde{\rho_A}(t) - 2a\widetilde{\rho_A}(t)a^\dagger) \quad (9.16)$$
$$\gamma \equiv 2(A - B) = 2\pi g(\omega)^2 D(\omega) .$$

9.2 General Properties

Equation (9.11) is called the **master equation for the Damped Harmonic Oscillator** and has the following properties:

1) Hermiticity. It is simple to verify that the Hermitian conjugate of the master equation gives back the same equation.
2) Normalization. It is not obvious that after tracing over the bath variables and making the Markov approximation, ρ_A is still normalized. However, it is quite simple to prove that if $Tr_A \rho_A(0) = 1$, then $Tr_A \rho_A(t) = 1$, for all times.

9.3 Expectation Values of Relevant Physical Quantities

When calculating, for example $\langle a^\dagger \rangle(t)$, this can be done in any picture, because they all give the same answer. However, we must be careful to calculate both $a^\dagger$ and ρ_A in the same picture

$$\langle a^\dagger \rangle(t) = Tr_A(\widetilde{\rho_A}(t)\widetilde{a^\dagger}(t)) , \qquad (9.17)$$

and

$$\widetilde{a^\dagger}(t) = \exp(i\omega a^\dagger a t)a^\dagger \exp(-i\omega a^\dagger a t) = a^\dagger \exp(i\omega t) .$$

Now we differentiate (9.17) with respect to time. We readily get

$$\frac{d}{dt}\langle a^\dagger \rangle(t) = \frac{d}{dt}\left\{ Tr_A \left[\widetilde{\rho_A}(t)a^\dagger \right] \exp(i\omega t) \right\} \qquad (9.18)$$

$$= Tr_A \left[\frac{\partial \widetilde{\rho_A}(t)}{\partial t}a^\dagger \exp(i\omega t) + i\omega Tr_A(\widetilde{\rho_A}(t)a^\dagger \exp(i\omega t)) \right] ,$$

and making use of the master equation (9.11), it is simple to obtain

$$\frac{d}{dt}\langle a^\dagger \rangle(t) = i(\omega + \Delta\omega)\langle a^\dagger \rangle(t) - (A - B)\langle a^\dagger \rangle(t) . \qquad (9.19)$$

The solution of (9.19) is

$$\langle a^\dagger \rangle(t) = \langle a^\dagger \rangle(0) \exp\left[i(\omega + \Delta\omega)t \right] \exp\left(-\frac{\gamma}{2}t \right) , \qquad (9.20)$$

$$\gamma \equiv 2(A - B) = 2\pi g(\omega)^2 D(\omega) .$$

Using the same procedure described above, one can also find $\langle a^\dagger a \rangle(t)$. The differential equation and solution are

$$\frac{d}{dt}\langle a^\dagger a \rangle(t) = -\gamma(\langle a^\dagger a \rangle(t) - \langle n(\omega) \rangle) , \qquad (9.21)$$

$$\langle a^\dagger a \rangle(t) = \langle a^\dagger a \rangle(0) \exp(-\gamma t) + \langle n(\omega) \rangle \left[1 - \exp(-\gamma t) \right] .$$

An interesting property is that $\langle a^\dagger a \rangle(t) \mid_{t \to \infty} \to \langle n(\omega) \rangle$, which has a simple interpretation. After a long time, the oscillator in contact with a heat bath gets thermalized, with the same average photon number as the thermal average, at the oscillator's frequency.

Once we know $\langle a^\dagger \rangle(t)$ and $\langle a \rangle(t)$, we can calculate the average position and momentum of the oscillator. The results are

$$\langle q \rangle(t) = \left(\frac{\hbar}{2\omega} \right)^{\frac{1}{2}} \left[\langle a \rangle(t) + \langle a^\dagger \rangle(t) \right] , \tag{9.22}$$

$$\langle p \rangle(t) = i \left(\frac{\hbar\omega}{2} \right)^{\frac{1}{2}} \left[-\langle a \rangle(t) + \langle a^\dagger \rangle(t) \right] ,$$

which can be written as

$$\langle q \rangle(t) = \exp\left(-\frac{\gamma}{2}t \right) \left[\langle q \rangle(0) \cos \omega t + \frac{\langle p \rangle(0)}{\omega} \sin \omega t \right] , \tag{9.23}$$

$$\langle p \rangle(t) = \exp\left(-\frac{\gamma}{2}t \right) \left[-\omega \langle q \rangle(0) \sin \omega t + \langle p \rangle(0) \cos \omega t \right] , \Delta\omega = 0 .$$

Finally, if we assume an initial minimum uncertainty state (MUS), it is simple to show that

$$\Delta p \Delta q = \frac{\hbar}{2} \left\{ 1 + 2\langle n(\omega) \rangle \left[1 - \exp(-\gamma t) \right] \right\} . \tag{9.24}$$

The result given by (9.24) shows again that for t→ ∞, $\Delta p \Delta q \rightarrow \frac{\hbar}{2} \left[1 + 2\langle n(\omega) \rangle \right]$, which is the uncertainty product for a thermal photon at frequency ω.

We notice here that the particular example of the damped harmonic oscillator is a simple one, in the sense that the expectation values we found are only coupled with moments of the same order and not higher. This is not generally the case, where the moment equations are coupled with higher orders, and one has to use some kind of approximate truncation scheme.

9.4 Time Evolution of the Density Matrix Elements

We are interested in the time evolution of $(\widetilde{\rho_A}(t))_{nn} \equiv p_n(t)$.

If we assume $\Delta\omega = 0$ and take the matrix elements of (9.11), we readily get

$$\frac{dp_n}{dt} = 2A(n+1)p_{n+1} + 2Bnp_{n-1} - p_n[2An + 2B(n+1)] . \tag{9.25}$$

Each term in the r.h.s. of (9.25) has an interpretation in terms of energy transitions, as described in the Fig. 9.1, where the arrows arriving to one of the energy levels increase $\frac{dp_n}{dt}$ and the ones leaving the level decrease the rate.

A common technique to solve this type of difference-differential equation is that of the generating function. We define

$$Q(x,t) = \sum_{n=0}^{\infty} (1 - x)^n p_n(t) , \tag{9.26}$$

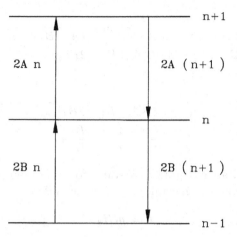

Fig. 9.1. Graphical representation of the time evolution equation for p_n, where each arrow pointing towards the n-th level increases p_n and viceversa

where p_n can be easily calculated as

$$p_n = \frac{(-1)^n}{n!} \frac{\partial^n}{\partial x^n} Q(x,t) \mid_{x=1} . \tag{9.27}$$

Multiplying (9.25) by $(1-x)^n$ and summing over n, we get

$$\frac{\partial Q}{\partial t} = -2BxQ - 2(A-B)x\frac{\partial Q}{\partial x} - 2Bx^2\frac{\partial Q}{\partial x} . \tag{9.28}$$

In deriving (9.28), we made use of relations such as

$$\sum p_n(n+1)(1-x)^n = -\frac{\partial Q}{\partial x} , \tag{9.29}$$

$$\sum p_n(n)(1-x)^n = -(1-x)^n\frac{\partial Q}{\partial x}, \text{etc} .$$

To solve (9.28), we make use of the method of characteristics. [3] (See Appendix B)

Equation (9.28) can be conveniently written as

$$\frac{dt}{1} = \frac{dx}{2\left[(A-B)+Bx\right]x} , \tag{9.30}$$

$$\frac{dx}{2\left[(A-B)+Bx\right]} = -\frac{dQ}{2BQ} .$$

The simultaneous solution of the system of differential equations is:

$$t + \frac{\ln H}{2(A - B)} = \frac{1}{2(A - B)} \ln \left[\frac{x}{2(A - B) + 2Bx} \right] , \tag{9.31}$$

$$- \ln Q + \ln K = \ln \left[2(A - B) + 2Bx \right] ,$$

or explicitly

$$K = Q \left[2(A - B) + 2Bx \right] , \tag{9.32}$$

$$H = \frac{x \exp \left[-2(A - B)t \right]}{2(A - B) + 2Bx} \tag{9.33}$$

where K and H are integration constants to be determined from the initial conditions, satisfying the general integral of the differential equation

$$K = g(H) , \tag{9.34}$$

g being an arbitrary function.

Now we consider the initial state of the oscillator to be a coherent state, that is

$$p_n(0) = \frac{\overline{n}^n}{n!} \exp(-\overline{n}) , \tag{9.35}$$

so that

$$Q(x, 0) = \sum_{n=0}^{\infty} (1 - x)^n \frac{\overline{n}^n}{n!} \exp -\overline{n} = \exp -\overline{n}x . \tag{9.36}$$

Making use of (9.32, 9.33, 9.34), one can write

$$Q(x, t) = \frac{1}{2(A - B) + 2Bx} g \left(\frac{x \exp \left[-2(A - B)t \right]}{2(A - B) + 2Bx} \right) , \tag{9.37}$$

and making use of the initial condition (9.36), we obtain

$$Q(x, t) = \frac{1}{1 + \frac{B}{A-B} x \left\{ 1 - \exp \left[-2(A - B)t \right] \right\}} \tag{9.38}$$

$$\exp \left[-x \frac{\overline{n} \exp \left[-2(A - B)t \right]}{1 + \frac{B}{A-B} x \left\{ 1 - \exp \left[-2(A - B)t \right] \right\}} \right] \tag{9.39}$$

In the limit $t \to \infty, Q(x, \infty) = \frac{1}{1 + \frac{B}{A-B} x} = \frac{1}{1 + x \langle n(\omega) \rangle} = \sum_{n=0}^{\infty} (1 - x)^n$ $p_n(\infty)$, and with the use of (9.27), we get

$$p_n(\infty) = \frac{\langle n(\omega) \rangle^n}{[1 + \langle n(\omega) \rangle]^{n+1}} . \tag{9.40}$$

As we can see, the oscillator, initially in a coherent state get's thermalized when $t \to \infty$, or

$$| \alpha \rangle \Rightarrow \textit{one more element of the reservoir}$$

9.5 The Glauber–Sudarshan Representation, and the Fokker–Planck Equation

It is convenient, for the purposes of this section, to define the **Bargmann** [4] states as

$$\| \alpha \rangle = \exp\left(\frac{|\alpha|^2}{2}\right) |\alpha\rangle . \tag{9.41}$$

It is simple to show from the definition of the coherent states that

$$a^\dagger \| \alpha \rangle = \frac{\partial}{\partial \alpha} \| \alpha \rangle, \tag{9.42}$$

$$\langle \alpha \| a = \frac{\partial}{\partial \alpha^*} \langle \alpha \| .$$

We now write the density operator in the P-representation

$$\rho = \int d^2\alpha \, \| \alpha \rangle \langle \alpha \| \exp(-|\alpha|^2) P(\alpha) , \tag{9.43}$$

then

$$a^\dagger \rho = \int d^2\alpha \frac{\partial}{\partial \alpha} (\| \alpha \rangle) \langle \alpha \| \exp(-|\alpha|^2) P(\alpha), \tag{9.44}$$

$$= \int d^2\alpha (\| \alpha \rangle \langle \alpha \|) \exp(-|\alpha|^2)(\alpha^* - \frac{\partial}{\partial \alpha}) P(\alpha) .$$

Thus, we have the following correspondence

$$a^\dagger \rho \rightarrow \left(\alpha^* - \frac{\partial}{\partial \alpha}\right) P(\alpha) , \tag{9.45}$$

$$\rho a \rightarrow \left(\alpha - \frac{\partial}{\partial \alpha^*}\right) P(\alpha) ,$$

$$\rho a^\dagger \rightarrow \alpha^* P(\alpha),$$

$$a\rho \rightarrow \alpha P(\alpha) ,$$

$$a^\dagger a\rho \rightarrow \left(\alpha^* - \frac{\partial}{\partial \alpha}\right) \alpha P(\alpha) ,$$

$$a a^\dagger \rho \rightarrow \alpha \left(\alpha^* - \frac{\partial}{\partial \alpha}\right) P(\alpha)$$

$$\rho a^\dagger a \rightarrow \left(\alpha - \frac{\partial}{\partial \alpha^*}\right) \alpha^* P(\alpha) ,$$

$$\rho a a^\dagger \rightarrow \alpha^* \left(\alpha - \frac{\partial}{\partial \alpha^*}\right) P(\alpha) .$$

If we now apply the above rules to the master equation 9.15, for $\Delta\omega = 0$, we get the **Fokker–Planck equation for the damped harmonic oscillator**:

$$\frac{\partial P(\alpha,\alpha^*,t)}{\partial t} = \frac{\gamma}{2}\left(\frac{\partial}{\partial\alpha}\alpha P + \frac{\partial}{\partial\alpha^*}\alpha^* P\right) + \gamma\langle n\rangle\frac{\partial^2 P}{\partial\alpha\partial\alpha^*} . \tag{9.46}$$

9.6 Time-Dependent Solution: The Method of the Eigenfunctions

We try to solve the Fokker–Planck equation (9.46), using the following ansatz [5, 6]

$$P(\alpha,\alpha^*,t) = \exp(-\lambda t)Q(\alpha,\alpha^*) . \tag{9.47}$$

By replacing the above ansatz in the Fokker–Planck Equation, we get the following result

$$LQ = -\lambda Q, \tag{9.48}$$

$$L \equiv \frac{\gamma}{2}\left(\alpha\frac{\partial}{\partial\alpha} + \alpha^*\frac{\partial}{\partial\alpha^*} + 2\right) + \gamma\langle n\rangle\frac{\partial^2}{\partial\alpha\partial\alpha^*} .$$

Now we perform the following change of variables

$$\alpha = \sqrt{\langle n\rangle}(x + iy), \tag{9.49}$$

$$Q(\alpha) = \exp-\frac{1}{2}\left(x^2 + y^2\right)N(x,y) .$$

With the above transformation, the eigenvalue equation (9.48) becomes just the well-known Schrödinger equation for the two-dimensional isotropic harmonic oscillator. The eigenfunctions and eigenvalues are

$$N_{n_x,n_y} = K_{n_x,n_y}\exp\left[-\frac{1}{2}\left(x^2 + y^2\right)\right]H_{n_x}(x)H_{n_y}(y) , \tag{9.50}$$

$$\epsilon = \frac{4\lambda}{\gamma} + 2 = 2n_x + 1 + 2n_y + 1 ,$$

where n_x and n_y are integer numbers, so the solution now becomes

$$P_{n_x,n_y} = K_{n_x,n_y}\exp\left\{-(x^2 + y^2)\right\}H_{n_x}(x)H_{n_y}(y)\exp\left[-\frac{\gamma}{2}\left(n_x + n_y\right)t\right] . \tag{9.51}$$

We calculate the normalization constant K_{n_x,n_y} from the normalization of the P distribution

$$\int\frac{\mathrm{d}^2\alpha}{\pi}P(\alpha,t) = 1,$$

and the properties of the Hermite polynomials

$$\int_{-\infty}^{\infty} dx H_n(x) H_m(x) \exp(-x^2) = \delta_{nm} 2^n n! \sqrt{\pi}, \tag{9.52}$$

$$H_0(x) = 1 ,$$

we write

$$\frac{1}{\pi} K_{n_x, n_y} \langle n \rangle \int dx \exp(-x^2) H_{n_x}(x) \int dy H_{n_y}(y) \exp(-y^2) \exp\left[-\frac{\gamma}{2}(n_x + n_y)t\right]$$

$$= \frac{1}{\pi} K_{n_x, n_y} \langle n \rangle \delta_{n_x,0} \delta_{n_y,0} \exp\left[-\frac{\gamma}{2}(n_x + n_y)t\right] \sqrt{\pi}\sqrt{\pi} = 1 ,$$

so that $K_{00} = \frac{1}{\langle n \rangle}$ and $P_{00} = \frac{1}{\langle n \rangle} \exp\left[-\frac{1}{2}(x^2 + y^2)\right]$, which correspond to the time independent steady-state solution with the $\lambda = 0$ eigenvalue. Going back to the α variables

$$\lim_{t \to \infty} P(\alpha, \alpha^*, t) = P_{00} = \frac{1}{\langle n \rangle} \exp\left[-\frac{|\alpha|^2}{\langle n \rangle}\right] ,$$

which corresponds, as it should, to the P-representation of the thermal density matrix.

9.6.1 General Solution

We now concentrate on the general solution of the Fokker–Planck equation, using conditional probability densities such as $P(\alpha, t \mid \alpha', 0)$ with the initial condition

$$P(\alpha, 0 \mid \alpha', 0) = \delta^2(\alpha - \alpha') . \tag{9.53}$$

Of course we notice that because

$$P(\alpha, t) = \int \frac{d^2\alpha'}{\pi} P(\alpha, t \mid \alpha', 0) P(\alpha', 0) ,$$

the solution will have the form

$$P(\alpha, t \mid \alpha', 0) = \sum_{n_x n_y} \frac{P_{n_x n_y}(x, y, t) P_{n_x n_y}(x', y', 0)}{\langle n \rangle P_{00}(x', y')} \tag{9.54}$$

Now we explain why $P(\alpha, t \mid \alpha', 0)$ given by (9.54) is a general solution of the Fokker–Planck equation (9.46).

In the first place, it consists of linear superposition of $P_{n_x n_y}(x, y, t)$ solutions. So we have only to verify that it satisfies the initial conditions.

Making use of another property of the Hermite polynomials

$$\sum_{m=0}^{\infty} \frac{1}{2^m m! \sqrt{\pi}} \exp(-x^2) H_m(x) H_m(x') = \delta(x - x') ,$$

thus

$$K_{n_x, n_y} = \frac{1}{\sqrt{2^{n_x} n_x! \sqrt{\pi} 2^{n_y} n_y! \sqrt{\pi}}} ,$$

and the initial condition (9.53) is satisfied.

9.7 Langevin's Equations

We now study the damped harmonic oscillator problem, in the Heisenberg picture, with the initial reservoir operators interpreted as the quantum version of a stochastic or Langevin force.

From the Hamiltonian (9.2), the Heisenberg's equations for a and b_j are [7]

$$\frac{da(t)}{dt} = -i\omega a(t) - i\sum_j g_j b_j(t) , \qquad (9.55)$$

$$\frac{db_j(t)}{dt} = -i\omega_j b_j(t) - ig_j^* a(t) .$$

Integrating formally the second of (9.55), we get

$$b_j(t) = b_j(t_0)\exp\left[-i\omega_j(t-t_0)\right] - ig_j^*\int_{t_0}^t dt' a(t')\exp\left[-i\omega_j(t-t')\right] . \quad (9.56)$$

We insert (9.56) back in (9.55), to get

$$\frac{da(t)}{dt} = -i\omega a(t) - i\sum_j g_j b_j(t_0)\exp\left[-i\omega_j(t-t_0)\right] \qquad (9.57)$$

$$- \sum_j \mid g_j \mid^2 \int_{t_0}^t dt' a(t')\exp\left[-i\omega_j(t-t')\right] .$$

The second term of the r.h.s. of (9.57) represents the fluctuating term, independent from the system oscillator variable and the third term is a back reaction to the oscillator.

We now want to eliminate the fast time varying terms and introduce a rotating frame by defining $A(t) = a(t)\exp(i\omega t)$ with $\left[A(t), A(t)^\dagger\right] = 1$, so (9.57) can be written now as

$$\frac{dA(t)}{dt} = -\sum_j \mid g_j \mid^2 \int_{t_0}^t dt' A(t')\exp\left[-i(\omega_j-\omega)(t-t')\right] + F(t),$$

$$F(t) = \textbf{Noise Operator} = -i\sum_j g_j b_j(t_0)\exp\left[-i\omega_j(t-t_0)\right]\exp i\omega t .$$

As, for a thermal bath $\langle b_j\rangle = 0$, therefore $\langle F(t)\rangle_B = 0$.

We follow the same procedure as in the previous section, that is, make the Markov approximation, convert the discrete sum into an integral and arrive to the following result

$$\frac{dA(t)}{dt} = -\frac{\gamma}{2}A(t) + F(t) . \qquad (9.58)$$

On the average (9.58) behaves classically

$$\frac{d\langle A(t)\rangle}{dt} = -\frac{\gamma}{2}\langle A(t)\rangle . \tag{9.59}$$

Equation (9.58) is called generically the **Langevin equation for the damped harmonic oscillator.** The original idea came from the study of Brownian motion, where the particles in a liquid suffer from the rapid impact of the liquid molecules on the Brownian particle, having as a net effect, a rapidly time varying force.

We have to point out that, although averaging (9.58) over the bath variables make the fluctuating force disappear, $F(t)$ plays a very important role of preserving the commutation relations, in other words if the equation $\frac{dA(t)}{dt} = -\frac{\gamma}{2}A(t)$ were true, then we would have $A(t) = A(t_0)\exp{-\frac{\gamma}{2}t}$ and therefore $[A(t), A(t)^{\dagger}] \to 0$, thus violating quantum mechanics.

9.7.1 Calculation of the Correlation Function $\langle F(t')F(t'')^{\dagger}\rangle_B$

According to the definition of $F(t)$, we can write

$$\langle F(t')F(t'')^{\dagger}\rangle_B = \sum_i \sum_j g_i g_j^* \langle b_i b_j^{\dagger}\rangle \exp{[i\omega(t'-t'')]}\exp{[i(\omega_j t'' - \omega_i t')]} (9.60)$$

$$= \sum_i \mid g_i \mid^2 \langle b_i b_i^{\dagger}\rangle \exp{[i\omega(t'-t'')]}\exp{[-i\omega_i(t'-t'')]}$$

$$= \int d\omega' D(\omega') \mid g(\omega') \mid^2 [\langle n(\omega')\rangle + 1]\exp{[i(\omega - \omega')(t'-t'')]}$$

$$= \gamma[\langle n(\omega)\rangle + 1]\delta(t'-t'') .$$

In a similar calculation, one gets

$$\langle F(t')^{\dagger}F(t'')\rangle_B = \gamma\langle n(\omega)\rangle\delta(t'-t'') . \tag{9.61}$$

9.7.2 Differential Equation for the Photon Number

The Heisenberg equation for the photon number can be written as

$$\frac{dA^{\dagger}A}{dt} = -\sum_j \mid g_j \mid^2 A^{\dagger}(t)\int_{t_0}^t dt' A(t')\exp{[i(\omega - \omega_j)(t-t')]} \quad (9.62)$$

$$-i\sum_j g_j A^{\dagger}(t)b_j(t_0)\exp{[i(\omega - \omega_j)(t-t_0)]} + adj .$$

Making the Markov approximation , transforming the sum into an integral, we get the following Langevin equation

$$\frac{dA^{\dagger}A}{dt} = -\gamma A^{\dagger}A + F'_{A^{\dagger}A}, \tag{9.63}$$

$$F'_{A^\dagger A} = i \sum_j g_j^* A(t) b_j^\dagger(t_0) \exp\left[-i(\omega - \omega_j)(t - t_0)\right] + adj \ .$$

It is simple to show that $\langle F'_{A^\dagger A} \rangle = \gamma \langle n \rangle$, so that is more convenient to define a noise operator with zero average $F_{A^\dagger A} = F'_{A^\dagger A} - \gamma \langle n \rangle$, so that

$$\frac{\mathrm{d}A^\dagger A}{\mathrm{d}t} = -\gamma A^\dagger A + \gamma \langle n \rangle + F'_{A^\dagger A} \ , \tag{9.64}$$

9.8 Other Master Equations

9.8.1 Two-Level Atom in a Thermal Bath

We attack here a different problem, but formally similar to the damped harmonic oscillator.

Consider a two-level atom interacting with a reservoir of harmonic oscillators. Physically, this could, for example, represent an atom decaying irreversibly when interacting with infinite vacuum modes of the electromagnetic field.

The Hamiltonian of this problem is

$$H = \frac{\hbar\omega}{2}\sigma_z + \hbar \sum_l \omega_l b_l^\dagger b_l + \hbar \sum_l (g_l b_l^\dagger \sigma + g_l^* b_l \sigma^\dagger) \ . \tag{9.65}$$

The derivation of the master equation for the atom can be derived exactly as in the case of the damped harmonic oscillator. The result gives back (9.15), with the substitution

$$a \to \sigma$$
$$a^\dagger \to \sigma^\dagger \ .$$

It then reads

$$\frac{\mathrm{d}\widetilde{\rho_A}}{\mathrm{d}t} = -\frac{\gamma}{2}(1 + \langle n(\omega) \rangle)(\widetilde{\rho_A}(t)\sigma^\dagger\sigma + \sigma^\dagger\sigma\widetilde{\rho_A}(t) - 2\sigma\widetilde{\rho_A}(t)\sigma^\dagger) \tag{9.66}$$
$$-\frac{\gamma}{2}\langle n(\omega) \rangle(\widetilde{\rho_A}(t)\sigma\sigma^\dagger + \sigma\sigma^\dagger\widetilde{\rho_A}(t) - 2\sigma^\dagger\widetilde{\rho_A}(t)\sigma) \ .$$

Making use of the Pauli matrix properties

$$\sigma\sigma^\dagger = \frac{1}{2}(1 - \sigma_z) \ ,$$

$$\sigma^\dagger\sigma = \frac{1}{2}(1 + \sigma_z) \ ,$$

one can write a different version of the master equation (9.66)

$$\frac{d\widetilde{\rho_A}}{dt} = \gamma(1 + \langle n(\omega)\rangle)\sigma\widetilde{\rho_A}(t)\sigma^\dagger + \gamma\langle n(\omega)\rangle\sigma^\dagger\widetilde{\rho_A}(t)\sigma \qquad (9.67)$$

$$-\widetilde{\rho_A}\gamma\left[\frac{1}{2} + \langle n(\omega)\rangle\right] - \frac{\gamma}{4}(\widetilde{\rho_A}\sigma_z + \sigma_z\widetilde{\rho_A})$$

The master equation written in the form (9.67) is suitable to calculate $\langle\sigma_z\rangle$.

$$\frac{d\langle\sigma_z\rangle}{dt} = \frac{d}{dt}Tr(\widetilde{\rho_A}\sigma_z) = Tr\left(\frac{d\widetilde{\rho_A}}{dt}\sigma_z\right), \qquad (9.68)$$

where we used the fact that $\widetilde{\sigma_z} = \sigma_z$.

By introducing the master equation in (9.68), and after some simple algebra, one arrives to

$$\frac{d\langle\sigma_z\rangle}{dt} = -\gamma - 2\gamma\left[\frac{1}{2} + \langle n(\omega)\rangle\right]\langle\sigma_z\rangle . \qquad (9.69)$$

The solution of (9.68) is

$$\langle\sigma_z\rangle(t) = \langle\sigma_z\rangle(0)\exp(-2\gamma t)\left[\frac{1}{2} + \langle n(\omega)\rangle\right] \qquad (9.70)$$

$$-\frac{1}{2\left[\frac{1}{2} + \langle n(\omega)\rangle\right]}\left\{1 - \exp(-2\gamma t)\left[\frac{1}{2} + \langle n(\omega)\rangle\right]\right\} .$$

We now take the aa matrix element of (9.66) and get

$$\frac{d}{dt}P_a(t) = \frac{d}{dt}\rho_{aa}(t) \qquad (9.71)$$

$$= -\gamma P_a(t) ,$$

and the solution is $P_a(t) = P_a(0)\exp(-\gamma t)$, which is the well-known result of E. Wigner and V. Weisskopf [8] , predicting an exponential decay of an atom initially in the excited state. We notice that here the single atom is interacting with an infinite reservoir of electromagnetic modes in vacuum, and the predicted behavior here is very different from the Rabi flopping of the atom **interacting with a single mode in a good cavity.**

9.8.2 Damped Harmonic Oscillator in a Squeezed Bath

We start with (9.8), and write it out in detail, using the Hamiltonian of (9.5), thus getting

$$\frac{d\rho_A}{dt} = -\int_0^t dt' Tr_B\left\{[a^\dagger G(t) + G^\dagger(t)a]\left[a^\dagger G(t') + G^\dagger(t')a\right]\rho_B \otimes \rho(t')\right.$$

$$\left. - [a^\dagger G(t) + G^\dagger(t)a]\rho_B \otimes \rho(t')\left[a^\dagger G(t') + G^\dagger(t')a\right]\right. \qquad (9.72)$$

$$- \left[a^\dagger G(t') + G^\dagger(t')a\right] \rho_B \otimes \rho(t') \left[a^\dagger G(t) + G^\dagger(t)a\right]$$
$$+ \rho_B \otimes \rho(t') \left[a^\dagger G(t') + G^\dagger(t')a\right] \left[a^\dagger G(t) + G^\dagger(t)a\right]\} \; .$$

Using the property

$$\langle G(t')G(t)\rangle = \langle G(t)G(t')\rangle \; , \tag{9.73}$$

$$\langle G^\dagger(t')G(t)\rangle = \langle G^\dagger(t)G(t')\rangle^* \; ,$$

$$\langle G(t')G^\dagger(t)\rangle = \langle G(t)G^\dagger(t')\rangle^* \; ,$$

$$\langle G^\dagger(t')G^\dagger(t)\rangle = \langle G^\dagger(t)G^\dagger(t')\rangle \; ,$$

we can write

$$\frac{\mathrm{d}\rho_A(t)}{\mathrm{d}t} = I_1 \left[a^\dagger a^\dagger \rho_A + \rho_A a^\dagger a^\dagger - 2a^\dagger \rho_A a^\dagger\right] \tag{9.74}$$
$$+ I_2 \left[aa\rho_A + \rho_A aa - 2a\rho_A a\right]$$
$$+ I_3 \left[a^\dagger a\rho_A + \rho_A a^\dagger a - 2a\rho_A a^\dagger\right]$$
$$+ I_4 \left[aa^\dagger \rho_A + \rho_A aa^\dagger - 2a^\dagger \rho_A a\right]$$

where we defined

$$I_1 = \int \mathrm{d}t' \langle G(t)G(t')\rangle \; , \tag{9.75}$$

$$I_2 = \int \mathrm{d}t' \langle G^\dagger(t)G^\dagger(t')\rangle \; ,$$

$$I_3 = \int \mathrm{d}t' \langle G(t)G^\dagger(t')\rangle \; ,$$

$$I_4 = \int \mathrm{d}t' \langle G^\dagger(t)G(t')\rangle \; .$$

We will calculate in detail one of them, for example I_1

$$I_1 = \int \mathrm{d}t' \langle G(t)G(t')\rangle \; , \tag{9.76}$$

$$= \int \mathrm{d}t' \sum_{k,k'} g_k g_{k'} \exp\left[i(\omega - \omega_k)t\right] \exp\left[i(\omega - \omega_{k'})t'\right] \langle b_k b_{k'}\rangle \; ,$$

$$= \int_0^t \mathrm{d}t' \sum_{k,k'} g_k g_{k'} \exp\left[i\omega(t+t')\right] \exp\left[-i(\omega_k t + \omega_{k'}t')\right] \langle b_k b_{k'}\rangle \; .$$

Now, we consider the two-mode squeezing correlation

$$\langle b_k b_{k'}\rangle = \langle b_k b_{-k}\rangle \delta_{k',-k} = M \tag{9.77}$$

placed symmetrically around the system frequency ω, so

$$\omega_{\pm k} = \omega \pm \Omega,$$

then, converting the sum into a frequency integral, with a density D, we can write

$$I_1 = M \int_0^\infty d\Omega D(\omega + \Omega) g^2(\omega + \Omega) \int_0^t dt' \exp\left[-i\Omega(t - t')\right]$$

If we assume that the functions D and g are slowly varying with frequency, then the time-dependent result after the frequency integration is sharply peaked and we can set $t \to \infty$, without much error, getting $\delta(\Omega)$ and a small frequency shift. The result then reads

$$I_1 = \frac{\gamma}{2}M + i\delta_1 , \tag{9.78}$$

with

$$\gamma = 2\pi D(\omega)g(\omega)^2, \tag{9.79}$$
$$\delta_1 = P \int_{-\infty}^\infty \frac{d\Omega}{\Omega} D(\Omega + \omega) g^2(\Omega + \omega) M$$

In a similar way, we obtain the other integrals

$$I_2 = \frac{\gamma}{2}M^* - i\delta_1 , \tag{9.80}$$

$$I_3 = \frac{\gamma}{2}(N + 1) + i\delta_2 , \tag{9.81}$$

$$I_4 = \frac{\gamma}{2}N - i\delta_2 , \tag{9.82}$$

with

$$\langle b_k^\dagger b_{k'} \rangle = N\delta_{k,k'} . \tag{9.83}$$

If we compare the above results with the average corresponding to the two-mode squeezed vacuum, we readily get

$$N = \sinh^2 r, \tag{9.84}$$
$$M = -\exp(i\theta) \sinh r \cosh r ,$$

obeying the relation

$$\sqrt{N(N + 1)} = \sqrt{\sinh^2 r \cosh^2 r} = |M| . \tag{9.85}$$

Ignoring the small frequency shifts δ_1 and δ_2, we finally write the master equation for the damped harmonic oscillator in a squeezed vacuum

$$\frac{d\rho}{dt} = \frac{\gamma}{2}(N + 1)\left(2a\rho a^\dagger - a^\dagger a\rho - \rho a^\dagger a\right) \tag{9.86}$$

$$+\frac{\gamma}{2}(N)\left(2a^\dagger\rho a - aa^\dagger\rho - \rho aa^\dagger\right)$$

$$+\frac{\gamma}{2}(M)\left[2a^\dagger\rho a^\dagger - a^\dagger a^\dagger\rho - \rho a^\dagger a^\dagger\right]$$

$$+\frac{\gamma}{2}(M^*)\left(2a\rho a - aa\rho - \rho aa\right).$$

In the particular case of a thermal reservoir $N \to \langle n \rangle, M \to 0$, and we recover (9.15).

This master equation is expected to give a correct description of a system driven by noise that comes from the squeezed vacuum, provided the squeezing is reasonably constant over the bandwidth of the system.

The above condition has to be satisfied, because if D, k, M are slowly varying functions of Ω, then the frequency integral approaches a δ-function in time and we are justified to set the upper limit to ∞.

9.8.3 Application: Spontaneous Decay in a Squeezed Vaccum

Once more, in the master equation [9, 10], we replace

$$a \to \sigma, a^\dagger \to \sigma^\dagger, \tag{9.87}$$

to study atomic decay, getting

$$\frac{d\rho}{dt} = \frac{\gamma}{2}\cosh^2 r\left(2\sigma\rho\sigma^\dagger - \sigma^\dagger\sigma\rho - \rho\sigma^\dagger\sigma\right) \tag{9.88}$$

$$+\frac{\gamma}{2}\sinh^2 r\left(2\sigma^\dagger\rho\sigma - \sigma\sigma^\dagger\rho - \rho\sigma\sigma^\dagger\right)$$

$$-\gamma\exp(i\theta)\sinh r\cosh r(\sigma^\dagger\rho\sigma^\dagger)$$

$$-\gamma\exp(-i\theta)\sinh r\cosh r(\sigma\rho\sigma).$$

In the above equation, we used the property $\sigma^2 = \sigma^{\dagger 2} = 0$.
If we take the expectation value of

$$\sigma_x = \frac{\sigma + \sigma^\dagger}{2}, \tag{9.89}$$

$$\sigma_x = \frac{\sigma - \sigma^\dagger}{2i},$$

$$\sigma_z = (2\sigma^\dagger\sigma - 1),$$

we get the following differential equations

$$\langle\dot\sigma_x\rangle = -\frac{\gamma}{2}\exp(2r)\langle\sigma_x\rangle \equiv -\gamma_x\langle\sigma_x\rangle, \tag{9.90}$$

$$\langle\dot\sigma_y\rangle = -\frac{\gamma}{2}\exp(-2r)\langle\sigma_y\rangle \equiv -\gamma_y\langle\sigma_x\rangle, \tag{9.91}$$

$$\langle\dot\sigma_z\rangle = -\gamma(2\sinh^2 r + 1)\langle\sigma_z\rangle - \gamma \equiv -\gamma_z\langle\sigma_z\rangle - \gamma. \tag{9.92}$$

As we can see, for large squeezing, both γ_x and γ_z become large, and γ_y very small.

On a time scale short compared to γ_y^{-1}, but larger than γ_x^{-1} and γ_z^{-1}, we get

$$\langle \sigma_x \rangle \rightarrow 0, \langle \sigma_z \rangle \rightarrow -\frac{1}{2\sinh^2 r + 1}. \tag{9.93}$$

Problems

9.1. The Dirac delta function can be defined as

$$\delta(\omega_0 - \omega) = \frac{2}{\pi}\mathrm{Lim}_{t\to\infty}\frac{\sin^2\left[\frac{(\omega_0-\omega)t}{2}\right]}{(\omega_0-\omega)^2 t}.$$

Prove that

a)

$$\int_{-\infty}^{\infty}\delta(\omega_0-\omega)d\omega = 1$$

b)

$$\int_{\omega_1}^{\omega_2}\delta(\omega_0-\omega)(\omega)d\omega = f(\omega_0),$$

with $\omega_1 < \omega_0 < \omega_2$.

9.2. Prove that the following are acceptable definitions or representations of the Dirac delta function

a)

$$\delta(\omega_0-\omega) = \frac{1}{2\pi}\mathrm{Lim}_{T_1,T_2\to\infty}\int_{-T_1}^{T_2}\exp[i(\omega_0-\omega)t]dt,$$

b)

$$\delta(\omega_0-\omega) = \mathrm{Lim}_{T\to\infty}\left[\frac{\sin(\omega_0-\omega)T}{(\omega_0-\omega)\pi}\right],$$

c)

$$\delta(\omega_0-\omega) = \frac{1}{\pi}\mathrm{Lim}_{\varepsilon\to 0}\frac{\varepsilon}{(\omega_0-\omega)^2+\varepsilon^2}.$$

9.3. Write down the Fokker–Planck equation of the damped harmonic oscillator in a squeezed bath, starting from (9.86).

9.4. The master equation

$$\dot{\rho}= \frac{A}{2}(2a^\dagger\rho a - aa^\dagger\rho - \rho aa^\dagger) - \frac{C}{2}(a^\dagger a\rho + \rho a^\dagger a - 2a\rho a^\dagger)$$

represents the laser theory in the lowest order approximation, where A is the gain and C the cavity loss.

Prove that the corresponding Fokker–Planck equation is

$$\frac{\partial P}{\partial t} = -\frac{A-C}{2}\left(\frac{\partial}{\partial \alpha}\alpha + \frac{\partial}{\partial \alpha^*}\alpha^*\right)P + A\frac{\partial^2 P}{\partial \alpha \partial \alpha^*}.$$

9.5. Generalizing the results of the problem 9.4, we take now the second-order Laser theory. For the following master equation

$$\frac{d\rho_f}{dt} = -\frac{A}{2}\{aa^\dagger\left[\rho_f - \left(\frac{g}{\gamma}\right)^2(aa^\dagger\rho_f + 3\rho_f aa^\dagger)\right]$$

$$+ \left[\rho_f - \left(\frac{g}{\gamma}\right)^2(\rho_f aa^\dagger + 3aa^\dagger\rho_f)\right]aa^\dagger$$

$$-2a^\dagger\left[\rho_f - 2\left(\frac{g}{\gamma}\right)^2(aa^\dagger\rho_f + \rho_f aa^\dagger)a\right]\}$$

$$-\frac{C}{2}\left(a^\dagger a\rho_f + \rho_f a^\dagger a - 2a\rho_f a^\dagger\right),$$

show that the corresponding Fokker–Planck equation is

$$\frac{\partial P}{\partial t} = -\frac{1}{2}\frac{\partial}{\partial \alpha}\left\{\left[(A-C) - 4A\frac{g^2}{\gamma^2}\mid \alpha\mid^2\right]\alpha P - A\frac{\partial P}{\partial \alpha^*}\right\}$$

$$-\frac{1}{2}\frac{\partial}{\partial \alpha^*}\left\{\left[(A-C) - 4A\frac{g^2}{\gamma^2}\mid \alpha\mid^2\right]\alpha^* P - A\frac{\partial P}{\partial \alpha}\right\}.$$

Notice that we regain the results of the Prob 9.4 when we neglect the g^2 terms.

9.6. Show for the damped harmonic oscillator that if one assumes initially a minimum uncertainty state, then at $t = t$

$$\Delta q(t)\Delta p(t) = \frac{\hbar}{2}[1 + 2\langle n(\omega)\rangle(1 - \exp(-\gamma t))]$$

References

1. Louisell, W.H.: Quantum Statistical Properties of Radiation. John Wiley, New York (1973)
2. Narducci, L.M.: WPI Lecture Notes (unpublished) (1973)
3. Sneddon, I.: Elements of Partial Differential Equations. Mc-Graw Hill, New York (1957)
4. Bargmann, V.: Commun. Pure. Appl. Math., **14**, 187 (1962)
5. Risken, H.: The Fokker-Planck Equation. Sringer Verlag, Berlin (1984)
6. Gardiner, C.W.: Handbook of Stochastic Methods. Springer Verlag, Berlin (1983).

7. Meystre, P., SargentIII, M.: Elements of Quantum Optics. Springer Verlag, Berlin (1990)
8. Weisskopf, V., Wigner, E.z.: Phys. 63, 54 (1930)
9. Collett, M.J., Gardiner, C.W.: Phys. Rev. A, **30**, 1386 (1984)
10. Gardiner, C.W.: Phys. Rev. Lett., **56**, 1917 (1986)

Further Reading

- Lax, M., Phys. Rev., **145**, 110 (1966)
- Haken, H., Laser Theory Springer Verlag, Berlin (1970)
- SargentIII, M., Scully, M.O., Lamb, W.E.: Laser Physics. AddisonWesley, Mass. USA (1974)
- Scully, M.O., Zubairy, M.S.: Quantum Optics. Cambridge University Press, Cambridge (1997)

10. Resonance Fluorescence

In this chapter, we study the scattered light from a two-level atom illuminated with a continuous field.

The theory of spontaneous emission was originally developed by **Weisskopf** and **Wigner** [1]. However, the sidebands for coherent single-mode fields were found by **Mollow** in the 1960s. [2]

The physical origin of the sidebands in the resonance fluorescence spectrum can be nicely seen using the dressed states [3].

10.1 Background

We assume a two-level atom driven by a continuous monochromatic field. The atom scatters the light in all directions, as shown in the Fig. 10.1.

For a weak incident field, the spectrum, as we shall see in some detail, exhibits a single peak much narrower that the natural linewidth of the atomic transition. As we increase the field, this spectrum splits into three peaks consisting in a central peak at the laser frequency and two sidebands symmetrically placed at $\pm R_n$ with respect to the center.

When the light intensity is very weak and the atom is initially in the ground state $|b\rangle$, it absorbs and scatters a single photon, whose frequency is identical to the laser frequency ω_L, in other words

$$S(\omega) = S_0 \delta(\omega - \omega_L) ,\qquad (10.1)$$

which is just the **Rayleigh scattering**.

In practice, however, the laser is not perfectly monochromatic, so that the scattered spectrum will have a narrow but finite (non-zero) linewidth.

Now, if we increase the laser power, the sidebands that appear in the spectrum can be understood intuitively, using the dressed picture description of the atom–field interaction, as we did in the Chap. 8.

In the resonant case, we saw that

$$E_{1n} = \hbar(n + \frac{1}{2}) + \hbar g \sqrt{n + 1} ,\qquad (10.2)$$

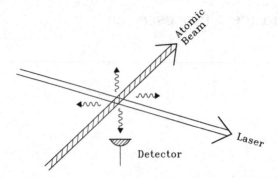

Fig. 10.1. Diagram showing the interaction between an atomic beam and a Laser-generating resonance fluorescence

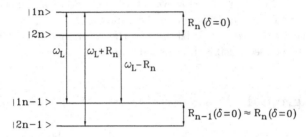

Fig. 10.2. Energy diagram showing the origin of the three frequencies in resonance fluorescence, corresponding to the central peak and the two sidebands

$$E_{2n} = \hbar\left(n + \frac{1}{2}\right) - \hbar g\sqrt{n+1} \,,$$

thus, the energy separation between the two levels is

$$R_n(\delta = 0) = 2g\sqrt{n+1} \,. \tag{10.3}$$

This is described in the Fig. 10.2.

As we can see from the dressed level picture, besides the transition at the laser frequency ω_L, it is also possible to have transitions at $\omega = \omega_L \pm R_n$, which correspond to the two sidebands.

The details of the heights and widths of the peaks requires a full quantum theory, including atomic losses.

10.2 Heisenberg's Equations

We begin with the Hamiltonian: [4]

$$H = H_A + H_r - e\mathbf{r} \cdot \mathbf{E} \,, \tag{10.4}$$

describing the interaction of the radiation field with a single electron atom.

If we consider many modes for the field, we can generalize the Jaynes–Cummings model, described in the Chap. 8, as

$$H = H_r + \frac{\hbar\omega_{ab}}{2}\sigma_z - i\hbar \sum_{\mathbf{k},\lambda} g_{\mathbf{k},\lambda}(\sigma_+ + \sigma_-)(a_{\mathbf{k},\lambda} - a_{\mathbf{k},\lambda}^\dagger) , \qquad (10.5)$$

where H_r is the free multimode radiation energy, and the interaction terms are a bit different now, since we are taking traveling waves rather than standing waves. Thus, we use the quantized field $\mathbf{E}$ described in the Chap. 3. Also, $\mathbf{k}$ is the wave-vector and λ the polarization, and

$$g_{\mathbf{k},\lambda} \equiv \sqrt{\frac{\omega_{\mathbf{k}}}{2\hbar\varepsilon_0 v}} e\mathbf{r}_{ab} \cdot \mathbf{e}_{\mathbf{k},\lambda} , \qquad (10.6)$$

with

$$[\sigma_+, \sigma_-] = \sigma_z , \qquad (10.7)$$

$$[\sigma_z, \sigma_\pm] = \pm 2\sigma_\pm . \qquad (10.8)$$

Also, one has

$$\left[a_{\mathbf{k},\lambda}, a_{\mathbf{k'},\lambda'}^\dagger\right] = \delta_{\mathbf{k},\mathbf{k'}}^{(3)} \delta_{\lambda,\lambda'} . \qquad (10.9)$$

The Heisenberg Equations of motion for the operators are

$$\dot{a}_{\mathbf{k},\lambda} = -i\omega_{\mathbf{k}} a_{\mathbf{k},\lambda} + g_{\mathbf{k},\lambda}(\sigma_+ + \sigma_-) , \qquad (10.10)$$

$$\dot{\sigma}_- = -i\omega_{ab}\sigma_- + \sum_{\mathbf{k},\lambda} g_{\mathbf{k},\lambda}\sigma_z(a_{\mathbf{k},\lambda} - a_{\mathbf{k},\lambda}^\dagger) , \qquad (10.11)$$

$$\dot{\sigma}_z = 2 \sum_{\mathbf{k},\lambda} g_{\mathbf{k},\lambda}(\sigma_- - \sigma_+)(a_{\mathbf{k},\lambda} - a_{\mathbf{k},\lambda}^\dagger) . \qquad (10.12)$$

Both $a_{\mathbf{k},\lambda}$ and σ_- have basically a positive frequency temporal dependence and σ_z varies slowly in time, thus making use of the rotating wave approximation, we can neglect the σ_+ and $a_{\mathbf{k},\lambda}^\dagger$ terms from the first two equations, respectively, and the counter-rotating terms $\sigma_- a_{\mathbf{k},\lambda}$ and $\sigma_+ a_{\mathbf{k},\lambda}^\dagger$ from the third one, writing a simplified version of (10.10, 10.11, 10.12)

$$\dot{a}_{\mathbf{k},\lambda} = -i\omega_{\mathbf{k}} a_{\mathbf{k},\lambda} + g_{\mathbf{k},\lambda}\sigma_- , \qquad (10.13)$$

$$\dot{\sigma}_- = -i\omega_{ab}\sigma_- + \sum_{\mathbf{k},\lambda} g_{\mathbf{k},\lambda}\sigma_z(a_{\mathbf{k},\lambda}) , \qquad (10.14)$$

$$\dot{\sigma}_z = -2 \sum_{\mathbf{k},\lambda} g_{\mathbf{k},\lambda}(\sigma_- a_{\mathbf{k},\lambda}^\dagger + \sigma_+ a_{\mathbf{k},\lambda}) . \qquad (10.15)$$

We can rewrite the last two equations as

$$\dot{\sigma}_- = -i\omega_{ab}\sigma_- - \frac{i}{\hbar}\mathbf{d}\cdot\sigma_z\mathbf{E}^+(\mathbf{t}) \,, \tag{10.16}$$

$$\dot{\sigma}_z = -\frac{2i}{\hbar}\mathbf{d}\cdot\left[\sigma_-\mathbf{E}^-(\mathbf{t})-\sigma_+\mathbf{E}^+(\mathbf{t})\right] \,, \tag{10.17}$$

with

$$\mathbf{E}^+(\mathbf{t}) = i\sum_{\mathbf{k},\lambda}\sqrt{\frac{\hbar\omega_{\mathbf{k}}}{2\varepsilon_0 v}}a_{\mathbf{k},\lambda}(t)\mathbf{e}_{\mathbf{k},\lambda} \,, \tag{10.18}$$

$$\mathbf{d} = e\mathbf{r}_{ab}.$$

Integrating (10.13), we get

$$a_{\mathbf{k},\lambda}(t) = a_{\mathbf{k},\lambda}(0)\exp(-i\omega_{\mathbf{k}}t)$$
$$+g_{\mathbf{k},\lambda}\int_0^t dt_1\sigma_-(t_1)\exp i\omega_{\mathbf{k}}(t_1-t) \,. \tag{10.19}$$

Thus, replacing the above result in (10.18), we get

$$\mathbf{E}^+(\mathbf{t}) = \mathbf{E}_0^+(\mathbf{t}) + \mathbf{E}_{RR}^+(\mathbf{t}) \,, \tag{10.20}$$

where

$$\mathbf{E}_0^+(\mathbf{t}) = i\sum_{\mathbf{k},\lambda}\sqrt{\frac{\hbar\omega_{\mathbf{k}}}{2\varepsilon_0 v}}a_{\mathbf{k},\lambda}(0)\exp(-i\omega_{\mathbf{k}}t)\mathbf{e}_{\mathbf{k},\lambda} \,, \tag{10.21}$$

$$\mathbf{E}_{RR}^+(\mathbf{t}) = i\sum_{\mathbf{k},\lambda}\sqrt{\frac{\hbar\omega_{\mathbf{k}}}{2\varepsilon_0 v}}g_{\mathbf{k},\lambda}\int_0^t dt_1\sigma_-(t_1)\exp\left[i\omega_{\mathbf{k}}(t_1-t)\right]\mathbf{e}_{\mathbf{k},\lambda} \,. \tag{10.22}$$

The first term $\mathbf{E}_0^+(\mathbf{t})$ is the solution of the homogeneous Maxwell equation and corresponds to the field at the position of the atom as if the atom was not there. On the other hand, the second term $\mathbf{E}_{RR}^+(\mathbf{t})$ represents the influence of the atom or, said in different words, the radiation reaction field of the electric dipole point.

Substituting (10.21, 10.22) in (10.20), we get

$$\mathbf{E}^+(\mathbf{t}) = i\sum_{\mathbf{k},\lambda}\sqrt{\frac{\hbar\omega_{\mathbf{k}}}{2\varepsilon_0 v}}a_{\mathbf{k},\lambda}(0)\exp(-i\omega_{\mathbf{k}}t)\mathbf{e}_{\mathbf{k},\lambda} \tag{10.23}$$

$$+i\sum_{\mathbf{k},\lambda}\sqrt{\frac{\hbar\omega_{\mathbf{k}}}{2\varepsilon_0 v}}g_{\mathbf{k},\lambda}\int_0^t dt_1\sigma_-(t_1)\exp\left[i\omega_{\mathbf{k}}(t_1-t)\right]\mathbf{e}_{\mathbf{k},\lambda} \,,$$

but, using (3.36), that is changing the sum into an integral with the corresponding density of modes

$$\sum_{\mathbf{k},\lambda} \rightarrow \sum_{\lambda=1,2} \int dk k^2 \int \sin\theta d\theta \int d\varphi \frac{v}{(2\pi)^3} , \qquad (10.24)$$

we can write

$$(\mathbf{E}_{RR}^+(t))_z = i \sum_{\mathbf{k},\lambda} \frac{\omega_\mathbf{k}}{2\varepsilon_0 v} (\mathbf{d} \cdot \mathbf{e}_{\mathbf{k},\lambda}) \mathbf{e}_{\mathbf{k},\lambda} \int_0^t dt_1 \sigma_-(t_1) \exp\left[i\omega_\mathbf{k}(t_1 - t)\right] \qquad (10.25)$$

$$= 2i \int dk k^2 \int d\varphi \frac{\omega_\mathbf{k}}{2\varepsilon_0 v} \frac{v}{(2\pi)^3} \mid \mathbf{d} \mid \int \cos^2\theta \sin\theta d\theta$$

$$\cdot \int_0^t dt_1 \sigma_-(t_1) \exp\left[i\omega_\mathbf{k}(t_1 - t)\right] ,$$

where we choose the dipole moment along the z-axis, so that $\mathbf{d} \cdot \mathbf{e}_{\mathbf{k},\lambda} = \mid \mathbf{d} \mid \cos\theta$, as shown in Fig. 10.3.

If we take the lowest-order approximation for $\sigma_-(t)$ and performing the θ integral, we get

$$\left[\mathbf{E}_{RR}^+(t)\right]_z = \frac{2i \mid \mathbf{d} \mid \sigma_-(t) \exp(-i\omega_{ab}t)}{3(2\pi)^2 \varepsilon_0 c^3} \int d\omega \omega^3 \int dt_1 \exp i(\omega - \omega_{ab})(t_1 - t) .$$
$$(10.26)$$

We calculate now the above time integral, with the change of variable $-\tau = t_1 - t$:

$$\int dt_1 \exp\left[i(\omega - \omega_{ab})(t_1 - t)\right] \qquad (10.27)$$

$$= \int_0^t d\tau \exp\left[i(\omega - \omega_{ab})\tau\right] t \rightarrow \infty \rightarrow \pi\delta(\omega - \omega_{ab}) .$$

The final result for the $\left[\mathbf{E}_{RR}^+(t)\right]_z$ field is

$$\left[\mathbf{E}_{RR}^+(t)\right]_z = \frac{1}{4\pi\varepsilon_0} \frac{2i \mid \mathbf{d} \mid \omega_{ab}^3}{3c^3} \sigma_-(t) , \qquad (10.28)$$

(in CGS units $\frac{1}{4\pi\varepsilon_0} = 1$), and the differential equations for the Heisenberg operators now become

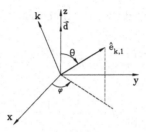

Fig. 10.3. The atomic dipole and k vector of the electric field in polar coordinates

$$\dot{\sigma}_- = -i(\omega_{ab} - i\beta)\sigma_- - \frac{i}{\hbar}\mathbf{d} \cdot \sigma_z \mathbf{E}_0^+(t) , \tag{10.29}$$

$$\dot{\sigma}_z = -2\beta(1 + \sigma_z) - \frac{2i}{\hbar}\mathbf{d} \cdot \left[\sigma_- \mathbf{E}_0^-(t) - \sigma_+ \mathbf{E}_0^+(t)\right] , \tag{10.30}$$

with

$$\beta \equiv \frac{1}{4\pi\varepsilon_0} \frac{2\,|\,\mathbf{d}\,|^2\,\omega_{ab}^3}{3\hbar c^3} . \tag{10.31}$$

10.3 Spectral Density, and the Wiener–Khinchine Theorem

If one has a random process $y(t)$, an interesting characteristic of this process is its spectrum.

We may define a Fourier Transform

$$y(t) = \int_{-\infty}^{\infty} \widetilde{y}\,(\omega) \exp(-i\omega t) \mathrm{d}\omega , \tag{10.32}$$

which, in principle, can be inverted as

$$\widetilde{y}\,(\omega) = \frac{1}{2\pi} \int_{-\infty}^{\infty} y(t) \exp(i\omega t) \mathrm{d}t . \tag{10.33}$$

We also define a spectral density or just plain spectrum, as the expectation value of $|\widetilde{y}\,(\omega)\,|^2$

$$S(\omega) = \langle |\widetilde{y}\,(\omega)\,|^2 \rangle , \tag{10.34}$$

whose physical meaning is the fluctuation strength associated with a definite frequency component.

However, a word of caution. If one is dealing with a stationary process, invariant under a translation of the time origin, then $y(t)$ does not go to zero for $t \to \pm\infty$, and therefore, this function is not square integrable and the Fourier transform does not exist in the usual sense. We will consider them as symbolic formulas, which can be given rigorous mathematical meaning in an enlarged functional space.

However, Wiener observed that the functions

$$\Gamma(\tau) = \underset{T \to \infty}{Lim} \frac{1}{2T} \int_{-T}^{T} y^*(t)y(t + \tau)\mathrm{d}t \tag{10.35}$$

and

$$\sigma(\omega) = \frac{1}{2\pi} \int_{-\infty}^{\infty} \Gamma(\tau) \frac{\exp(i\omega\tau) - 1}{i\tau} \mathrm{d}\tau \tag{10.36}$$

do exist.

One can define an alternative spectrum as

$$S(\omega) = \frac{d\sigma(\omega)}{d\omega} = \frac{1}{2\pi} \int_{-\infty}^{\infty} \exp(i\omega\tau)\Gamma(\tau)d\tau \ . \tag{10.37}$$

One can also invert the above formula to get

$$\Gamma(\tau) = \int \exp(-i\omega\tau)S(\omega)d\omega \ . \tag{10.38}$$

In order to understand the relation between the two definitions of spectra (10.34 and 10.37), we calculate the ensemble average of $\widetilde{y}^{*}(\omega)\,\widetilde{y}(\omega')$

$$\langle \widetilde{y}^{*}(\omega)\,\widetilde{y}(\omega')\rangle = \frac{1}{(2\pi)^2} \int\!\!\int_{-\infty}^{\infty} \langle y^{*}(t)y(t')\rangle \exp i(\omega' t' - \omega t)dtdt' , \tag{10.39}$$

and since $y(t)$ is stationary and ergodic (the time and ensemble averages are equal)

$$\langle y^{*}(t)y(t')\rangle = \Gamma(t'-t) \ , \tag{10.40}$$

where Γ is the two-time correlation function.

Substituting (10.40) in (10.39), and changing variable: $t'-t = \tau$, we get

$$\langle \widetilde{y}^{*}(\omega)\,\widetilde{y}(\omega')\rangle = \frac{1}{(2\pi)^2} \int_{-\infty}^{\infty} \exp it(\omega'-\omega)dt \int_{-\infty}^{\infty} d\tau\Gamma(\tau)\exp i\omega'\tau \ , \tag{10.41}$$

which implies

$$\langle \widetilde{y}^{*}(\omega)\,\widetilde{y}(\omega')\rangle = \widetilde{\Gamma}(\omega)\delta(\omega-\omega') \ , \tag{10.42}$$

with

$$\widetilde{\Gamma}(\omega) = \frac{1}{(2\pi)} \int_{-\infty}^{\infty} d\tau\Gamma(\tau)\exp(i\omega\tau) \ . \tag{10.43}$$

We notice that (10.42) is telling us that the generalized Fourier components are uncorrelated.

On the other hand, (10.43) is giving us a measure of the strength of the fluctuations at a given frequency; thus, we may identify it with the spectrum

$$\widetilde{\Gamma}(\omega) = S(\omega) \ . \tag{10.44}$$

We also notice that the singularity in (10.42) can be easily removed by integrating, over a small range containing ω, in ω'

$$S(\omega) = \underset{\Delta\omega\to 0}{Lim} \int_{\omega-\frac{\Delta\omega}{2}}^{\omega+\frac{\Delta\omega}{2}} d\omega' \langle \widetilde{y}^{*}(\omega)\,\widetilde{y}(\omega')\rangle \ , \tag{10.45}$$

which is equivalent to (10.34).

The pair of formulas (10.37) and (10.38) are known as the Wiener–Khintchine theorem, which tells us that for a stationary random process, the autocorrelation function and the power spectrum are a Fourier pair.

If we apply these ideas to the correlation of a quantum field, we get

$$S(\omega) = \frac{1}{2\pi} \int_{-\infty}^{\infty} \exp(i\omega\tau) \langle \mathbf{E}^-(t)\mathbf{E}^+(t+\tau)\rangle_{ss} d\tau . \tag{10.46}$$

10.4 Emission Spectra from Strongly Driven Two-Level Atoms

The power delivered to the field can be expressed as energy per unit time

$$P(t) = \frac{\mathrm{d}Energy}{\mathrm{d}t} = \frac{\mathrm{d}}{\mathrm{d}t} \sum_{\mathbf{k},\lambda} \hbar\omega_{\mathbf{k}} \langle a_{\mathbf{k},\lambda}^\dagger(t)a_{\mathbf{k},\lambda}(t)\rangle \tag{10.47}$$

$$= \sum_{\mathbf{k},\lambda} g_{\mathbf{k},\lambda} \hbar\omega_{\mathbf{k}} \left[\langle a_{\mathbf{k},\lambda}^\dagger(0)\sigma_-(t)\rangle \exp(i\omega_{\mathbf{k}}t) + \langle \sigma_+(t)a_{\mathbf{k},\lambda}(0)\rangle \exp(-i\omega_{\mathbf{k}}t) \right]$$

$$+ 2\,\mathrm{Re} \left\{ \sum_{\mathbf{k},\lambda} g_{\mathbf{k},\lambda}^2 \hbar\omega_{\mathbf{k}} \int_0^t \mathrm{d}t_1 \langle \sigma_+(t_1)\sigma_-(t)\rangle \exp\left[-i\omega_{\mathbf{k}}(t_1-t)\right] . \right\}$$

The first two terms in the above expression refer to the change in the energy of the field because of stimulated energy and absorption. On the other hand, the last term is the scattered radiation out of the incident beam as resonance fluorescence.

We shall concentrate in this last term:

$$= 2\,\mathrm{Re} \left\{ \sum_{\mathbf{k},\lambda} \frac{\omega_{\mathbf{k}} \cos^2\theta \,|\,\mathbf{d}\,|^2}{2v\varepsilon_0} \int_0^t \mathrm{d}t_1 \langle \sigma_+(t_1)\sigma_-(t)\rangle \exp\left[i\omega_{\mathbf{k}}(t-t_1)\right] \right\} . \tag{10.48}$$

$$= \frac{|\,\mathbf{d}\,|^2}{3\pi^2 c^3 \varepsilon_0}\,\mathrm{Re} \int \omega^4 \mathrm{d}\omega \int_0^t \mathrm{d}t_1 \langle \sigma_+(t_1)\sigma_-(t)\rangle \exp\left[i\omega(t-t_1)\right] ,$$

where, in the last term, we took the limit $v \to \infty$.
Defining

$$S(t) \equiv \sigma_-(t) \exp i\omega_L t , \tag{10.49}$$

we can write

$$P_s(t) = \frac{1}{4\pi\varepsilon_0} \frac{4\,|\,\mathbf{d}\,|^2}{3\pi c^3}\,\mathrm{Re} \left\{ \int_0^\infty \omega^4 \mathrm{d}\omega \int_0^t \mathrm{d}t_1 \langle S_+(t_1)S_-(t)\rangle \exp\left[i(\omega-\omega_L)(t-t_1)\right] \right\} ,$$

and assuming that the spectrum will be concentrated near $\omega \cong \omega_L$, we approximate

$$P_s(\infty) = \frac{1}{4\pi\varepsilon_0} \frac{4 |\mathbf{d}|^2 \omega^4}{3\pi c^3} \operatorname{Re}\left\{ \int_0^\infty d\omega \int_0^\infty d\tau \langle S_+(t_0)S_-(t_0 + \tau)\rangle \exp\left[i(\omega - \omega_L)\tau\right] \right\} .$$
$$(10.50)$$

In the last step, we also took the stationary limit $t \to \infty$, and $\tau \equiv t - t_1$, so that the correlation function $\langle S_+(t_0)S_-(t_0 + \tau)\rangle$ only depends on τ, where t_0 is a time much longer than the radiative lifetime.

The power spectrum of the resonance fluorescence is defined as

$$S(\omega) = 2\operatorname{Re}\left\{ \int_0^\infty d\tau \langle S_+(t_0)S_-(t_0 + \tau)\rangle \exp\left[i(\omega - \omega_L)\tau\right] \right\} . \quad (10.51)$$

As we see, basically, we have to calculate the atomic correlation function [4]

$$g(\tau) = \langle S_+(t_0)S_-(t_0 + \tau)\rangle . \quad (10.52)$$

On the other hand, we can write the Heisenberg´s equations for the atomic operators, using the above definition of $S(t)$, and assuming, for simplicity, that the incident field is linearly polarized parallel to the dipole $\mathbf{d}$

$$\dot{S}(t) = -i(\omega_{ab} - \omega_L - i\beta)S(t) - \frac{id}{\hbar}\sigma_z(t)E_0^+(t)\exp(i\omega_L t) , \quad (10.53)$$

$$\dot{\sigma}_z(t) = -2\beta\left[1 + \langle\sigma_z(t)\rangle\right] - \frac{2i}{\hbar}d\left[S(t)E_0^-(t)\exp(-i\omega_L t) - S_+(t)E_0^+(t)\exp(i\omega_L t)\right] . \quad (10.54)$$

Making use of (10.54), we can find a differential equation for $g(\tau)$. It is simple to verify that

$$\frac{dg(\tau)}{d\tau} = -i((\omega_{ab} - \omega_L - i\beta))g(\tau)$$
$$+ \frac{dE_0}{2\hbar}\langle S(t_0)\sigma_z(t_0 + \tau)\rangle , \quad (10.55)$$

where we took initially a coherent state for the driving field, so

$$E_0^+(t) \mid \psi\rangle = \frac{i}{2}E_0\exp(-i\omega_L t) \mid \psi\rangle . \quad (10.56)$$

In order to have a closed set of equations, we further define

$$h(\tau) = \langle S_+(t_0)\sigma_z(t_0 + \tau)\rangle \quad (10.57)$$

and

$$f(\tau) = \langle S_+(t_0)S_+(t_0 + \tau)\rangle . \quad (10.58)$$

The equations satisfied by these three functions g, h, f are [4]

$$\left[\frac{d}{d\tau} + i(\Delta - i\beta)\right] g(\tau) = \frac{\Omega}{2} h(\tau) , \tag{10.59}$$

$$\left(\frac{d}{d\tau} + 2\beta\right) h(\tau) = -2\beta\langle S_+(t_0)\rangle - \Omega g(\tau) - \Omega f(\tau) , \tag{10.60}$$

$$\left[\frac{d}{d\tau} - i(\Delta + i\beta)\right] f(\tau) = \frac{\Omega}{2} h(\tau) , \tag{10.61}$$

with

$$\Delta = \omega_{ab} - \omega_L, \tag{10.62}$$

$$\Omega = \frac{dE_0}{\hbar} . \tag{10.63}$$

In the derivation of the above equations, we made use of the approximation [4]

$$\left[S_+(t_0), E_0^-(t+\tau)\right] \approx 0 . \tag{10.64}$$

We have to solve the above equations, with the following initial conditions

$$g(0) = \langle S_+(t_0)S(t_0)\rangle = \frac{1}{2}\left[1 + \langle\sigma_z(t_0)\rangle\right] , \tag{10.65}$$

$$h(0) = \langle S_+(t_0)\sigma_z(t_0)\rangle = -\langle S_+(t_0)\rangle , \tag{10.66}$$

$$f(0) = \langle S_+(t_0)S_+(t_0)\rangle = 0 , \tag{10.67}$$

where we used the properties of the Pauli-spin matrices.

The initial values $\langle\sigma_z(t_0)\rangle$ and $\langle S_+(t_0)\rangle$ can be easily obtained from the steady-state solutions of the equations

$$\langle \dot{S}(t)\rangle = -i(\Delta - i\beta)\langle S(t)\rangle + \frac{i\Omega}{2}\langle\sigma_z(t)\rangle , \tag{10.68}$$

$$\langle \dot{\sigma}_z(t)\rangle = -2\beta(1 + \langle\sigma_z(t)\rangle) - \Omega\left(\langle S_+(t)\rangle + \langle S(t)\rangle\right) , \tag{10.69}$$

obtained by taking the expectation values of the (10.53) and (10.54).

Equations (10.68, 10.69) are usually referred to as the Optical Bloch equations.

By setting the time derivatives to zero, we readily get

$$g(0) = \frac{\Omega^2}{(4\Delta^2 + 4\beta^2 + 2\Omega^2)} , \tag{10.70}$$

$$h(0) = \frac{\Omega(\beta + i\Delta)}{(2\Delta^2 + 2\beta^2 + \Omega^2)} . \tag{10.71}$$

In the particular case of high intensity ($\Omega \gg \beta$) and zero detuning, we get

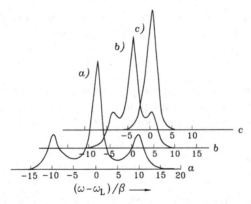

Fig. 10.4. Resonance Fluorescence Spectrum for the follwing parameters: $\Omega = 10\beta$ (**a**); $\Omega = 5\beta$ (**b**); $\Omega = 2\beta$ (**c**); In all cases we took $\Delta = 0$ (after [4])

$$g(\tau) = \frac{1}{2}\left[\exp(-\beta\tau) + \exp(-\frac{3\beta\tau}{2})\cos(\Omega\tau)\right] + (\frac{\beta}{\Omega})^2 , \qquad (10.72)$$

and for the spectrum

$$S(\omega) = 2\pi(\frac{\beta}{\Omega})^2\delta(\omega - \omega_L) + \frac{\frac{\beta}{2}}{(\omega - \omega_L)^2 + \beta^2}$$

$$+ \frac{\frac{3\beta}{8}}{(\omega - \omega_L - \Omega)^2 + \frac{9\beta^2}{4}} + \frac{\frac{3\beta}{8}}{(\omega - \omega_L + \Omega)^2 + \frac{9\beta^2}{4}} . \qquad (10.73)$$

These results were obtained by Burshtein [6], Newstein [5] and Mollow [2].

The more general case, for any Δ and Ω can be also obtained either analytically or numerically.

In Fig. 10.4, we show the resonance fluorescence spectrum, omitting the δ-component for a) $\Omega = 10\beta$, b) $\Omega = 5\beta$, c) $\Omega = 2\beta$. In all cases we took $\Delta = 0$. It is quite striking how the sidebands appear as we increase the laser intensity at the positions $\omega_L \pm \Omega$, with their heights in a ratio 3:1, with respect to the central peak

Experimentally, **the three peaked spectrum** was observed by **Schuda** et al [8], **Walther** et al [9], **Ezekiel** et al [7], and a good agreement between theory and experiment was found.

10.5 Intensity Correlations

As we saw already, from (10.28), the field radiated by a two-level atom is, to a good approximation, proportional to the atomic lowering operator at a retarded time

$$E^+(\mathbf{r}, t) \sim \sigma(t - \frac{r}{c}) , \qquad (10.74)$$

and, if for the moment, we ignore the vector character of the field, the spectrum is related to the first-order correlation function:

$$G^{(1)}(\mathbf{r}, t; \mathbf{r}, t + \tau) = \langle E^-(\mathbf{r}, t) E^+(\mathbf{r}, t + \tau) \rangle . \tag{10.75}$$

Now, we may consider the intensity measurements registered by photodetectors at two different space–time points.

In the Chap. 6, we defined the joint probability density

$$p(r, t_1; r, t_2) \propto G^{(2)}(\mathbf{r}_1, t_1; \mathbf{r}_2, t_2) \tag{10.76}$$

$$= \langle E^-(\mathbf{r}_1, t_1) E - (\mathbf{r}_2, t_2) E^+(\mathbf{r}_2, t_2) E^+(\mathbf{r}_1, t_1) \rangle .$$

A Classical experiment dealing with this type of correlation is the Brown–Twiss experiment [10]. We show the experimental setup in the Fig. 10.5.

A half-silvered mirror divides the incident beam in two identical beams, whose intensities are recorded in the photodetectors P_1 and P_2. One takes the product of $I_1(t)$ and $I_2(t + \tau)$ and averages over a time t, keeping the value of τ fixed. Then, one takes many different values for τ.

If the two beams are independent, this average should be independent of τ. However, the observation of a small bump in the experimental curve, as shown in Fig. 10.6, indicates that the photons have a distinct tendency to arrive in pairs, or photon-bunching effect.

One could regard the effect as a result of the boson nature of the photons.

However, this is not quite true. In Chap. 6 we already discussed the negative Brown–Twiss effect or photon-antibunching effect displaying an anticorrelation for $\tau = 0$.

Carmichael and **Walls** [11] predicted for the first time that photon antibunching should be an observable effect in resonance fluorescence.

In order to study this effect, we consider the joint probability density of photodetection at a same point but two different times

$$p(\mathbf{r}, t; \mathbf{r}, t + \tau) = p(\mathbf{r}, t, t + \tau) \propto G^{(2)}(\mathbf{r}, t; \mathbf{r}, t + \tau) . \tag{10.77}$$

Once more, in the stationary regime, $p(\mathbf{r}, t; \mathbf{r}, t + \tau)$ is independent of t and is proportional to the second-order atomic dipole correlation function

$$g^{(2)}(\tau) = \langle \sigma_+(t_0) \sigma_+(t_0 + \tau) \sigma_-(t_0 + \tau) \sigma_-(t_0) \rangle ,$$

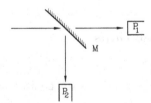

Fig. 10.5. Experimental setup to measure intensity correlations

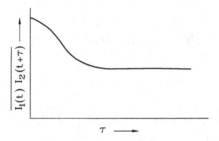

Fig. 10.6. Photon bunching effect near $\tau = 0$

$$= \langle S_+(t_0)S_+(t_0 + \tau)S(t_0 + \tau)S(t_0)\rangle . \qquad (10.78)$$

On the other hand, from the properties of Pauli's spin matrices

$$S_+(t_0 + \tau)S(t_0 + \tau) = \qquad (10.79)$$

$$\sigma_+(t_0 + \tau)\sigma_-(t_0 + \tau) = \frac{1}{2} + \sigma_z(t_0 + \tau) ,$$

so, we can write

$$g^{(2)}(\tau) = \frac{1}{2}\langle S_+(t_0)S(t_0)\rangle + \frac{1}{2}\langle S_+(t_0)\sigma_z(t_0 + \tau)S(t_0)\rangle .$$

$$= \frac{1}{2}g(0) + \frac{1}{2}G(\tau) , \qquad (10.80)$$

where

$$G(\tau) \equiv \langle S_+(t_0)\sigma_z(t_0 + \tau)S(t_0)\rangle . \qquad (10.81)$$

We now proceed in a similar way to the last section, namely, we calculate differential equations for $G(\tau)$ and other correlation functions. Define

$$F(\tau) \equiv \langle S_+(t_0)S_+(t_0 + \tau)S(t_0)\rangle , \qquad (10.82)$$

$$H(\tau) \equiv \langle S_+(t_0)S(t_0 + \tau)S(t_0)\rangle . \qquad (10.83)$$

Using the Heisenberg equations for the atomic operator, we readily find three differential equations for G, F, H

$$\frac{dG(\tau)}{d\tau} = -2\beta g(0) - 2\beta G(\tau) - \Omega(F(\tau) + H(\tau)) , \qquad (10.84)$$

$$\frac{dF(\tau)}{d\tau} = i(\Delta + i\beta)F(\tau) + \frac{\Omega}{2}G(\tau) , \qquad (10.85)$$

$$\frac{dH(\tau)}{d\tau} = -i(\Delta - i\beta)F(\tau) + \frac{\Omega}{2}G(\tau) , \qquad (10.86)$$

with the initial conditions

$$G(0) = \langle S_+(t_0)\sigma_z(t_0)S(t_0)\rangle = -\langle S_+(t_0)S(t_0)\rangle = -g(0) , \qquad (10.87)$$

$$F(0) \equiv \langle S_+(t_0)S_+(t_0)S(t_0) \rangle = 0 , \tag{10.88}$$

$$H(0) \equiv \langle S_+(t_0)S(t_0)S(t_0) \rangle = 0 . \tag{10.89}$$

From (10.80), we immediately notice that

$$g^{(2)}(\tau = 0) = 0 , \tag{10.90}$$

thus, we see that there is an antibunching effect.
The exact solution for $\Delta = 0$ is

$$g^{(2)}(\tau) = g(0)^2 \left[1 - \exp(-\frac{3\beta\tau}{2})(\cos \Omega'\tau + \frac{2\beta}{2\Omega'}\sin \Omega'\tau) \right] , \tag{10.91}$$

with

$$\Omega' \equiv \sqrt{\Omega^2 - \frac{\beta^2}{4}} . \tag{10.92}$$

The behaviour of

$$\frac{g^{(2)}(\tau)}{g(0)^2} = (g^{(2)}(\tau))_{\text{norm}} \tag{10.93}$$

is shown in Fig. 10.7, where the solid line corresponds to $\Omega = 5$ and the dotted line to $\Omega = 0.5$. In both cases we took $\beta = 0.5$.

On the other hand, $\left[g^{(2)}(\tau) \right]_{\text{norm}}$ can also be expressed as a photon number correlation:

$$\left[g^{(2)}(\tau) \right]_{\text{norm}} = \frac{\langle : n(t)n(t + \tau) : \rangle}{\langle n(t) \rangle^2} . \tag{10.94}$$

For chaotic light, $\left[g^{(2)}(\tau) \right]_{\text{norm}} = 2$, that is the correlation is twice the random background correlation, showing the tendency of photons to bunch together.

On the other hand, as we mentioned before, in resonance fluorescence and for $\tau = 0$, antibunching occurs.

The physical interpretation of this effect is quite simple. Right after the detection of the first photon, the atom is in the lower state and it requires a

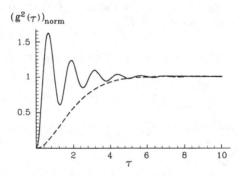

Fig. 10.7. Second-order intensity correlation $(g^{(2)}(\tau))_{\text{norm}}$ versus time, for $\Omega = 5.0$ (solid line), $\Omega = 0.5$ (dashed line). In both cases, $\beta = 0.5$

finite (non-zero) amount of time to get back to the excited state and be able to emit the second photon.

Experiments verifying the antibunching effect in resonance fluorescence were performed by **Kimble** et al. [12] and **Leuchs** et al. [13].

Lately, the influence of quantum interference on the resonance fluorescent spectrum was studied when the system was in contact with coloured or white noise [14, 15, 16, 17, 18, 19, 20, 21, 22]

Problems

10.1. Prove (10.72).
10.2. Prove (10.73).
10.3. Prove (10.91) for $\Delta = 0$.

References

1. Weisskopf, V., Wigner, E.: Z. Phys., **63**, 54 (1930)
2. Mollow, B.R., Phys. Rev., **188**, 1969 (1969)
3. Cohen-Tannoudji, C., Reynaud, S.: J. Phys. B, **10**, 3451 (1976)
4. Knight, P.L., Milonni, P.W.: Phys. Rep., **66**, 21 (1980)
5. Newstein, A.L.: Phys. Rev., **167**, 89 (1968)
6. Burshtein, A.L.: Sov. Phys. J.E.T.P, **21**, 567 (1965), *ibid* **22**, 939 (1966)
7. Yu, F.Y., Grove, R.E., Ezekiel, S.: Phys. Rev. Lett., **35**, 1426 (1975) Also:R.E.Grove, F.Y.Yu, S.Ezekiel, Phys.Rev.A, **15**, 227 (1977).
8. Schuda, F., Stroud, C.R., Hercher, M.: J. Phys. B, Atom. Mol. Phys, **7**, L198 (1975)
9. Walther, H.: Bull. Am. Phys. Soc, II20, 1467 (1975), also: Hartig, W., Rasmussen, W., Schieder, R., Walther, H.: Z. Physik A, **278**, 205 (1976)
10. Handbury Brown R., Twiss, R.Q.: Nature, **177**, 27 (1956)
11. Carmichael, H.J., Walls, D.F.: J. Phys. B, Mol. Phys, **9**, 1199 (1976)
12. Kimble, H.J., Dagenais, M., Mandel, L.: Phys. Rev. Lett., **39**, 691 (1977), *ibid* Phys. Rev. A, **18**, 201 (1978)
13. Cresser, J.D., Hager, J., Leuchs, G., Rateike, F.M., Walther H.: In: Dissipative Systems in Quantum Optics. Bonifacio, R. (ed.) Springer Verlag (1982)
14. Zhu, S.Y., Narducci, L.M., Scully, M.O.: Phys. Rev. A, **52**, 4971 (1995)
15. Xia, H.R., Ye, C.Y., Zhu, S.Y.: Phys. Rev. Lett., **77**, 1031 (1996)
16. Zhu, S.Y., Scully, M.O.: Phys. Rev. Lett., **76**, 388 (1996)
17. Zhou, P., Swain, S.: Phys. Rev. Lett., **77**, 3995 (1996)
18. Zhou, P., Swain, S.: Phys. Rev. Lett., **78**, 832 (1997)
19. Paspalakis, E., Keitel, C.H., Knight, P.L.: Phys. Rev. A, **58**, 4868 (1998)
20. Smyth, W.S., Swain, S.: Phys. Rev. A, **59**, R2583 (1999)
21. Zhou, P., Swain, S.: Phys. Rev. A, **59**, 1603 (1999)
22. Li, F., Zhu, S.Y.: Phys. Rev. A, **59**, 2330 (1999)

Further Reading

Line Narrowing in Fluorescence

- Narducci, L.M., Scully, M.O., Oppo, G.L., Ru, P., Tredice, J.R.: Phys. Rev. A, **42**, 1630 (1990)
- Keitel, C.H., Narducci, L.M., Scully, M.O.: App. Phys. B, **60**, 5153 (1995)
- Keitel, C.H.: J. Mod. Opt, **43**, 1555 (1996)

Discussion on the Physical Spectrum

- Eberly, J.H., Wodkiewicz, K.: JOSA B, **67**, 1252 (1977)

11. Quantum Laser Theory: Master Equation Approach

In this chapter, we study the Laser Theory, with the master equation approach. We include the influence of pump statistics.

For large fields interacting with atoms, the Semiclassical description, that is considering the atoms quantum mechanically and the field classical, seems to be adequate, to describe the most classical features, such as threshold, steady-state intensity, etc. However, whenever quantum fluctuations are to be considered, like to determine the laser linewidth, photon statistics, etc. we require the fully quantized field.

The Hamiltonian of a two-level atom interacting with a single mode (cavity mode) of the field is described by the Jaynes–Cummings Hamiltonian

$$H = \frac{\hbar\omega}{2}\sigma_z + \hbar\Omega a^\dagger a + \hbar g(\sigma^+ a + \sigma^- a^\dagger) , \tag{11.1}$$

within the dipole and rotating-wave approximations. It is convenient to split the Hamiltonian into two terms

$$H = H_1 + H_2 , \tag{11.2}$$

where

$$H_1 = \frac{\hbar\Omega}{2}\sigma_z + \hbar\Omega a^\dagger a , \tag{11.3}$$

$$H_2 = \hbar\frac{\Delta\Omega}{2}\sigma_z + \hbar g(\sigma^+ a + \sigma^- a^\dagger),$$

with

$$\Delta\Omega \equiv \omega - \Omega . \tag{11.4}$$

It is simple to see that $[H_1, H_2] = 0$, and when we go to the interaction picture, the dynamics is governed by $H_2 \equiv V$. The time evolution operator can be exactly computed as

$$U(\tau) = \exp(-i\frac{V\tau}{\hbar}) = \sum_{n=0}^{\infty} \frac{(\frac{-i\tau}{\hbar})^n}{n!} V^n = \sum_{n=0}^{\infty} \frac{(-i\tau)^n}{n!} \begin{bmatrix} \frac{\Delta\Omega}{2} & ga \\ ga^\dagger & -\frac{\Delta\Omega}{2} \end{bmatrix}^n . \tag{11.5}$$

It is simple to show that

$$\begin{bmatrix} \frac{\Delta\Omega}{2} & ga \\ ga^\dagger & -\frac{\Delta\Omega}{2} \end{bmatrix}^{2m} = \begin{bmatrix} (\varphi + g^2)^m & 0 \\ 0 & (\varphi)^m \end{bmatrix}, \tag{11.6}$$

$$\begin{bmatrix} \frac{\Delta\Omega}{2} & ga \\ ga^\dagger & -\frac{\Delta\Omega}{2} \end{bmatrix}^{2m+1} = \begin{bmatrix} \frac{\Delta\Omega}{2}(\varphi + g^2)^m & g(\varphi + g^2)^m a \\ ga^\dagger(\varphi + g^2)^m & -\frac{\Delta\Omega}{2}(\varphi)^m \end{bmatrix}$$

where $\varphi \equiv g^2 a^\dagger a + (\frac{\Delta\Omega}{2})^2$. It then follows that

$$U(\tau) = \begin{bmatrix} \cos(\tau\sqrt{\varphi + g^2}) - \frac{i\Delta\Omega}{2}\frac{\sin(\tau\sqrt{\varphi+g^2})}{\sqrt{\varphi+g^2}} & -ig\frac{\sin(\tau\sqrt{\varphi+g^2})}{\sqrt{\varphi+g^2}}a \\ -iga^\dagger\frac{\sin(\tau\sqrt{\varphi+g^2})}{\sqrt{\varphi+g^2}} & \cos(\tau\sqrt{\varphi}) + \frac{i\Delta\Omega}{2}\frac{\sin(\tau\sqrt{\varphi})}{\sqrt{\varphi}} \end{bmatrix}.$$
$$\tag{11.7}$$

If the initial atom-field density operator is $\rho(0)$, after an interaction time t, will be given by $\rho(\tau) = U(\tau)\rho(0)U^\dagger(\tau) = U(\tau)\rho_f(0)\begin{bmatrix} 1 & 0 \\ 0 & 0 \end{bmatrix}U^\dagger(\tau)$, assuming that initially the atom is in the upper state factorizes with the field. By performing the matrix product and tracing over the atom, that is adding up the diagonal elements, we find

$$\rho_f(\tau) = \cos(\lambda\tau)\rho_f(0)\cos(\lambda\tau) + g^2 a^\dagger \left[\frac{\sin(\lambda\tau)}{\lambda}\right]\rho_f(0)\left[\frac{\sin(\lambda\tau)}{\lambda}\right]a \equiv M(\tau)\rho_f(0),$$
$$\tag{11.8}$$

with $\lambda \equiv g\sqrt{a^\dagger a + 1}$, and we assumed zero detuning. M is the gain superoperator acting on ρ.

11.1 Heuristic Discussion of Injection Statistics

We assume that a dense flux of atoms goes through an excitation region, and each atom has a probability p of being excited from the ground state c to the upper level a (See Fig. 11.1).

We further assume that the levels a and b are involved in the laser or maser transition, and that the level b remains unpopulated. We also assume that the beam has a regular distribution before arriving to the excitation region, so the number K of the atoms which cross that region, during a time Δt, is given by

$$K = R\Delta t, \tag{11.9}$$

where R is the injection rate and Δt is much larger than the time interval between consecutive atoms.

This model may describe a system called a **micromaser**, in which case, one has a beam of highly excited or Rydberg atoms crossing a high-quality microwave cavity, with a couple of energy levels resonant with the microwave field inside the cavity. If we assume that τ is the interaction time of each atom with the cavity field, we may use the same model to describe the excitation

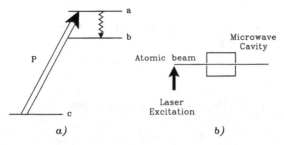

Fig. 11.1. (a) An atom is excited from the level c to the upper level. The lasing transition occurs between the a and b levels. (b) An atomic beam arrives to the excitation region where Rydberg states (micromaser) or excited states (laser) are generated prior entering the cavity (micromaser) or participating in the laser action (laser)

process of a laser, in which case τ is related to the atomic lifetime of the lasing levels. In the past, people neglected the effects of pump statistics. However, some recent experiments in micromasers and lasers showed that by controlling the pump noise, one could get a large reduction in the photon number fluctuations. The probability for k atoms to be excited during a time Δt is given by

$$P(k, K) = \begin{bmatrix} K \\ k \end{bmatrix} p^k (1 - p)^{K-k} . \tag{11.10}$$

The average number of excited atoms and the variance is given by

$$\overline{k} = \sum_{k=0}^{k=K} kP(k, K) = pK = r\Delta t , \tag{11.11}$$

$$\overline{\Delta k^2} = (1 - p)\overline{k},.$$

with $r \equiv Rp$.

11.2 Master Equation for Generalized Pump Satistics

Let the interaction time between the atom and the field be τ. Also, we assume that the j-th atom is "injected" at the time t_j. So the field, after interacting with the j-th atom, can be written as (we skip the field subindex)

$$\rho(t_j + \tau) = M(\tau)\rho(t_j) . \tag{11.12}$$

Now, if k atoms are excited, then

$$\rho^{(k)}(t) = M^k(\tau)\rho(0) . \tag{11.13}$$

Of course, if we do not know the number of atoms, but only probabilistically by (11.10), we then have

$$\rho(t) = \sum_{k=0}^{K} \binom{K}{k} p^k (1-p)^{K-k} M^k(\tau)\rho(0) \qquad (11.14)$$

$$= [1 + p(M-1)]^K \rho(0) ,$$

with K=Rt.

Differentiating (11.14), with respect to time, we get

$$\frac{d\rho(t)}{dt} = \frac{r}{p} \ln\left[1 + p(M-1)\right]\rho(t) + L\rho(t) . \qquad (11.15)$$

Equation (11.15) is our Generalized master equation. [1, 2]

In (11.15), we have added the cavity loss term denoted by $L\rho(t)$. This term, can be borrowed from the quantum theory of damping, where the oscillator is our single mode field interacting with a reservoir, at zero temperature. Thus,

$$L\rho(t) = \frac{C}{2}(2a\rho a^\dagger - a^\dagger a\rho - \rho a^\dagger a) , \qquad (11.16)$$

$$C = \frac{1}{t_{cav}} = \frac{\Omega}{Q} ,$$

Q being the cavity quality factor and t_{cav}, the photon's lifetime inside the cavity.

If the average photon number is sufficiently large and the distribution narrow, one can expand (11.15) and get

$$\frac{d\rho}{dt} = r(M-1)\rho(t) - \frac{1}{2}rp(M-1)^2\rho(t) + L\rho(t) . \qquad (11.17)$$

If we now use the expression for M given by (11.8), we get

$$\frac{d\rho}{dt} = r(1+p)\left\{\cos(\lambda\tau)\rho\cos(\lambda\tau) + g^2 a^\dagger\left[\frac{\sin(\lambda\tau)}{\lambda}\right]\rho\left[\frac{\sin(\lambda\tau)}{\lambda}\right]a\right\} - r(1+\frac{p}{2})\rho$$

$$\qquad (11.18)$$

$$-\frac{rp}{2}\{\cos^2(\lambda\tau)\rho\cos^2(\lambda\tau) + g^2\cos(\lambda\tau)a^\dagger\left[\frac{\sin(\lambda\tau)}{\lambda}\right]\rho\left[\frac{\sin(\lambda\tau)}{\lambda}\right]a\cos(\lambda\tau)$$

$$+g^2 a^\dagger\left[\frac{\sin(\lambda\tau)}{\lambda}\right]\cos(\lambda\tau)\rho\cos(\lambda\tau)\left[\frac{\sin(\lambda\tau)}{\lambda}\right]a$$

$$+g^4 a^\dagger\left[\frac{\sin(\lambda\tau)}{\lambda}\right]a^\dagger\left[\frac{\sin(\lambda\tau)}{\lambda}\right]\rho\left[\frac{\sin(\lambda\tau)}{\lambda}\right]a\left[\frac{\sin(\lambda\tau)}{\lambda}\right]a\} + L\rho .$$

From the expression above, one can calculate ρ_{nn} and $\rho_{n,n+1}$ that will give us the photon statistics and laser linewidth, respectively.

Finally, we notice that the two extreme cases are $p = 0$, while pR = constant, in which case the Bernoulli distribution becomes a Poissonian distribution, corresponding to the random injection case and the usual Scully–Lamb laser theory, whereas $p = 1$ corresponds to a regular injection of atoms.

The generalized master (11.15) was derived under a couple of approximations. First, there is the course graining approximation $\frac{\Delta \rho}{\Delta t} \approx \frac{d\rho}{dt}$, such that during Δt there were many atoms injected into the interaction region, which is normally a good approximation for lasers but could run into problems when dealing with a small photon number, for example in a micromaser. The other approximation is that the loss is independent of the gain, which may also lead to erroneous results when $p \neq 0$ [3]. This latter assumption is exact in the Poissonian case.

11.3 The Quantum Theory of the Laser: Random Injection $(p = 0)$

In the case of zero detuning, the time evolution operator 11.7 becomes

$$U(\tau) = \begin{bmatrix} \cos g\tau \sqrt{aa^\dagger} & -i \left[\frac{\sin g\tau \sqrt{aa^\dagger}}{\sqrt{aa^\dagger}} \right] a \\ -ia^\dagger \left[\frac{\sin g\tau \sqrt{aa^\dagger}}{\sqrt{aa^\dagger}} \right] & \cos g\tau \sqrt{a^\dagger a} \end{bmatrix}, \tag{11.19}$$

and one can write

$$\rho_f(t + \tau) = \cos(\sqrt{aa^\dagger} g\tau) \rho_f(t) \cos(\sqrt{aa^\dagger} g\tau)$$
$$+ a^\dagger \left[\frac{\sin(\sqrt{aa^\dagger} g\tau)}{\sqrt{aa^\dagger}} \right] \rho(t) \left[\frac{\sin(\sqrt{aa^\dagger} g\tau)}{\sqrt{aa^\dagger}} \right] a. \tag{11.20}$$

To make a realistic model of a laser, we assume that the atoms have a distribution of time they spend in the cavity. In the case of the two-level atom model, as in this theory, the two levels decay at a rate γ, and the time distribution is

$$P(\tau) = \gamma \exp - \gamma \tau. \tag{11.21}$$

Now, defining again a course time grain $\Delta t \gg \langle \tau \rangle$, we can write

$$\left(\frac{d\rho_f}{dt} \right)_{\text{gain}} \approx \frac{\rho_f(t + \Delta t) - \rho_f(t)}{\Delta t} = r \left[\rho_f(t + \Delta t) - \rho_f(t) \right] \tag{11.22}$$

$$= -r\rho_f(t) + r \int_0^{\Delta t \to \infty} d\tau \exp - \gamma \tau$$
$$\times \left[\begin{array}{c} \cos(\sqrt{aa^\dagger} g\tau) \rho_f(t) \cos(\sqrt{aa^\dagger} g\tau) \\ + a^\dagger \left[\frac{\sin(\sqrt{aa^\dagger} g\tau)}{\sqrt{aa^\dagger}} \right] \rho(t) \left[\frac{\sin(\sqrt{aa^\dagger} g\tau)}{\sqrt{aa^\dagger}} \right] a \end{array} \right]$$

For a typical laser, the arguments of the sine and cosine are small, and one can expand (up to 4-th order in g)

$$\cos g\tau \sqrt{aa^\dagger} \approx 1 - \frac{g^2 \tau^2}{2} aa^\dagger + \frac{1}{24} g^4 \tau^4 aa^\dagger aa^\dagger..,$$

$$\left[\frac{\sin g\tau\sqrt{aa^\dagger}}{\sqrt{aa^\dagger}}\right] \approx g\tau - \frac{g^3\tau^3 aa^\dagger}{6}\dots$$

and replace into the master equation 11.22, getting

$$\frac{d\rho_f}{dt} = -\frac{\mathcal{A}}{2}\{aa^\dagger\left[\rho_f - (\frac{g}{\gamma})^2\left(aa^\dagger\rho_f + 3\rho_f aa^\dagger\right)\right] \tag{11.23}$$

$$+\left[\rho_f - (\frac{g}{\gamma})^2\left(\rho_f aa^\dagger + 3aa^\dagger\rho_f\right)\right]aa^\dagger$$

$$-2a^\dagger\left[\rho_f - 2(\frac{g}{\gamma})^2\left(aa^\dagger\rho_f + \rho_f aa^\dagger\right)\right]a\}$$

$$-\frac{C}{2}\left(a^\dagger a\rho_f + \rho_f a^\dagger a - 2a\rho_f a^\dagger\right),$$

where

$$\mathcal{A} \equiv \frac{2rg^2}{\gamma^2}, \tag{11.24}$$

$$\mathcal{B} \equiv \frac{4g^2}{\gamma^2}\mathcal{A},$$

$$C = \frac{\Omega}{Q}.$$

and to derive (11.23), we used the following integrals

$$\int \tau^2\gamma\exp(-\gamma\tau)d\tau = \frac{2}{\gamma^2},$$

$$\int \tau^4\gamma\exp(-\gamma\tau)d\tau = \frac{24}{\gamma^4}.$$

The coefficient $\mathcal{A}$ is the gain, $\mathcal{B}$ the saturation and C the cavity loss.

11.3.1 Photon Statistics

We go back to the full non-linear theory.

We take the nm matrix elements of (11.22) and perform the time integrals, using

$$\gamma\int_0^\infty d\tau\exp(-\gamma\tau)\left(\begin{array}{c}\cos g\tau\sqrt{n+1}\cos g\tau\sqrt{m+1}\\ \sin g\tau\sqrt{n+1}\sin g\tau\sqrt{m+1}\end{array}\right) \tag{11.25}$$

$$= \frac{\left(\begin{array}{c}1 + (\frac{g}{\gamma})^2(n+m+2)\\ 2(\frac{g}{\gamma})^2\sqrt{(n+1)(m+1)}\end{array}\right)}{1 + 2(\frac{g}{\gamma})^2(n+m+2) + (\frac{g}{\gamma})^4(n-m)^2},$$

getting

$$\left(\frac{d\rho}{dt}\right)_{nm} = -\frac{\mathcal{N}'_{nm}\mathcal{A}}{1 + \mathcal{N}_{nm}\frac{\mathcal{B}}{\mathcal{A}}}\rho_{nm} + \frac{\sqrt{nm}\mathcal{A}}{1 + \mathcal{N}_{n-1,m-1}\frac{\mathcal{B}}{\mathcal{A}}}\rho_{n-1,m-1} \quad (11.26)$$
$$-\frac{C}{2}(n+m)\rho_{nm} + C\sqrt{(n+1)(m+1)}\rho_{n+1,m+1},$$

where $\mathcal{A}$ and $\mathcal{B}$ have already been defined and

$$\mathcal{N}'_{nm} = \frac{1}{2}(n+m+2) + \frac{\frac{1}{8}(n-m)^2\mathcal{B}}{\mathcal{A}}, \quad (11.27)$$
$$\mathcal{N}_{nm} = \frac{1}{2}(n+m+2) + \frac{\frac{1}{16}(n-m)^2\mathcal{B}}{\mathcal{A}},$$

and we included, as usual, the cavity losses. This is the Scully–Lamb laser theory. [4]

For the photon statistics, we take the diagonal element of (11.27), getting

$$\left(\frac{d\rho}{dt}\right)_{nn} = -\frac{\mathcal{A}(n+1)}{1 + (n+1)\frac{\mathcal{B}}{\mathcal{A}}}\rho_{nn} + \frac{n\mathcal{A}}{1 + n\frac{\mathcal{B}}{\mathcal{A}}}\rho_{n-1,n-1} \quad (11.28)$$
$$-C(n)\rho_{nn} + C(n+1)\rho_{n+1,n+1}.$$

The term $\left[\frac{\mathcal{A}(n+1)}{1+(n+1)\frac{\mathcal{B}}{\mathcal{A}}}\right]\rho_{nn}$ represents the gain for $\rho_{n+1,n+1}$, due to stimulated emission caused by the presence of the amplifying atoms, and the term $C(n+1)\rho_{n+1,n+1}$ is the loss on this level.

In steady state $(\frac{d\rho}{dt})_{nn} = 0$, and we get

$$\rho_{n+1,n+1} = \frac{\frac{\mathcal{A}}{C}}{1 + (n+1)\frac{\mathcal{B}}{\mathcal{A}}}\rho_{nn}, \quad (11.29)$$

and the solution is

$$\rho_{nn} = \rho_{00}\left(\frac{\mathcal{A}}{C}\right)^n \prod_{k=0}^{n}\left(1 + \frac{k\mathcal{B}}{\mathcal{A}}\right)^{-1}. \quad (11.30)$$

When $\frac{\mathcal{A}}{C} < 1$, the laser is below threshold, since ρ_{nn} is decreasing monotonically with n and the normalization condition gives us: $(\frac{k\mathcal{B}}{\mathcal{A}} \ll 1)$

$$\sum_n \rho_{nn} = 1 = \rho_{00}\sum_n\left(\frac{\mathcal{A}}{C}\right)^n = \frac{\rho_{00}}{1 - \frac{\mathcal{A}}{C}}, \quad (11.31)$$

and if we define $\frac{\mathcal{A}}{C} = \exp(-\frac{\hbar\Omega}{K_B T})$, then

$$\rho_{nn} = \left(1 - \frac{\mathcal{A}}{C}\right)\left(\frac{\mathcal{A}}{C}\right)^n, \quad (11.32)$$

becomes the Bose–Einstein statistics for the black-body radiation. That is, the laser below threshold behaves like an incandescent lamp, with a given temperature.

For the case $\frac{A}{C} > 1$, we use the exact formula (11.29) and get

$$\rho_{nn} = \rho_{00} \left(\frac{A}{C}\right)^n \prod_{k=1}^{n} \left(1 + \frac{kB}{A}\right)^{-1} . = \rho_{00} \left(\frac{A^2}{BC}\right)^n \prod_{k=1}^{n} \left(\frac{A}{B} + k\right)^{-1} \tag{11.33}$$

We calculate the average photon number

$$\langle n \rangle = \sum_n n\rho_{nn} = \rho_{00} \sum_n \left(n + \frac{A}{B} - \frac{A}{B}\right) \left(\frac{A^2}{BC}\right)^n \prod_{k=1}^{k=n} \left[\frac{1}{\left(k + \frac{A}{B}\right)}\right] = \tag{11.34}$$

$$\rho_{00} \frac{A^2}{BC} \sum_n \left(\frac{A^2}{BC}\right)^{n-1} \prod_{k=1}^{k=n-1} \left[\frac{1}{\left(k + \frac{A}{B}\right)}\right] - \frac{A}{B} \sum_{n=1}^{\infty} \rho_{nn}$$

$$= \frac{A^2}{BC} - \frac{A}{B}(1 - \rho_{00}) = \frac{A}{B}\left(\frac{A}{C} - 1\right) + \frac{A}{B}\rho_{00}.$$

Well over the threshold $\langle n \rangle \to \frac{A^2}{BC}$ and the photon statistics

$$\rho_{nn} \approx \frac{(\exp-\langle n \rangle)\,(\langle n \rangle)^n}{n!} \tag{11.35}$$

that corresponds to a Poisson statistics.

The change in photon statistics below and above threshold is illustrated in the Fig. 11.2.

We notice that for $\frac{B}{A} \ll 1$,

$$\rho_{nn} = \rho_{00} \prod_{k=1}^{n} \left[\frac{A^2}{BC\left(\frac{A}{B} + k\right)}\right],$$

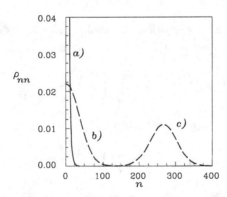

Fig. 11.2. Steady-state photon statistics versus n, for the cases below (**a**), at (**b**) and above (**c**) threshold. (After [4])

$$\rho_{nn} = \rho_{00} \prod_{k=1}^{n} \left[\frac{\mathcal{A}}{C(1 + \frac{kB}{\mathcal{A}})} \right],$$

$$\rho_{nn} \approx \rho_{00} \prod_{k=1}^{n} \left(\frac{\mathcal{A} - kB}{C} \right).$$

11.3.2 The Fokker–Planck Equation: Laser Linewidth

We start with the approximate master (11.23), and neglecting the non-linear terms, we can use the rules described in (9.45), to write a Fokker–Planck Equation in terms of the Glauber's P distribution as

$$\frac{\partial P}{\partial t} = -\frac{\mathcal{A} - C}{2} \left(\frac{\partial}{\partial \alpha} \alpha + \frac{\partial}{\partial \alpha^*} \alpha^* \right) P + \mathcal{A} \frac{\partial^2 P}{\partial \alpha \partial \alpha^*}. \tag{11.36}$$

The non-linear term can be included in the Fokker–Planck equation, using a clever trick [5] of observing that (11.36) is modified, according to (11.23) by replacing $P \to P \left[1 - \frac{4g^2}{\gamma^2} \mid \alpha \mid^2 \right]$, getting

$$\frac{\partial P}{\partial t} = -\frac{1}{2} \frac{\partial}{\partial \alpha} \left\{ \left[(\mathcal{A} - C) - 4\mathcal{A} \frac{g^2}{\gamma^2} \mid \alpha \mid^2 \right] \alpha P - \mathcal{A} \frac{\partial P}{\partial \alpha^*} \right\} \tag{11.37}$$
$$- \frac{1}{2} \frac{\partial}{\partial \alpha^*} \left\{ \left[(\mathcal{A} - C) - 4\mathcal{A} \frac{g^2}{\gamma^2} \mid \alpha \mid^2 \right] \alpha^* P - \mathcal{A} \frac{\partial P}{\partial \alpha} \right\}.$$

This equation has been extensively studied, particularly in connection with the no lasing–lasing phase transition [6, 7].

For the linewidth purposes, the linear Fokker–Planck Equation (11.36) is sufficient. We go to polar coordinates

$$\alpha = r \exp i\varphi$$

and get, neglecting the radial variations,

$$\frac{\partial P(\theta, t)}{\partial t} = \frac{\mathcal{A}}{4\langle n \rangle} \frac{\partial^2}{\partial \theta^2} P(\theta, t) \equiv \frac{D}{2} \frac{\partial^2}{\partial \theta^2} P(\theta, t). \tag{11.38}$$

D is the phase diffusion constant and corresponds, as we shall see from the Langevin theory, precisely to the Schawlow–Townes laser linewidth, when $A \approx C$, that is not far from the threshold. The present fourth-order expansion loses its validity well above threshold. It is interesting to observe that C is the empty cavity linewidth, thus the formula

$$D = \frac{C}{2\langle n \rangle} \tag{11.39}$$

shows that the linewidth is decreased by a factor $\langle n \rangle^{-1}$.

These results were also calculated by Lax [8], Gordon [9] and Haken [10].

11.3.3 Alternative Derivation of the Laser Linewidth

We present, in this section, a different approach to the laser linewidth, which is related to the off-diagonal elements of the field density matrix.

The expressions obtained here will be useful in calculating the quantum phase fluctuations in a laser (Chap. 15).

We begin by rewriting the equation of motion of the off-diagonal elements $\rho_{n,n+k}$, in a form related to the diagonal elements. We introduce the following notation

$$\rho_{n,n'} = \rho_{n,n+k} \equiv \phi_n(k,t) , \tag{11.40}$$

where k is the distance from the main diagonal. From (11.26), and expanding to the lowest order in $\frac{\mathcal{B}}{\mathcal{A}}$, we get

$$\dot{\phi}_n(k,t) = -\frac{k^2}{8}\mathcal{B}\phi_n(k,t) - \left(n+1+\frac{k}{2}\right)\left[\mathcal{A} - \mathcal{B}\left(n+1+\frac{k}{2}\right)\right]\phi_n(k,t) \tag{11.41}$$

$$+[n(n+k)]^{\frac{1}{2}}\left[\mathcal{A} - \mathcal{B}\left(n+\frac{k}{2}\right)\right]\phi_{n-1}(k,t)$$

$$-\mathcal{C}\left(n+\frac{k}{2}\right)\phi_n(k,t)$$

$$+[(n+1)(n+k+1)]^{\frac{1}{2}}\phi_{n+1}(k,t) .$$

The first three terms represent the gain and the last two the loss in the cavity.

We assume a general solution of the form

$$\mu_s^{(k)} = \sum_s \varphi_s(n,k)\exp[-\mu_s^{(k)}t] , \tag{11.42}$$

where the $\mu_s^{(k)}$ are the eigenvalues and $\varphi_s(n,k)$ the eigenvectors. As we will see, the eigenvalues are either positive or zero.

In the case $k = 0$ (diagonal elements), the fact that the steady-state solution exists implies that

$$\mu_0^{(0)} = 0 . \tag{11.43}$$

For the off-diagonal elements, one finds that $\mu_s^{(k)} > 0$, implying that the solution (for $k \neq 0$)

$$\mu_s^{(k)} \underset{t\to\infty}{\to} 0 . \tag{11.44}$$

Now, for the laser far above threshold, the lowest eigenvalue $\mu_0^{(k)}$ will be small, and we look for a solution of the form

$$\phi_n(k,t) = N_k\left[\prod_{l=0}^{n}\left(\frac{\mathcal{A}-l\mathcal{B}}{\mathcal{C}}\right)\prod_{m=0}^{n+k}\left(\frac{\mathcal{A}-m\mathcal{B}}{\mathcal{C}}\right)\right]^{\frac{1}{2}}\exp(-\mu_0^{(k)}t) \tag{11.45}$$

$$\approx \sqrt{\rho_{n,n}\rho_{n+k,n+k}} \exp(-\mu_0^{(k)}t).$$

One finds [4] that to a very good approximation, the above differential equation is satisfied for

$$\mu_0^{(k)} = \frac{1}{2}k^2 D , \qquad (11.46)$$

for

$$D = \frac{C}{2\langle n \rangle} . \qquad (11.47)$$

Thus, the off-diagonal elements of the laser field are given by

$$\rho_{n,n+k}(t) = \rho_{n,n+k}(0) \exp(-\mu_0^{(k)}t) , \qquad (11.48)$$

which back in the Schrödinger picture becomes

$$\rho_{n,n+k}(t) = \rho_{n,n+k}(0) \exp(-i\Omega t - \mu_0^{(k)}t) . \qquad (11.49)$$

As we did in Chap. 8, a one-mode stationary field is written as

$$E(t) = \varepsilon(a + a^\dagger) \sin kz , \qquad (11.50)$$

so the statistical average will be

$$E(t) = \varepsilon \sum_n (\rho_{n,n+1}(t)\sqrt{n+1} + cc) \sin kz$$

$$= \varepsilon \sum_n (\rho_{n,n+1}(0)\sqrt{n+1} \exp(-i\Omega t - \mu_0^{(k)}t) + cc) \sin kz$$

$$= E_0 \cos(\Omega t + \varphi) \exp(-\frac{D}{2}t) , \qquad (11.51)$$

where

$$E_0 = 2\varepsilon A , \qquad (11.52)$$

and

$$\sum_n \rho_{n,n+1}(0)\sqrt{n+1} = A \exp(-i\varphi) . \qquad (11.53)$$

The decay of the electric field can be understood as a result of a random walk of the ensemble average electric field because of the stochastic process that influences the system. After a certain amount of time, the phase of the field will have diffused to cover uniformly the whole 2π range. This will be seen in more detail in Chap. 15, when dealing with the quantum phase.

Finally, the Fourier transform of the average electric field gives the laser spectrum

$$\mid E(\omega) \mid^2 = \mid \int_0^\infty dt \exp(-i\omega t) E_0 \cos(\Omega t + \varphi) \exp\left(-\frac{Dt}{2}\right) \mid^2$$

$$= \frac{E_0^2}{4} \left[\frac{1}{(\omega - \Omega)^2 + (\frac{D}{2})^2} \right] , \tag{11.54}$$

which is a Lorenzian with a full width at half maximum of D.

As a final note on this issue, the stochasticity of the phase of the laser field has its origin on the spontaneous emission. Thus, if one could somehow control this spontaneous emission, it would be possible to decrease the linewidth by a substantial amount.

This subject will be treated in some detail in the Chap. 13, when dealing with the correlated emission laser and the methods of quenching the phase diffusion in a laser.

11.4 Quantum Theory of the Micromaser: Random injection ($p = 0$)

11.4.1 Generalities

As we have seen in the previous chapter, one of the simplest and most fundamental systems to study radiation–matter coupling is a single two-level atom interacting with the one-mode electromagnetic field. For a long time, this model remained only a theoretical scheme, as it was not possible to test experimentally the effects predicted by the model. These effects, among others, are for example the modification of the spontaneous emission rates of a single atom, in a resonant cavity, the oscillatory exchange of energy between the atom and the field, the disappearance and quantum revival of the Rabi notation.

However, the situation has changed over the last decade mainly because of two important factors. The introduction of highly tunable Dye Lasers, which can excite large populations of **highly excited atomic states, with a high principal quantum number n, called Rydberg states. Or these atoms are referred to as the Rydberg atoms.**

Such excited atoms are very suitable for the atom-radiation experiments because they are very strongly coupled to the radiation field, as the transition rates between neighbouring levels scale with n^4. Also, these transitions are in the microwave region, where photons can live longer, thus allowing longer interaction times. Finally, Rydberg atoms have long lifetimes, with respect to spontaneous decay [11, 12, 13]. The strong coupling of the Rydberg atom with the field can be physically understood because the dipole moment scales with the atomic radius that scales with n^2, thus when dealing with $n \sim 70$, we are talking about very large dipole moments.

To understand how the spontaneous emission rate is modified by a cavity, one has to study the effects of the cavity walls on the mode density. The continuum is replaced by a discrete set of modes, one of which may be resonant with the atom. In this case, the spontaneous decay rate is enhanced by a factor

$$\frac{\gamma_c}{\gamma_f} = \frac{2\pi Q}{v_c \omega_{ab}^3} , \qquad (11.55)$$

with γ_c, γ_f being the spontaneous decay rate with the cavity and in free space, respectively. However, when the cavity is detuned, the decay rate will decrease. The atom cannot emit, as it is not resonant with the cavity. Both effects of reduction and enhancement of spontaneous emission have been observed.

The reduction was observed by Drexhage et al. (For a review, see Drexhage, 1974 [14]), where the fluorescence of an active medium is observed, near a mirror. Also, similar effects were observed by De-Martini et al. (1987) [15] and Gabrielse and Dehmelt (1985) [16].

11.4.2 The Micromaser

A one-atom maser is described in the Fig. 11.3.

A collimated beam of Rubidium atoms is passed through a velocity selector. Before entering a high Q superconducting microwave cavity, the atom is excited to a high n-level and converted in a Rydberg atom.

Micromaser cavities are made of Niobium and cooled down to a small fraction of a degree Kelvin.

The Rydberg atoms are detected in the upper or lower maser levels by two field ionization detectors, and the fields are adjusted so that in the first detector, only the atoms in the upper state are ionized.

Maser operation was demonstrated by tuning the cavity to the maser transition and recording, simultaneously, the flux of atoms in the excited state [11].

As shown in the Fig. 11.4, on resonance, a reduction of the signal is observed, for relatively small atomic fluxes (1750 at\s). Higher fluxes produce power broadening and a small frequency shif.

Also, the two-photon micromaser was experimentally demonstrated [17].

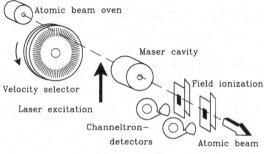

Fig. 11.3. The experimental setup of a Micromaser. (After [10])

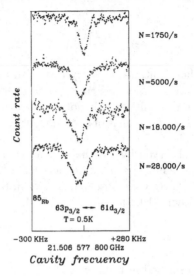

Fig. 11.4. Maser action of one-atom maser is manifested as a decrease of the number of atoms in the excited state. As the atomi flux is increased, there is a small frequency shift and power broadening. (After [10])

In the quantum theory of the micromaser [18], the atomic spontaneous emission rate into the free space modes is neglected. Also, because high-quality cavities have been achieved, one assumes that the photon lifetime is much longer than the transit time through the cavity, implying that we may neglect the cavity damping while the atom is in the cavity. Also, as the flux is kept low, the time interval between the atoms is much longer than the flight time and hence the cavity is empty most of the time.

Mathematically, the time evolution of the field in the micromaser is given by (11.18). For simplicity, we study the case $p = 0$. In this case, for the diagonal matrix elements, the master equation becomes ($\rho_{nn}(t) = p(n, t)$)

$$\frac{dp(n, t)}{dt} = r \left[- \sin^2(g\sqrt{n+1}\tau)p(n, t) + \sin^2(g\sqrt{n}\tau)p(n - 1, t) \right] \quad (11.56)$$

$$+ C(\langle n \rangle_{th} + 1) \left[(n + 1)p(n + 1, t) - np(n, t) \right] \quad (11.57)$$

$$+ C\langle n \rangle_{th} \left[np(n - 1, t) - (n + 1)p(n, t) \right],$$

where $\langle n \rangle_{th}$ is the average thermal photon number.

At steady state, and considering detailed balance

$$\left[r \sin^2(g\sqrt{n+1}\tau) + C\langle n \rangle_{th}(n + 1) \right] p_o(n) \quad (11.58)$$

$$= C(\langle n \rangle_{th} + 1)(n + 1)p_o(n + 1),$$

where $p_o(n)$ is the steady-state solution of (11.56).

The solution of the recursion relation (11.58) is [17]

$$p_o(n) = p_o(0) \left(\frac{\langle n \rangle_{\text{th}}}{\langle n \rangle_{\text{th}} + 1} \right)^n \prod_{k=1}^{n} \left[1 + \frac{n_{\text{ex}} \sin^2(\theta \sqrt{\frac{k}{n_{\text{ex}}}})}{k \langle n \rangle} \right], \qquad (11.59)$$

where $p_o(0)$ is the normalization constant determined by the condition

$$\sum_{n=0}^{\infty} p_o(n) = 1,$$

$$\theta \equiv \sqrt{n_{\text{ex}}} g \tau,$$

$$n_{\text{ex}} \equiv \frac{r}{C}.$$

A typical photon number distribution is shown in the Fig. 11.5.

Also, with $p_o(n)$, we can get the variance, that is shown in the Fig. 11.6.

The sub-Poissonian regions are due to the multipeak structure of the photon distribution.

11.4.3 Trapping States

Under the Jaynes–Cummings dynamics, the time evolution of the atom-field coupling is given, in the interaction picture (on resonance) by [18]:

$$\sum_n S_n \mid n \rangle (\alpha \mid a \rangle + \beta \mid b \rangle) \rightarrow \qquad (11.60)$$

$$\sum_n S_n U(\tau) \mid n \rangle (\alpha \mid a \rangle + \beta \mid b \rangle)$$

$$= \sum_n S_n \begin{bmatrix} \cos(g\tau\sqrt{aa^\dagger}) & -i \left[\frac{\sin(g\tau\sqrt{aa^\dagger})}{\sqrt{aa^\dagger}} \right] a \\ -ia^\dagger \left[\frac{\sin(g\tau\sqrt{aa^\dagger})}{\sqrt{aa^\dagger}} \right] & \cos(g\tau\sqrt{a^\dagger a}) \end{bmatrix} \begin{bmatrix} \alpha \\ \beta \end{bmatrix} \mid n \rangle$$

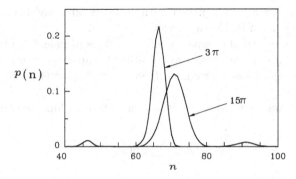

Fig. 11.5. Steady-state photon statistics, for $n_{\text{ex}} = 200$, $\langle n \rangle_{\text{th}} = 0.1$ and $\theta = 3\pi, 15\pi$. (After [17])

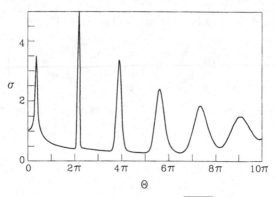

Fig. 11.6. Normalized standard deviation $\sigma = \sqrt{\frac{(\Delta n)^2}{\langle n \rangle}}$ as a function of θ. A Poissonian distribution corresponds to $\sigma = 1, n_{\text{ex}} = 200$. (After [17])

$$= \mid a \rangle \sum_n S_n \left\{ \alpha \cos(g\tau\sqrt{n+1}) \mid n \rangle - i\beta\sqrt{n} \left[\frac{\sin(g\tau\sqrt{n})}{\sqrt{n}} \right] \mid n-1 \rangle \right\}$$

$$+ \mid b \rangle \sum_n S_n \left[\beta \cos g\tau\sqrt{n} \mid n \rangle - i\alpha \sin(g\tau\sqrt{n+1}) \mid n+1 \rangle \right].$$

$$\equiv \mid f_a \rangle \mid a \rangle + \mid f_b \rangle \mid b \rangle.$$

If, for some $n = N$, we have $g\tau\sqrt{N} = q\pi$, q=integer, then the downward coupling from $\mid N \rangle \rightarrow \mid N-1 \rangle$ vanishes, and the regions above and below $\mid N \rangle$ are disconnected. This state is referred to as the downtrapping state.

Similarly, for an integer number p, the condition

$$\sqrt{M+1}g\tau = p\pi \tag{11.61}$$

corresponds to the upwards trapping condition.

Trapping states separate the whole Fock space in disconnected blocks, and, if the initial states of the field is within one of this blocks, the whole dynamics will take place within the block.

However, this is a simple picture without losses, and having only no atoms or, at most one atom inside the cavity.

Both cooperative and dissipative effects will result in leaks in the trapping blocks, which, after some time, are not able to trap any longer.

Pure states can be generated, at steady state, within these blocks, by the field mode

To see this, we assume for the field a pure state within the trapping block

$$\mid f \rangle = \sum_{n=N}^{n=M} S_n \mid n \rangle, \tag{11.62}$$

where the number states $\mid N \rangle$ and $\mid M \rangle$ states are the lower and upper bounds of the trapping block.

Now, we look for possible steady states for the field. This means that after the interaction with the next atom crossing the cavity, the state of the field will be the same, up to some global phase factor

$$| f\rangle(\alpha \mid a\rangle + \beta \mid b\rangle) \Longrightarrow \exp i\phi \mid f\rangle(\alpha' \mid a\rangle + \beta' \mid b\rangle) , \qquad (11.63)$$

and making use of (11.60) we get

$$\alpha' \exp i\phi \mid f\rangle = \mid f_a\rangle , \qquad (11.64)$$
$$\beta' \exp i\phi \mid f\rangle = \mid f_b\rangle,$$

which yields the following recursion relations

$$S_n = \frac{i\beta \sin(g\tau\sqrt{n+1})}{\alpha' \exp i\phi - \alpha \cos(g\tau\sqrt{n+1})} S_{n+1} , \qquad (11.65)$$
$$S_n = -i\frac{\beta' \exp i\phi - \beta \cos(g\tau\sqrt{n+1})}{\alpha \sin(g\tau\sqrt{n+1})} S_{n+1}.$$

Equations (11.65) have to be satisfied simultaneously, for all n within the block.

These relations are satisfied under two possible sets of conditions:

(a)

$$\exp i\phi = \pm 1 , \qquad (11.66)$$
$$\alpha' = -\alpha,$$
$$\beta' = \beta.$$

(b)

$$\exp i\phi = \pm 1 , \qquad (11.67)$$
$$\alpha' = \alpha,$$
$$\beta' = -\beta.$$

The conditions a and b lead to different recursion relations. In the case a

$$S_n = i\frac{\alpha}{\beta} \cot\left(\frac{g\tau\sqrt{n}}{2}\right) S_{n-1} , \qquad (11.68)$$

and in the case b

$$S_n = -i\frac{\alpha}{\beta} \tan\left(\frac{g\tau\sqrt{n}}{2}\right) S_{n-1} . \qquad (11.69)$$

The two states are referred to as cotangent and tangent states, respectively.

It is simple to verify that if one takes into account the boundaries of the block in phase space, for the cotangent states

$$\sqrt{N}g\tau = q\pi, \text{for } q \text{ even},$$ (11.70)

$$\sqrt{M+1}g\tau = p\pi, \text{ for } p \text{ odd},$$

and the reverse is true for the tangent states.

From the recursion relation (11.68), a cotangent state can be written as

$$| \cot\rangle = \sum_{n=N}^{n=M} S_n \mid n\rangle$$ (11.71)

$$= C(i)^n(\frac{\alpha}{\beta})^n \sqcap_{j=1}^n \cot\left(\frac{g\tau\sqrt{j}}{2}\right).$$

One of the most interesting properties of the cotangent states is that they are squeezed [19].

As, in practice, the initial conditions include the vacuum state, the cotangent states are the more interesting ones. We take $N = 0$ ($q = 0$) and some odd integer number for p, representing a $2\pi p$ rotation of an initially excited atom.

Because the state is exactly known, it is straightforward to compute the quadrature fluctuations.

In the Fig. 11.7, we show the variation of the two quadratures for a cotangent state for $N = 0$, $M = 20$ and $p = 1$(a) and $p = 3$(b), as a function of the probability of the upper state $\mid \alpha \mid^2$. The maximum squeezing is obtained for a large α. Also, the squeezing increases with the size of the trapping block.

Trapping states provide a way to build up Fock states or superposition of Fock states in cavities. It is important, though, to remark that our discussion is somewhat idealized, even in the case of the trapped vacuum state, that does not require the assumption of zero dissipation.

Actually, the atoms coming out from the oven have a Poissonian arrival distribution, so that there is always a finite probability to have two atoms simultaneously in the cavity. This cooperative effect produces a great disruption in the trapping states, even for a very low-density beam, with a probability of having two atoms less that 1%. [20, 21].

Finally, recent experiments show that one can also have microlasers, that is laser oscillations with one atom in the optical region [22, 23].

In these experiments, a beam of Ba atoms is first excited by a laser pulse, before entering the optical cavity. Laser oscillations were observed, with an average photon number ranging from a fraction of a photon to 11. The rapid increase of the photon number with the pump, when the average number of atoms approaches 1, departing from the linear regime, may be attributed to cooperative effects.

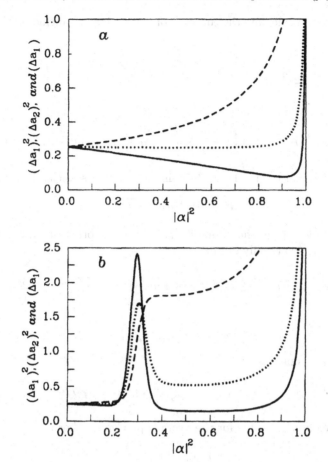

Fig. 11.7. Squeezing of a cotangent state with $p = 1$(**a**) and $p = 3$(**b**) bound between $N = 0$ and $M = 20$, as a function of $|\alpha|^2$. The solid line corresponds to $(\Delta a_1)^2$, the dashed line to $(\Delta a_2)^2$ and the dotted line to the product $(\Delta a_1)(\Delta a_2)$. $\phi = \frac{\pi}{2}$

11.5 Quantum Theory of the Laser and the Micromaser with Pump Statistics ($p \neq 0$)

We take the diagonal matrix elements of (11.18) and get

$$\frac{d\rho_{nn}}{dt} = r\left[-\sin^2(g\sqrt{n+1}\tau)\rho_{nn} + \sin^2(g\sqrt{n}\tau)\rho_{n-1,n-1}\right] \qquad (11.72)$$

$$+\frac{rp}{2}[\{\sin^4(g\sqrt{n}\tau) + \sin^2(g\sqrt{n}\tau)\sin^2(g\sqrt{n+1}\tau)\}\rho_{n-1,n-1}$$

$$-\sin^2(g\sqrt{n}\tau)\sin^2(g\sqrt{n-1}\tau)\rho_{n-2,n-2}$$

$$-\sin^4(g\sqrt{n+1}\tau)\rho_{nn}] + (L\rho)_{nn}.$$

Now, we can calculate the equation of motion for $\langle n \rangle$ (including the cavity loss term)

$$\frac{d\langle n \rangle}{dt} = \sum_n n \frac{d\rho_{nn}}{dt} = r \sum_n \alpha_n \rho_{nn} - C\langle n \rangle = r\langle \alpha_n \rangle - C\langle n \rangle , \qquad (11.73)$$

with

$$\alpha_n = \sin^2(g\sqrt{n+1}\,\tau) \left\{ 1 + \frac{p}{2} \left[\sin^2(g\sqrt{n+1}\,\tau) - \sin^2(g\sqrt{n+2}\,\tau) \right] \right\} . \qquad (11.74)$$

We notice that for $n \gg 1, g\tau$, the gain α_n is just its semiclassical expression

$$\alpha_n = \sin^2(g\sqrt{n+1}\,\tau) , \qquad (11.75)$$

and independent from the pump statistics. On the other hand, the variance $v = \langle n^2 \rangle - \langle n \rangle^2$ is readily obtained as

$$\frac{dv}{dt} = 2r\langle \alpha_n \Delta n \rangle + r\langle (\alpha_n - p\sin^2(g\sqrt{n+1}\,\tau)\sin^2(g\sqrt{n+2}\,\tau)) \rangle - 2Cv + C\langle n \rangle , \qquad (11.76)$$

where

$$\Delta n \equiv n - \langle n \rangle .$$

For $n \gg 1$, one has

$$\frac{dv}{dt} = 2r\langle \alpha_n \Delta n \rangle + r\langle (\alpha_n - p\alpha_n^2) \rangle - 2Cv + C\langle n \rangle . \qquad (11.77)$$

The steady-state value for the average photon number is

$$r\langle \alpha_n \rangle = Cn_{ss} , \qquad (11.78)$$

which can be seen graphically as the intersection of the gain curve with the loss (straight line). See Fig. 11.8.

It is not difficult to see that the steady-state value n_{ss} is only stable if the slope of the gain curve is smaller than the slope of the loss, that is

$$r \left(\frac{d\alpha_n}{dn} \right)_{n_{ss}} < C . \qquad (11.79)$$

To see this, we expand the gain around the intersection point

$$\alpha_n \approx (\alpha_n)_{n_{ss}} + \left(\frac{d\alpha_n}{dn} \right)_{n_{ss}} (n - n_{ss}) + \dots . \qquad (11.80)$$

Now, if n is slightly larger than n_{ss}

$$n \geq n_{ss} , \qquad (11.81)$$

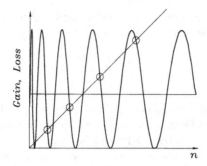

Fig. 11.8. The steady-state photon number found as an intersection of the gain curve with the loss straight line

$$\Delta n = n - n_{ss} > 0,$$

then one can write

$$\frac{\mathrm{d}\langle n \rangle}{\mathrm{d}t} = r\left[(\alpha_n)_{n_{ss}} + \left(\frac{\mathrm{d}\alpha_n}{\mathrm{d}n}\right)_{n_{ss}} \Delta n \right] - C\left[n_{ss} + \Delta n \right] \qquad (11.82)$$

$$= \left[r\left(\frac{\mathrm{d}\alpha_n}{\mathrm{d}n}\right)_{n_{ss}} - C \right] \Delta n \,,$$

and if the equilibrium is stable, $\frac{\mathrm{d}\langle n \rangle}{\mathrm{d}t}$ should be negative in order to go back to the equilibrium point, so the condition given by (11.79) is satisfied.

Defining a normalized photon number $\eta_n \equiv \frac{n}{n_{ex}}$, where $n_{ex} \equiv \frac{r}{C}$ is the number of excited atoms entering the cavity, during the cavity damping time, we can see from (11.78) that $0 \leq \eta \leq 1$, and the stability condition can be written now as

$$\left(\frac{\mathrm{d}\alpha_n}{\mathrm{d}\eta_n}\right)_{n=n_{ss}} < 1 \,. \qquad (11.83)$$

Now, we assume that the photon distribution is highly peaked around a single maximum, and we can expand α_n around n_{ss}, then we find (using (11.77) in steady state)

$$v = \left(\frac{\mathrm{d}\alpha_n}{\mathrm{d}\eta_n}\right)_{n=n_{ss}} v + \frac{n_{ex}}{2}\langle\alpha_n\rangle - \frac{pr}{2C}\langle\alpha_n^2\rangle + \frac{\langle n \rangle}{2} \,, \qquad (11.84)$$

or

$$v = \frac{1}{1 - \left(\frac{\mathrm{d}\alpha_n}{\mathrm{d}\eta_n}\right)_{n=n_{ss}}} \langle n - \frac{pn_{ex}}{2}\alpha_n^2 \rangle \,. \qquad (11.85)$$

The above expression exhibits the role of the pump statistics on the photon number noise. For the case of the micromaser, one gets a sub-Poissonian distribution, even if $p \to 0$, if $\left(\frac{\mathrm{d}\alpha_n}{\mathrm{d}\eta_n}\right)_{n=n_{ss}} < 0$. As p is increased, this behaviour is enhanced even further.

A somewhat simpler expression can be obtained for the variance, in the case $n_{ss} \gg 1$, in which case $\langle \alpha_n^2 \rangle \approx \langle \alpha_n \rangle^2$, and we get

$$v \cong \frac{n_{ss} \left(1 - \frac{p\eta_s}{2}\right)}{1 - \left(\frac{d\alpha_n}{d\eta_n}\right)_{n=n_{ss}}} , \tag{11.86}$$

where η_s is its value at steady state. Choosing $p = \eta_s = 1$, one has a 50% of photon noise reduction because of the regularity of the pump, when compared with the same micromaser with a Poissonian pump.

Now, we turn our attention to the laser case. Averaging over the atomic lifetimes, we get

$$\alpha_n^{\text{laser}} = \int_0^\infty d\tau \exp(-\gamma\tau) \sin^2(g\sqrt{n}\tau) = \frac{2g^2 n}{\gamma^2 + 4g^2 n} , \tag{11.87}$$

and according to (11.78)

$$\eta_s = \frac{2g^2 n_{ss}}{\gamma^2 + 4g^2 n_{ss}} , \tag{11.88}$$

thus η_s is always smaller than $\frac{1}{2}$, approaching this value for large n_{ss}. From (11.87), we get

$$\left(\frac{d\alpha_n}{d\eta_n}\right)_{n=n_{ss}} = 1 - 2\eta_s \geq 0 . \tag{11.89}$$

Replacing (11.89) into (11.86), we get

$$v = \frac{\left(1 - \frac{p\eta_s}{2}\right)}{2\eta_s} n_{ss} . \tag{11.90}$$

Now, we can discuss the influence of the pump fluctuations over the photon-number variance in the laser. In the case of a Poissonian distribution of the incoming atoms $(p = 0)$, the variance is always larger than the mean number of photons, only approaching this value well above threshold, and indeed, when $\eta_s \to \frac{1}{2}, v \to n_{ss}$ as it should. On the other hand, if we go to the regular pumping limit $(p = 1)$, we can get a considerable noise reduction well above threshold, when $\eta_s \to \frac{1}{2}$, getting $v = \frac{3}{4}n_{ss}$, thus getting up to 25% noise reduction.

Finally, it is possible to show [2] that the phase diffusion constant is not affected by the pump statistics.

Problems

11.1. Prove (11.7).
11.2. Prove (11.18).
11.3. Prove (11.23).
11.4. Prove (11.37) from (11.23).

References

1. Yu, M., Golubev, I.V., Sokolov, I.V.: Zh. Eksp. Teor. Fiz, **87**, 408 (1984); Sov. Phys. JETP, **60**, 234.
2. Bergou, J., Davidovich, L., Orszag, M., Benkert, C., Hillery, M., Scully, M.O., Opt. Comm., **72**, 82 (1989);Bergou, J., Davidovich, L., Orszag, M., Benkert, C., Hillery, M., Scully, M.O.: Phys. Rev. A, **40**, 5073 (1989); See also: Haake, F., Tan, S.M., Walls, D.: Phys. Rev. A, **40**, 7121 (1989)
3. An excellent discussion on this point, as well and on noise supression in quantum optical systems is found in: Davidovich, L.: Rev. Mod. Phys., **68**, 127 (1996)
4. Scully, M.O., Lamb, W.E.: Phys. Rev., **159**, 208(1967); See also: Sargent III, M., Scully, M.O., Lamb, W.E.: Laser Physics. Addison Wesley, Mass (1974); Scully, M.O., Lamb, W.E.: Phys. Rev., **159**, 208 (1967); Scully, M.O., Lamb, W.E.: Phys. Rev., **179**, 368 (1969)
5. Stenholm, S.: Phys. Rep., **6**, 1 (1973)
6. Degiorgio, V., Scully, M.O.: Phys. Rev., **2**, 1170 (1970)
7. Graham, R., Haken, H.: Z. Physik, **237**, 31 (1970)
8. Lax, M.: Phys. Rev., **157**, 213 (1967)
9. Gordon, J.P.: Phys. Rev., **161**, 367 (1967)
10. Haken, H.: Z. Physik, **190**, 327 (1966)
11. Walther, H.: Phys. Rep., **219**, 263 (1992)
12. Haroche, S., Raimond, J.M.: In: Bates, D., Benderson, B. (eds) Advances in Atomic and Molecular Physics. Volume 20, 350, Academic Press, New York (1985)
13. Gallas, J.A.C., Leuchs, G., Walther, H., Figger, H.: In: Bates, D., Benderson, B. (eds) Advances in Atomic and Molecular Physics. Volume 20, 413 Academic Press, New York (1985)
14. Drexhage, K.H.: In: Wolf, E. (ed.) Progress in Optics. Volume 12, North Holland, Amsterdam (1974)
15. De-Martini, F., Innocenti, G., Jacovitz, G., Mantolini D.: Phys. Rev. Lett., **29**, 2955 (1987)
16. Gabrielse, G., Dehmelt, H.: Phys. Rev. Lett., **55**, 67 (1985)
17. Brune, M.S., Raimond, J.M., Goy, P., Davidovich, L., Haroche, S.: Phys. Rev. Lett., **59**, 1899 (1987)
18. Filipowicz, P., Javanainen, J., Meystre, P., Phys. Rev. A, **34**, 3077 (1986)
19. Slosser, J.J., Meystre, P.: Phys. Rev. A, **41**, 3867 (1990)
20. Orszag, M., Ramirez, R., Retamal, J.C., Saavedra, C.: Phys. Rev. A, **49**, 2933 (1994)
21. Wehner, E., Seno, R., Sterpi, N., Englert, B.G., Walther, H.: Opt. Comm, **110**, 655 (1994)
22. An, K., Childs, J.J., Dasari, R.R., Feld, M.: Phys. Rev. Lett., **73**, 3375 (1994)
23. Weidinger, M., Varcoe, B.T.H., Heerlein, R., Walther, H.: Phys. Rev. Lett., **82**, 3795 (1999)

Further Reading

• Arecchi, F.T., Ricca, A.M.: Phys. Rev. A, **15**, 308 (1977)

- Casagrande, F., Lugiato, L.A.: Phys. Rev. A, **14**, 778 (1976)
- Graham, R.: Smith, W.A, Opt. Comm, **7**, 289 (1973)
- Lugiato, L.A.: Physics, **81A**, 565 (1976)
- Risken, H.: Z. Phys., **191**, 186 (1965)
- Risken, H., Vollmer, H.D.: Z.Phys., **201**, 323 (1967)
- Scully, M.O.: In: Glauber, R. (ed.) Proceedings of the International School of Physics "Enrico Fermi" Course XLII. Academic, New York (1969)
- Scully, M.O., Zubairy, M.S.: Quantum Optics. Cambridge University Press, Cambridge (1997)

12. Quantum Laser Theory: Langevin Approach

In this chapter, we study the Laser Theory using the Langevin Approach. We include the influence of pump statistics.

In the previous chapter, we studied the influence of the pump statistics on the amplitude and phase fluctuations of the laser radiation, making use of the Master Equation approach. We, thus, derived a generalized Master Equation in terms of a parameter p that represented the probability for an atom to be excited to the upper level, before entering into the cavity.

The two extreme cases were $p \to 0$ (Poisson statistics) and $p \to 1$ (regular statistics). What we found was that the pump statistics had no influence on the phase fluctuations or linewidth but had a strong influence on the photon number fluctuations.

In this chapter, we discuss the influence of the pump statistics from a different point of view. We use the Langevin formalism [1], including generalized noise operators as to include the effects of the pump noise [2].

We show here that, again, the photon number fluctuations can be reduced by simply reducing the pump fluctuations. Furthermore, we generalize the arguments of the previous chapter, allowing for different atomic decay constants from the two levels.

12.1 Quantum Langevin Equations

Our physical system is described, again in the Fig. 11.1, where the atoms are prepared initially in the upper level $\mid a \rangle$. The two levels $\mid a \rangle$ and $\mid b \rangle$ constitute the lasing transition, which is coupled to one mode of the radiation field, inside the cavity.

The Hamiltonian of this system, in the rotating wave approximation, is given by

$$H = \hbar \omega a^\dagger a + \sum_{j=1}^{N} (\epsilon_a \mid a \rangle \langle a \mid + \epsilon_b \mid b \rangle \langle b \mid + \epsilon_c \mid c \rangle \langle c \mid)_j \qquad (12.1)$$

$$+ \hbar g \sum_j \Theta(t - t_j)(a^\dagger \sigma^j + \sigma^{j\dagger} a),$$

where $\Theta(t)$ is the usual step function.

In the above Hamiltonian, $\sigma^j = (\mid b\rangle\langle a \mid)_j$ represents the polarization operator for the j-th atom. The cavity losses as well as the atomic decay are modelled in the usual way, coupling the system to heat reservoirs. We find the following equations of motion:

$$\dot{a} = -i\omega a - \frac{C}{2}a - ig\sum_j \Theta(t - t_j)\sigma^j + F_C, \tag{12.2}$$

$$\dot{\sigma}^j = -i\omega_{ab}\sigma^j - \gamma\sigma^j + ig\Theta(t - t_j)(\sigma_{aa}^j + \sigma_{bb}^j)a + F_{ba}^j, \tag{12.3}$$

$$\dot{\sigma_{aa}}^j = -\gamma\sigma_{aa}^j + ig\Theta(t - t_j)(a^\dagger\sigma^j - \sigma^{j\dagger}a) + F_{aa}^j, \tag{12.4}$$

$$\dot{\sigma}_{bb}^j = -\gamma\sigma_{bb}^j - ig\Theta(t - t_j)(a^\dagger\sigma^j + \sigma^{j\dagger}a) + F_{bb}^j, \tag{12.5}$$

where $\sigma_{aa}^j = (\mid a\rangle\langle a \mid)_j, \sigma_{bb}^j = (\mid b\rangle\langle b \mid)_j$ and $\omega_{ab} = \frac{\epsilon_a - \epsilon_b}{\hbar}$.

For now and in the sake of simplicity, we have assumed that the two atomic decay constants are equal with value γ. However, this assumption will be relaxed at the end.

Now, we look at the noise terms. From the damped harmonic oscillator, we saw already that

$$\langle F_C^\dagger(t)F_C(t')\rangle = C\langle n\rangle_{th}\delta(t - t') \tag{12.6}$$
$$= 0 \text{ at } T = 0,$$

$$\langle F_C(t)F_C^\dagger(t')\rangle = C(1 + \langle n\rangle_{th})\delta(t - t') \tag{12.7}$$
$$= C\delta(t - t') \text{ at } T = 0,$$

$$\langle F_C(t)\rangle = 0. \tag{12.8}$$

For the atomic noise correlation functions, we first derive the Einstein relations.

12.1.1 The Generalized Einstein's Relations

We write a quantum Langevin equation, in the absence of atom–field interaction: For a general discussion on stochastic processes, see appendix D.

$$\dot{A}_\mu = D_\mu(t) + F_\mu(t), \tag{12.9}$$

where $D_\mu(t)$ is the drift operator for $A_\mu(t)$ and $F_\mu(t)$ is the corresponding noise operator with zero reservoir average $\langle F_\mu(t)\rangle = 0$.

We also write the two time average for the noise operator as

$$\langle F_\mu(t)F_\nu(t')\rangle = 2\langle D_{\mu\nu}\rangle\delta(t - t'). \tag{12.10}$$

Now, we start with the identity

$$A_\mu(t) = A_\mu(t - \Delta t) + \int_{t-\Delta t}^{t} dt' \, \dot{A}_\mu(t'),$$

to obtain the system noise correlation function

$$\langle A_\mu(t)F_\nu(t)\rangle = \langle A_\mu(t-\Delta t)F_\nu(t)\rangle + \int_{t-\Delta t}^{t} dt' \, \langle [D_\mu(t') + F_\mu(t')] \, F_\nu(t)\rangle,$$
(12.11)

and as $A_\mu(t - \Delta t)$ cannot be affected by a noise term $F_\nu(t)$ at a later time, the first term in (12.11) is zero. The same argument applies to the first term in the integral $\langle D_\mu(t') \, F_\nu(t)\rangle$, so we are left with only one term

$$\langle A_\mu(t)F_\nu(t)\rangle = \int_{t-\Delta t}^{t} dt' \, \langle F_\mu(t') \, F_\nu(t)\rangle = \frac{1}{2}\int_{-\infty}^{\infty} dt' \, \langle F_\mu(t') \, F_\nu(t)\rangle, \quad (12.12)$$

and substituting (12.10) into (12.12), we get

$$\langle A_\mu(t)F_\nu(t)\rangle = \langle D_{\mu\nu}\rangle. \tag{12.13}$$

It is simple to prove that, also

$$\langle F_\nu(t)A_\mu(t)\rangle = \langle D_{\mu\nu}\rangle. \tag{12.14}$$

Now we write

$$
\begin{aligned}
\frac{d}{dt}\langle A_\mu(t)A_\nu(t)\rangle &= \langle \dot{A}_\mu(t) \, A_\nu(t)\rangle + \langle A_\mu(t) \, \dot{A}_\nu(t)\rangle \\
&= \langle D_\mu(t) \, A_\nu(t)\rangle + \langle F_\mu(t) \, A_\nu(t)\rangle \\
&\quad + \langle A_\mu(t) \, D_\nu(t)\rangle + \langle A_\mu(t)F_\nu(t)\rangle,
\end{aligned}
$$

and using (12.13 and 12.14), we get the **Generalized Einstein's relations**

$$2\langle D_{\mu\nu}\rangle = -\langle A_\mu(t) \, D_\nu(t)\rangle - \langle D_\mu(t) \, A_\nu(t)\rangle + \frac{d}{dt}\langle A_\mu(t)A_\nu(t)\rangle. \tag{12.15}$$

12.1.2 The Atomic Noise Moments

From the Eistein's relations, one can easily calculate the atomic noise moments. Let us take as an example $\langle F_{ba}^\dagger(t)F_{ba}(t')\rangle$. so we take $A_\mu = \sigma^\dagger$ and $A_\nu = \sigma$. From (12.15), we get

$$
\begin{aligned}
2\langle D_{\sigma^\dagger\sigma}\rangle &= -\langle \sigma^\dagger \, (-\gamma\sigma)\rangle - \langle -\gamma\sigma^\dagger\sigma\rangle + \frac{d}{dt}\langle \sigma_{aa}\rangle \tag{12.16} \\
&= \gamma\langle \sigma_{aa}\rangle,
\end{aligned}
$$

where we used the property $\sigma^\dagger\sigma = \sigma_{aa}$ and (12.3). We leave as an exercise to the reader to verify the rest of the atomic correlations

$$\langle F_{ba}^{\dagger j}(t) F_{ba}^{j}(t') \rangle = \gamma \langle \sigma_{aa}^{j} \rangle \delta(t - t'), \tag{12.17}$$

$$\langle F_{ll}^{j}(t) F_{ll}^{j}(t') \rangle = \gamma \langle \sigma_{ll}^{j} \rangle \delta(t - t'), l = a, b \tag{12.18}$$

$$\langle F_{ba}^{\dagger j}(t) F_{aa}^{j}(t') \rangle = \gamma \langle \sigma^{j} \rangle \delta(t - t'), \tag{12.19}$$

$$\langle F_{ba}^{\dagger j}(t) F_{bb}^{j}(t') \rangle = \gamma \langle \sigma^{j\dagger} \rangle \delta(t - t'). \tag{12.20}$$

We now proceed to eliminate the fast varying terms in (12.2, 12.3, 12.4, 12.5). For simplicity, we assume resonance $\omega_{ab} = \omega$ and define

$$\widetilde{a(t)} = \exp(i\omega t)a(t), \widetilde{\sigma^{j}}(t) = \exp(i\omega t)\sigma^{j}(t). \tag{12.21}$$

It is quite evident that the equations of motion for $\widetilde{a}$ and $\widetilde{\sigma}^{j}$ are the same as for a and σ^{j} with the only difference that the terms proportional to ω_{ab} and ω are omitted.

The following step in this theory is to define the collective operators summed over all the atoms. This is very convenient with the adiabatic approximation we are going to make. Thus, we define

$$M(t) = -i \sum_{j} \Theta(t - t_j)\sigma^{j}(t), \tag{12.22}$$

$$N_a(t) = \sum_{j} \Theta(t - t_j)\sigma_{aa}^{j}(t), \tag{12.23}$$

$$N_b(t) = \sum_{j} \Theta(t - t_j)\sigma_{bb}^{j}(t). \tag{12.24}$$

The operators M, N_a and N_b represent the macroscopic polarization and the populations in the a and b levels, respectively.

With the above definitions, the equation of motion for the field now becomes

$$\dot{a} = -\frac{C}{2}a + Mg + F_\gamma. \tag{12.25}$$

On the other hand, for the atomic operator N_a, we differentiate (12.23) and substitute (12.4), getting

$$\dot{N}_a = \sum_{j} \Theta(t - t_j) \dot{\sigma}_{aa}^{j} + \delta(t - t_j)\sigma_{aa} \tag{12.26}$$

$$= \sum_{j} \delta(t - t_j)\sigma_{aa}^{j}(t_j) - \gamma N_a - g(a^{\dagger}M + M^{\dagger}a)$$

$$-i \sum_{j} \Theta(t - t_j) F_{ba}^{j}(t).$$

The first term in the r.h.s. of (12.26) corresponds to the pumping of the atoms to the excited state. This can be seen as follows:

$$\langle \sum_j \delta(t-t_j)\sigma_{aa}^j(t_j)\rangle = \langle \sum_j \delta(t-t_j)\langle\sigma_{aa}^j(t_j)\rangle\rangle_S \qquad (12.27)$$

$$= \langle \sum_j \delta(t-t_j)\rangle_S.$$

In the above result, we used the fact that the atoms are initially prepared in the upper state, so that $\langle\sigma_{aa}^j(t_j)\rangle = 1$. Also, there is a second bracket with a subscript S, showing that a statistical average has been performed over all the terms that depend on the random injection times t_j.

Assuming an average injection rate of atoms R, then one can write

$$\langle \sum_j \delta(t-t_j)\sigma_{aa}^j(t_j)\rangle = \langle \sum_j \delta(t-t_j)\rangle_S \qquad (12.28)$$

$$= R\int_{-\infty}^{\infty} dt_j \delta(t-t_j) = R.$$

To separate the drift from the noise terms, we add and substract R, getting

$$\dot{N_a} = R - \gamma N_a - g(a^\dagger M + M^\dagger a) + F_a, \qquad (12.29)$$

with

$$F_a(t) = \sum_j \Theta(t-t_j)F_{aa}^j(t) + \sum_j \delta(t-t_j)\sigma_{aa}^j(t_j) - R. \qquad (12.30)$$

In a similar way, one derives the other operator equations

$$\dot{N_b} = -\gamma N_b + g(a^\dagger M + M^\dagger a) + F_b, \qquad (12.31)$$

$$\dot{M} = -\gamma M + g(N_a - N_b)a + F_M, \qquad (12.32)$$

with

$$F_b(t) = \sum_j \Theta(t-t_j)F_{bb}^j(t) + \sum_j \delta(t-t_j)\sigma_{bb}^j(t_j), \qquad (12.33)$$

$$F_M(t) = -i\sum_j \Theta(t-t_j)F_{ba}^j(t) - i\sum_j \delta(t-t_j)\sigma^j(t_j). \qquad (12.34)$$

Now, we calculate the atomic noise correlation functions . As an example, we calculate $\langle F_a(t)F_a(t')\rangle$.

$$\langle F_a(t)F_a(t')\rangle = \langle \sum_{j,k} \Theta(t-t_j)\Theta(t'-t_k)\langle F_{aa}^j(t)F_{aa}^k(t')\rangle\rangle_S \qquad (12.35)$$

$$+\langle \sum_{j,k} \delta(t-t_j)\delta(t'-t_k)\langle\sigma_{aa}^j(t)\sigma_{aa}^k(t')\rangle\rangle_S - R^2\rho_{aa}^2.$$

We notice, once again, that in the above expression, two types of average have been considered.

On one hand, one has the usual quantum mechanical average over both the variables, and on the other hand, we have the statistical average, symbolized by the subscript S. We also, replaced the symbol $\langle \sigma_{aa}^j(t) \rangle$ by ρ_{aa}, just to differentiate the various terms appearing in the following analysis. At the end, we will set $\rho_{aa} = 1$, consistent with the initial preparation of the atoms.

In the first term of (12.35), only terms with $j = k$ contribute, because the atoms are independent of each other. Also, we separate the second term in two contributions, one with $j = k$ and the other one with $j \neq k$. In this second term, products of the type $\langle \sigma_{aa}^j(t_j)\sigma_{aa}^k(t_k) \rangle = \rho_{aa}^2$. We get the following result:

$$\langle F_a(t)F_a(t') \rangle = \langle \sum_{j,k} \Theta(t - t_j)\gamma \langle \sigma_{aa}^j(t) \rangle \rangle_S \delta(t - t')$$

$$+ \langle \sum_j \delta(t - t_j)\delta(t' - t_j)\rho_{aa} \rangle_S$$

$$+ \left[\langle \sum_{j \neq k} \delta(t - t_j)\delta(t' - t_k) \rangle_S - R^2 \right] \rho_{aa}^2,$$

or in the final form

$$\langle F_a(t)F_a(t') \rangle = \gamma \langle N_a \rangle \delta(t - t') + \langle \sum_j \delta(t - t_j) \rangle_S \rho_{aa}\delta(t - t') \quad (12.36)$$

$$+ \left[\langle \sum_{j \neq k} \delta(t - t_j)\delta(t' - t_k) \rangle_S - R^2 \right] \rho_{aa}^2.$$

In the Appendix (C), we show that

$$\left[\langle \sum_{j \neq k} \delta(t - t_j)\delta(t' - t_k) \rangle_S - R^2 \right] \rho_{aa}^2 = -pR\delta(t - t'), \quad (12.37)$$

so that

$$\langle F_a(t)F_a(t') \rangle = [\langle \gamma N_a \rangle + R(1 - p)] \delta(t - t'). \quad (12.38)$$

The rest of the correlation functions are calculated in a similar way. The result is, for the non-vanishing terms,

$$\langle F_M^\dagger(t)F_M(t') \rangle = [\langle \gamma N_a \rangle + R] \delta(t - t'). \quad (12.39)$$

$$\langle F_b(t)F_b(t') \rangle = \langle \gamma N_b \rangle \delta(t - t'). \quad (12.40)$$

$$\langle F_b(t)F_M(t') \rangle = \langle \gamma M \rangle \delta(t - t'). \quad (12.41)$$

The differential equations for the field and the atomic variables plus the noise correlation values completely describe the laser under an arbitrary pump statistics.

12.2 *C*-Number Langevin Equations

To solve the present problem , we have to convert the four-operator equations into c-number equations. To do that in a unique way, we have to define a prescribed ordering, the choice of which is completely arbitrary. We choose the following ordering: $a^\dagger, M^\dagger, N_a, N_b, M, a$. We now, derive the equations of motion for their c-number versions $\varepsilon, \mathcal{M}, \mathcal{N}_a, \mathcal{N}_b$. As the (12.25, 12.29, 12.31 and 12.33) are already in the chosen normal order, we can write directly

$$\dot{\varepsilon} = -\frac{C}{2}\varepsilon + g\mathcal{M} + \mathcal{F}_C, \tag{12.42}$$

$$\dot{\mathcal{M}} = -\gamma\mathcal{M} + g(\mathcal{N}_a - \mathcal{N}_b)\varepsilon + \mathcal{F}_\mathcal{M}, \tag{12.43}$$

$$\dot{\mathcal{N}}_a = R - \gamma\mathcal{N}_a - g(\varepsilon^*\mathcal{M} + \mathcal{M}^*\varepsilon) + \mathcal{F}_a, \tag{12.44}$$

$$\dot{\mathcal{N}}_b = -\gamma\mathcal{N}_b + g(\varepsilon^*\mathcal{M} + \mathcal{M}^*\varepsilon) + \mathcal{F}_b, \tag{12.45}$$

The Langevin forces $\mathcal{F}$ have the following properties:

$$\langle\mathcal{F}_k(t)\rangle = 0, \tag{12.46}$$

$$\langle\mathcal{F}_k(t)\mathcal{F}_l(t')\rangle = 2D_{kl}\delta(t - t'). \tag{12.47}$$

The diffusion coefficients D_{kl} are determined in such a way that the second moments calculated from the c-number equations agree with those calculated from the operator equations. We illustrate this procedure, we calculate $D_{\mathcal{M}\mathcal{M}}$.

From (12.32), we get

$$\frac{\mathrm{d}}{\mathrm{d}t}\langle M(t)M(t)\rangle = -2\gamma\langle MM\rangle + \langle MF_M\rangle \tag{12.48}$$
$$+ \langle F_M M\rangle + g\left[\langle(N_a - N_b)Ma\rangle + \langle M(N_a - N_b)a\rangle\right]$$

We notice that the second term in the square bracket is not in the normal form; therefore, we use the commutation relation

$$[M, N_a - N_b] = 2M,$$

also the second and third terms vanish, so

$$\frac{\mathrm{d}}{\mathrm{d}t}\langle M(t)M(t)\rangle = -2\gamma\langle MM\rangle + 2g\langle(N_a - N_b)Ma\rangle + 2g\langle Ma\rangle. \tag{12.49}$$

We now obtain the corresponding c-number equation

$$\frac{\mathrm{d}}{\mathrm{d}t}\langle\mathcal{M}\mathcal{M}\rangle = -2\gamma\langle\mathcal{M}\mathcal{M}\rangle + 2g\left[\langle(\mathcal{N}_a - \mathcal{N}_b)\mathcal{M}\varepsilon)\rangle\right] + 2D_{\mathcal{M}\mathcal{M}}. \tag{12.50}$$

By comparing the right-hand sides of (12.49 and 12.50), we readily get

$$2D_{\mathcal{M}\mathcal{M}} = 2g\langle \mathcal{M}\varepsilon\rangle. \tag{12.51}$$

In a similar way, we can calculate the rest of the diffusion coefficients. The results are given in the following table:

$2D_{\mathcal{M}^*\mathcal{M}} = \gamma\langle \mathcal{N}_a\rangle + R,$

$2D_{\mathcal{M}\mathcal{M}} = 2g\langle \mathcal{M}\varepsilon\rangle,$

$2D_{\mathcal{N}_b\mathcal{M}} = \gamma\langle \mathcal{M}\rangle,$

$2D_{\mathcal{N}_a\mathcal{N}_a} = \gamma\langle \mathcal{N}_a\rangle + R(1-p) - g(\langle \varepsilon^*\mathcal{M}+\varepsilon\mathcal{M}^*\rangle)$

$2D_{\mathcal{N}_b\mathcal{N}_b} = \gamma\langle \mathcal{N}_b\rangle - g(\langle \varepsilon^*\mathcal{M}+\varepsilon\mathcal{M}^*\rangle)$

$2D_{\mathcal{N}_a\mathcal{N}_b} = g(\langle \varepsilon^*\mathcal{M}+\varepsilon\mathcal{M}^*\rangle)$

12.2.1 Adiabatic Approximation

We want, now to solve (12.42, 12.43, 12.44, 12.45).Typically, in Laser problems, the atomic decay constant γ is much larger than the photon decay C, or in other words, we have two very different time scales in the problem: A short time corresponding to a typical variation of the atomic variables and a much longer time over which there is a sizeable variation of the field. Under these conditions, we can use the adiabatic approximation, where we neglect the time derivatives of the atomic variables, thus calculating $\mathcal{N}_a, \mathcal{N}_b, \mathcal{M}$, in terms of the field. The result is

$$\mathcal{M} = \frac{g}{\gamma}(\mathcal{N}_a - \mathcal{N}_b)\varepsilon + \frac{\mathcal{F}_\mathcal{M}}{\gamma}, \tag{12.52}$$

$$\mathcal{N}_a = \frac{\left[R\left(1+\frac{2g^2}{\gamma^2}I\right) + \left(1+\frac{2g^2}{\gamma^2}I\right)\mathcal{G}_a + \frac{2g^2}{\gamma^2}I\mathcal{G}_b\right]}{\left[\gamma(1+\frac{4g^2}{\gamma^2}I)\right]}, \tag{12.53}$$

$$\mathcal{N}_b = \frac{\left[R\frac{2g^2}{\gamma^2}I + \left(1+\frac{2g^2}{\gamma^2}I\right)\mathcal{G}_b + \frac{2g^2}{\gamma^2}I\mathcal{G}_a\right]}{\left[\gamma(1+\frac{4g^2}{\gamma^2}I)\right]}, \tag{12.54}$$

where $I \equiv \varepsilon\varepsilon^*$ is the intensity of the field, and the noise functions $\mathcal{G}_a$ and $\mathcal{G}_b$ are defined as

$$\mathcal{G}_a = \mathcal{F}_a - \frac{g}{\gamma}(\mathcal{F}_\mathcal{M}^*\varepsilon + \varepsilon^*\mathcal{F}_\mathcal{M}), \tag{12.55}$$

$$\mathcal{G}_b = \mathcal{F}_b + \frac{g}{\gamma}(\mathcal{F}_\mathcal{M}^*\varepsilon + \varepsilon^*\mathcal{F}_\mathcal{M}). \tag{12.56}$$

Now, if we replace the results of (12.52, 12.53, 12.54) in the equation of motion for ε (12.44), we get

$$\dot{\varepsilon} = -\frac{C}{2}\varepsilon + \frac{\mathcal{A}}{2}\left[\frac{1}{\left(1+\frac{\beta}{\mathcal{A}I}\right)}\right]\varepsilon + \mathcal{F}_\varepsilon \tag{12.57}$$

where the parameters $\mathcal{A}$ and $\mathcal{B}$ are the gain and saturation coefficient, defined as

$$\mathcal{A} = \frac{2g^2 R}{\gamma^2}, \mathcal{B} = \frac{8g^4 R}{\gamma^4}, \tag{12.58}$$

and

$$\mathcal{F}_\varepsilon = \mathcal{F}_C + \frac{g}{\gamma}\mathcal{F}_M + \frac{g^2}{\gamma^2}\left[\frac{1}{\left(1 + \frac{B}{AI}\right)}\right](\mathcal{G}_a - \mathcal{G}_b)\varepsilon. \tag{12.59}$$

The noise force $\mathcal{F}_\varepsilon$ is characterized by the correlation functions

$$\langle \mathcal{F}_\varepsilon(t) \rangle = 0 \tag{12.60}$$

$$\langle \mathcal{F}_\varepsilon^*(t)\mathcal{F}_\varepsilon(t') \rangle = 2\langle D_{\varepsilon^*\varepsilon} \rangle \delta(t - t'), \tag{12.61}$$

$$\langle \mathcal{F}_\varepsilon(t)\mathcal{F}_\varepsilon(t') \rangle = 2\langle D_{\varepsilon\varepsilon} \rangle \delta(t - t'). \tag{12.62}$$

The diffusion coefficients $D_{\varepsilon^*\varepsilon}$ and $D_{\varepsilon\varepsilon}$ determine the strength of the noise and can be calculated directly from the definition of $\mathcal{F}_\varepsilon(t)$.

We leave it to the reader to verify that

$$2D_{\varepsilon\varepsilon} = -\mathcal{A}\left[\frac{1}{\left(1 + \frac{B}{AI}\right)}\right]^2 \frac{\mathcal{B}\varepsilon^2}{4\mathcal{A}}\left[3 + \frac{p}{2} + \left(\frac{\mathcal{B}}{AI}\right)\right], \tag{12.63}$$

$$2D_{\varepsilon^*\varepsilon} = \mathcal{A}\left[\frac{1}{1 + \left(\frac{B}{AI}\right)}\right]^2 \left[1 + \left(\frac{\mathcal{B}I}{4\mathcal{A}}\right)\left(3 - \frac{p}{2} + \frac{\mathcal{B}}{AI}\right)\right]. \tag{12.64}$$

With the above results, we are now ready to calculate the phase and intensity fluctuations in the laser, in terms of the various parameters, including the pump statistics.

12.3 Phase and Intensity Fluctuations

In this section, we analyse the fluctuation properties of the phase and intensity of the field.

For this purpose, we with the field amplitude in polar coordinates, that is

$$\varepsilon = \sqrt{I}\exp i\varphi. \tag{12.65}$$

From (12.50), we now derive two differential equations for I and φ

$$\dot{\varphi} = F_\varphi, \tag{12.66}$$

$$\dot{I} = -CI + \frac{\mathcal{A}}{1 + \frac{B}{AI}}I + F_I. \tag{12.67}$$

In (12.67), we neglected the small contribution to the drift because of noise.

The diffusion coefficients corresponding to the noise forces F_I and F_φ are found to be

$$D_{\varphi\varphi} = \frac{\mathcal{A}}{4I} \left[\frac{1}{\left(1 + \frac{\mathcal{B}}{\mathcal{A}I}\right)} \right] \left(1 + \frac{\mathcal{B}}{2\mathcal{A}}I\right),$$

(12.68)

and

$$D_{II} = \frac{\mathcal{A}}{\left(1 + \frac{\mathcal{B}}{\mathcal{A}I}\right)^2} \left[1 - p\frac{\mathcal{B}}{4\mathcal{A}}I\right] I.$$

(12.69)

12.4 Discussion

The quantum mechanical description of the amplitude and phase and their measurement has turned out to be troublesome, and it is still a matter of discussion [3].

Early attempts to introduce the amplitude and phase operators in a quantum formalism goes way back to Dirac in 1927.

However, if the photon number is large, we can bypass the above complications and state that the phase in (12.65) is in excellent agreement to the measured phase of the electromagnetic field, and we can identify φ with the phase of the radiation field.

As far as the intensity I is concerned, we have to be careful, because this classical quantity was originally associated to normally ordered operators.

For the photon number average, there is no ordering problem, and one has

$$\langle n \rangle = \langle a^\dagger a \rangle = \langle I \rangle,$$

(12.70)

however, for the photon number fluctuations

$$\begin{aligned}
\langle (\Delta n)^2 \rangle &= \langle a^\dagger a a^\dagger a \rangle - \langle a^\dagger a \rangle^2 \\
&= \langle a^\dagger a^\dagger a a \rangle - \langle a^\dagger a \rangle^2 + \langle a^\dagger a \rangle \\
&= \langle I^2 \rangle - \langle I \rangle^2 + \langle I \rangle \\
&= \langle (\Delta I)^2 \rangle + \langle I \rangle.
\end{aligned}$$

(12.71)

The steady-state intensity is easily found by setting the drift term in (12.67) equal to zero, thus getting

$$I_o = \frac{\mathcal{A}(\mathcal{A} - C)}{\mathcal{B}C}.$$

(12.72)

We now turn our attention to the phase diffusion. Using (12.66), we can write

$$\frac{\mathrm{d}}{\mathrm{d}t}\langle \varphi^2 \rangle = \frac{\mathrm{d}}{\mathrm{d}t} \int_0^t \mathrm{d}t' \int_0^t \mathrm{d}t'' \langle \dot{\varphi}(t')\, \dot{\varphi}(t'') \rangle$$

(12.73)

$$= \frac{d}{dt} \int_0^t dt' \int_0^t dt'' \langle F_\varphi(t') F_\varphi(t'') \rangle$$

$$= \langle 2D_{\varphi\varphi} \rangle.$$

If we now substitute the expression for $D_{\varphi\varphi}$ into (12.73), we integrate, getting

$$\langle \varphi^2 \rangle = \frac{1}{4I_o} + \frac{A+C}{4I_o} t. \tag{12.74}$$

The integration constant $\frac{1}{4I_o}$ comes from the vacuum fluctuations and can be considered an added noise to the one generated by the spontaneous emission, which corresponds to the second term in (12.74). This last result is the well-known Schawlow–Townes result [4], which states that the laser phase diffuses linearly in time. An important observation is that **the phase diffusion is independent of the pump parameter** p. Therefore, the phase of the electromagnetic field is completely independent of the particular pump mechanism. On the other hand, the intensity diffusion depends on p, and hence, we expect the photon number fluctuations to also depend on the pump parameter.

Now, we proceed to study the intensity fluctuations, around its steady-state value. For that purpose, we define $\Delta I = I - I_o$, and linearizing (12.67), we get

$$\frac{d}{dt}(\Delta I) = -C \frac{A-C}{A} \Delta I + F_I. \tag{12.75}$$

Equation 12.75 describes a simple Markoff Process. This type of stochastic differential equation, with a linear drift term, is called Ornstein–Uhlenbeck differential equation. (For an introduction on the subject, see appendix D. [5, 6].)

The steady state variance of a Ornstein–Uhlenbeck process is given by

$$\langle (\Delta I)^2 \rangle = \frac{D_{II}}{C \frac{A-C}{A}}. \tag{12.76}$$

Combining (12.69, 12.71) and (12.76), we readily get

$$\langle (\Delta n)^2 \rangle = \left(\frac{A}{A-C} - \frac{p}{4} \right) n_o, \tag{12.77}$$

where $n_o = I_o$ is the average photon number inside the cavity. As pointed out also in the last chapter, the photon number fluctuations crucially depend on the particular pump mechanism. In the case of a Poisson statistics ($p = 0$), the variance of the photon number is always greater than the mean photon number, and in the high intensity limit, both quantities tend to be equal, which, as seen in the last chapter, corresponds to a Poisson distribution for the photons.

On the other hand, for a pump noise suppressed laser $(p>0)$, the photon number variance can be smaller than n_o, which corresponds to a sub-Poissonian photon statistics. The optimum noise reduction corresponds to the high intensity limit $(\mathcal{A} >> C)$ with a regular injection $(p = 1)$, in which case

$$\langle(\Delta n)^2\rangle = 0.75 n_o, \tag{12.78}$$

that corresponds to a 25% noise reduction in the photon number fluctuations, with respect to the Poissonian case.

In this theory, we assumed for simplicity that the atomic decay times of both levels were equal, $\gamma_a = \gamma_b = \gamma$.

This may not be true in some laser systems. For the more general case, the decay constants for N_a, N_b and M are γ_a, γ_b and Γ, respectively. A calculation completely analogous to the present one gives:

$$\langle(\Delta n)^2\rangle = \left[\frac{\mathcal{A}}{\mathcal{A} - C} - \frac{\gamma_a}{\gamma_b + \gamma_a}\frac{p}{4}\right] n_o. \tag{12.79}$$

As we can see from (12.79), in the high intensity limit and in the case $\gamma_b << \gamma_a$, if the injection is regular, we may reduce the photon number variance by 50% with respect to the Poissonian case. (some publications speak about 50% photon number squeezing; we prefer to use this word only in relation to the quadratures, to avoid confusion.) [7].

Finally, lasers with small active medium volume were recently developed, like vertical cavity surface emitting diode lasers [8, 9], heterostructure diode lasers [10], microdroplets [11], high-Q Fabry-Perot microcavity lasers [12]. A quantum theory of the thresholdless laser was developed by Protschenko et al. [13] based on the Heisenberg–Langevin equations of motion for the atomic and field operators.

Problems

12.1. Verify the atomic noise correlations given by (12.17, 12.18, 12.19, 12.20).
12.2. Prove (12.39, 12.40, 12.41).
12.3. Prove (12.63 and 12.64).
12.4. Prove (12.68 and 12.69).

References

1. Benkert, C., Scully, M.O., Bergou, J., Davidovich, L., Hillery, M., Orszag, M.: Phys. Rev. A, **41**, 2756 (1990)
2. For a different approach to pump noise, see: Machida, S., Yamamoto, Y., Itaya, Y., Phys. Rev. Lett., **58**, 100 (1987); Marte, M., Ritsch, H., Walls, D.F., Phys. Rev. Lett., **61**, 1093 (1988)

3. For many papers on the various interpretations of the quantum phase, see for example: In: Schleich, W.P., Barnett, S.M.W. (eds) Quantum Phase and Quantum Phase Measurements. Physica Scripta, **T48** (1993)
4. Schawlow, A.L., Townes, C.H.: Phys. Rev., **112**, 1940 (1958)
5. Van Kampen, N.G.: Stochastic Processes in Physics and Chemistry. North-Holland, Amsterdam (1981)
6. Gardiner, C.W.: Handbook of Stochastic Processes. Springer Verlag, Berlin (1985)
7. Fontenelle, M.T., Davidovich, L.: Phys. Rev. A, **51**, 2560 (1995)
8. Choa, F.S., Shih, M.H., Fan, J.Y., Simonis, G.J., Liu, P.L., Tanburn-Ek, T., Logan, R.A., Trang, W.T., Sargent, A.M.: App. Phys. Lett., **67**, 2777 (1995)
9. Huffaken, D.L., Shin, J., Deppe, D.G.: App. Phys. Lett., **66**, 1723 (1995); Huffaken, D.L., Deng, H., Deng, Q., Deppe, D.G.: App. Phys. Lett., **69**, 3477 (1997)
10. Feit, Z., McDonald, M., Woods, R.J., Archambault, V., Mak, P.: App. Phys. Lett., **68**, 738 (1996)
11. Taniguchi, H., Tomisawa, H., Kido, J.: App. Phys. Lett., **66**, 1578 (1995); Tanosaki, S., Taniguchi, H., Tsujita, K., Inaba, H.: App. Phys. Lett., **69**, 719 (1996)
12. An, K., Childs, J.J., Desari, R.R., Feld, M.S.: Phys. Rev. Lett., **73**, 3375 (1994)
13. Protschenko, I., Domokos, P., Lefevre-Seguin, L., Hare, J., Raimond, J.M., Davidovich, L.: Phys. Rev. A, **59**, 1667(1999)

Further Reading

- Gardiner, C.W.: Quantum Noise. Springer Verlag, Berlin (1991)
- Haken, H.: Laser Theory. Springer Verlag, Berlin (1970)
- Haken, H.: Light, vols. 1 and 2, Springer Verlag, Berlin (1981)
- Lax, M.: In: Kelley, P.L, Lax. B. Tannenwald, P.E. (eds), Physico of Quantum Electronics (MC Graw will, Newyork, 1966)
- Lax, M.: In: Chretien, M., Gross, E.P., Dreser, S. (eds) Statistical Physics, Phase Transition and Superconductivity. vol II, Gordon and Breach, New York (1968)
- Risken, H.: The Fokker Planck Equation. Springer Verlag, Berlin (1984)
- SargentIII, M., Scully, M.O., Lamb, W.E.: Laser Physics. Addison Wesley, Mass USA (1974)
- Scully, M.O., Zubairy, M.S.: Quantum Optics. Cambridge University Press, Cambridge (1997)

13. Quantum Noise Reduction 1

In this chapter, we study the correlated emission laser (CEL).

In many areas of modern physics, ultrasmall displacements are detected optically. The small displacement is converted into a change of optical path length in an interferometer.

This detection scheme is done usually in two ways:

1. In a passive detection scheme, where laser light, generated outside, is sent through a cavity, and the change in the path length results in a phase shift. The shift is then detected by homodyning the output beam with a reference beam. This phase shift is generally small, because the light spends only a finite time in the cavity, limited by the cavity lifetime. In this type of measurements, the limiting noise source is the photon-counting error or shot noise, reflecting the photon number fluctuations at the detector.

2. In the so-called active detection scheme, the laser light is generated inside the cavity and the operating frequency of the system changes due to the change in the path length, which results in a phase shift proportional to the measurement time, leading, in general to a bigger signal as compared to the first case. The shift is then detected by heterodyning the output light with that from a reference beam. The limiting quantum noise source, in this case, is the **spontaneous fluctuations of the relative phase between the two lasers, or in other words, the relative phase diffusion noise.**

The question posed in the first part of this chapter is the following one.

Can one possibly quench the spontaneous emission quantum noise from the relative phase of two lasers?

The whole subject of CEL is directed to answer this particular question.

A geometrical representation of the CEL is shown in the Fig. 13.1, where $\delta\varepsilon_1$ and $\delta\varepsilon_2$ are the contributions to the fields 1 and 2 by spontaneous emission of a photon in the two respective modes. To get an intuitive picture of the effect, consider three-level atoms in a double cavity, interacting with two quantum fields, E_1 and E_2, and a classical microwave field E_3, resonant with the upper two level and originating the necessary correlation between the spontaneous emission into the fields 1 and 2 (Fig. 13.2).

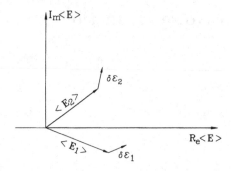

Fig. 13.1. Geometrical representation of the correlated emission laser. $\delta\varepsilon_1$ and $\delta\varepsilon_2$ are the contributions of a spontaneously emitted photon in the modes 1 and 2, respectively

In the case described above, the spontaneously emitted fields 1 and 2 are strongly correlated. To see this, consider a state vector given by [1]

$$| \psi\rangle = \alpha \exp(-i\phi_a)\,| a,0\rangle + \beta \exp(-i\phi_b)\,| b,0\rangle + \gamma_1\,| c,1_1\rangle + \gamma_2\,| c,1_2\rangle\,, \quad (13.1)$$

where $1_i, i = 1,2$ correspond to a photon emitted into the fields 1 and 2, respectively. Now, the expectation value of the fields

$$E_i = \epsilon_i a_i \exp i(\mathbf{k}_i\cdot\mathbf{r} - \nu_i t)\,, \quad (13.2)$$
$$i = 1,2\,,$$

vanishes, due to the orthogonality of the atomic states. **However, the crossed term does not vanish**

$$\langle\psi\,|\,E_1^\dagger E_2\,|\,\psi\rangle = \epsilon_1\cdot\epsilon_2\gamma_1^*\gamma_2\exp\left[-i(\mathbf{k}_1-\mathbf{k}_2)\cdot\mathbf{r} + i(\nu_1-\nu_2)t\right]\,, \quad (13.3)$$

thus giving a clear indication that the spontaneously emitted photons at frequencies ν_1 and ν_2 are correlated.

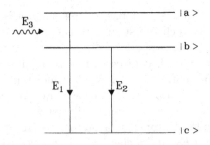

Fig. 13.2. In the three-level atom, the two upper levels a and b are coupled to a classical microwave field of frequency ν_3. The emissions from the $(b-c)$ and $(a-c)$ transitions are strongly correlated

Strongly motivated by the above arguments, we are led to investigate the diffusion of the relative phase angle between the two modes, in a system where the lasing three-level atoms are placed in a double cavity.

13.1 Correlated Emission Laser Systems

13.1.1 The Quantum Beat Laser

The Model

We consider the system described in the Fig. 13.2, in which three-level atoms are being pumped into a state $\mid a \rangle$ at a rate r_a. The external field at frequency ν_3 is characterized by a Rabi frequency Ω. The Hamiltonian is

$$H = H_0 + V , \tag{13.4}$$

where

$$H_0 = \sum_{i=a,b,c} \hbar\omega_i \mid i\rangle\langle i \mid +\hbar\nu_1 a_1^\dagger a_1 + \hbar\nu_2 a_2^\dagger a_2 , \tag{13.5}$$

and

$$
\begin{aligned}
V = {} & \hbar g_1(a_1 \mid a\rangle\langle c \mid +a_1^\dagger \mid c\rangle\langle a \mid) \\
& +\hbar g_2(a_2 \mid b\rangle\langle c \mid +a_2^\dagger \mid c\rangle\langle b \mid) \\
& -\frac{\hbar\Omega}{2}(\exp(-i\nu_3 t - i\phi) \mid a\rangle\langle b \mid + \exp(i\nu_3 t + i\phi) \mid b\rangle\langle a \mid) .
\end{aligned}
\tag{13.6}
$$

It is convenient to go to the interaction picture

$$V_I = \exp\left(\frac{i}{\hbar}H_0 t\right)(V)\exp(-\frac{i}{\hbar}H_0 t) . \tag{13.7}$$

After a direct calculation, one finds that

$$V_I = V_1 + V_2 , \tag{13.8}$$

with

$$V_1 = -\frac{\hbar\Omega}{2}\begin{pmatrix} 0 & \exp(-i\phi) & 0 \\ \exp(i\phi) & 0 & 0 \\ 0 & 0 & 0 \end{pmatrix} , \tag{13.9}$$

and

$$V_2 = \hbar\begin{bmatrix} 0 & 0 & g_1 a_1 \exp(i\Delta_1 t) \\ 0 & 0 & g_2 a_2 \exp(i\Delta_2 t) \\ g_1 a_1^\dagger \exp(-i\Delta_1 t) & g_2 a_2^\dagger \exp(-i\Delta_2 t) & 0 \end{bmatrix} . \tag{13.10}$$

In (13.10), we introduced the definitions

$$\Delta_1 = w_a - w_c - v_1; \quad \Delta_2 = w_b - w_c - v_2 . \qquad (13.11)$$

In the future, we will assume that $\Delta_1 = \Delta_2 = \Delta$ and that the driving field is resonant with the a–b transition.

Our next goal is to approximate our Hamiltonian model as to end up with a time-independent Hamiltonian after some sort of rotating wave approximation.

This is most easily achieved by introducing a second interaction picture [2]

$$V_{II} = \exp\left(\frac{i}{\hbar} V_1 t\right) V_2 \exp\left(-\frac{i}{\hbar} V_1 t\right) . \qquad (13.12)$$

One can easily check the following properties of V_1

$$(V_1)^{2n} = \left(\frac{\hbar\Omega}{2}\right)^{2n} \begin{bmatrix} 1 & 0 & 0 \\ 0 & 1 & 0 \\ 0 & 0 & 0 \end{bmatrix} , \qquad (13.13)$$

$$(V_1)^{2n+1} = \left(\frac{\hbar\Omega}{2}\right)^{2n} (V_1) .$$

With the above expressions, one can calculate the transformation explicitly:

$$\exp\left(\pm\frac{i}{\hbar} V_1 t\right) = \begin{bmatrix} \cos(\frac{\Omega}{2})t & \mp i\sin(\frac{\Omega}{2})t \exp(-i\phi) & 0 \\ \mp i\sin(\frac{\Omega}{2})t\exp(i\phi) & \cos(\frac{\Omega}{2})t & 0 \\ 0 & 0 & 1 \end{bmatrix} .$$
$$(13.14)$$

With the above expression, one can calculate explicitly the interaction Hamiltonian

$$V_{II} = \begin{pmatrix} 0 & 0 & V_{ac} \\ 0 & 0 & V_{bc} \\ V_{ac}^\dagger & V_{bc}^\dagger & 0 \end{pmatrix} , \qquad (13.15)$$

where

$$V_{ac} = \frac{1}{2}\left\{(\exp\left[i\left(\Delta + \frac{\Omega}{2}\right)t\right](g_1 a_1 - g_2 a_2 \exp(-i\phi)\right\} \qquad (13.16)$$
$$+ \exp\left[i\left(\Delta - \frac{\Omega}{2}\right)t\right](g_1 a_1 + g_2 a_2 \exp(-i\phi)) ,$$
$$V_{bc} = \frac{1}{2}\{-\exp\left[i\left(\Delta + \frac{\Omega}{2}\right)t\right](g_1 a_1 \exp(i\phi) - g_2 a_2)$$
$$+ \exp\left[i\left(\Delta - \frac{\Omega}{2}\right)t\right](g_1 a_1 \exp(i\phi) + g_2 a_2)\} .$$

The condition on CEL found in [3] was $\Delta = \frac{\Omega}{2}$. We immediately see that this condition is appealing, because one of the time dependences disappear and the other term $\propto \exp 2\Delta t$ can be neglected in a rotating wave approximation. The conditions of validity of this approximation will be discussed later.

We define, for convenience, a non-Hermitian operator A

$$A \equiv \frac{g_1 a_1 \exp(i\phi) + g_2 a_2 \exp(-i\phi)}{\sqrt{g_1^2 + g_2^2}} , \qquad (13.17)$$

so that it is easy to verify that

$$[A, A^\dagger] = 1 , \qquad (13.18)$$

and

$$V_{II} = \begin{pmatrix} 0 & 0 & A \exp(-i\frac{\phi}{2}) \\ 0 & 0 & A \exp(i\frac{\phi}{2}) \\ A^\dagger \exp(i\frac{\phi}{2}) & A^\dagger \exp(-i\frac{\phi}{2}) & 0 \end{pmatrix} , \qquad (13.19)$$

with $g \equiv \frac{1}{2}\sqrt{g_1^2 + g_2^2}$.

The Solution

We are now going to develop the non-linear theory of the quantum beat laser, whose Hamiltonian is given in (13.19).

The Schrödinger's equation in the second interaction picture is

$$i\hbar \frac{\partial \psi}{\partial t} = V_{II}\psi . \qquad (13.20)$$

Here, ψ is a column vector with three components, ψ_a, ψ_b, ψ_c. We get the following coupled equations

$$i\frac{\partial \psi_a}{\partial t} = gA\psi_c - i\frac{\gamma}{2}\psi_a , \qquad (13.21)$$

$$i\frac{\partial \psi_b}{\partial t} = gA\psi_c - i\frac{\gamma}{2}\psi_b ,$$

$$i\frac{\partial \psi_c}{\partial t} = gA^\dagger\psi_a + gA^\dagger\psi_b - i\frac{\gamma}{2}\psi_c .$$

In (13.20), ϕ was eliminated by a trivial transformation: $\psi_a \exp\frac{i\phi}{2} \rightarrow \psi_a, \psi_b \exp-\frac{i\phi}{2} \rightarrow \psi_b$. Also, we have introduced the phenomenological decay constant γ, for simplicity, the same for all three levels.

The solution is

$$\psi_a = \psi_b = \frac{1}{\sqrt{2}} \exp-\frac{\gamma}{2}(t - t_0) \cos\left[g\sqrt{2}AA^\dagger(t - t_0)\right] \psi_f(t_0) , \qquad (13.22)$$

$$\psi_c = -i \exp-\frac{\gamma}{2}(t - t_0)A^{-1}(AA^\dagger)^{\frac{1}{2}} \sin\left[g\sqrt{2}AA^\dagger(t - t_0)\right] \psi_f(t_0) ,$$

where we have taken the initial condition with the atom injected in the excited state, that is: $\psi_a(t_0) = \psi_f(t_0)$, a function of the field variables only.

The Master Equation

For the second interaction picture, one can write

$$\frac{d\rho}{dt} = -\frac{i}{\hbar} [V_{II}, \rho] .$$ (13.23)

The reduced density operator, for the field is $\rho_f = Tr\rho$. Making use of (13.19), one can write

$$\frac{d\rho_f}{dt} = -ig \left\{ [A, (\rho_{ca} + \rho_{cb})] + [A^\dagger, (\rho_{ac} + \rho_{bc})] \right\} + L ,$$ (13.24)

where L is the loss term to be specified later. We need to calculate $(\rho_{ac} + \rho_{bc})$. We adopt the following procedure [4]: we first calculate the one-atom contribution injected at time t_0 into the upper level $\mid a\rangle$ and then add all the contributions from $t - \gamma^{-1}$ to t

$$(\rho_{ac} + \rho_{bc}) = r_a \int_{t-\gamma^{-1}}^{t} dt_0 \, [\psi_a(t, t_0) - \psi_b(t, t_0)] \, \psi_c^\dagger(t, t_0)\psi$$

$$= \sqrt{2} i r_a \int_{t-\gamma^{-1}}^{t} dt_0 \exp -\gamma(t - t_0) \cos \left[g\sqrt{2}(AA^\dagger)^{\frac{1}{2}}(t - t_0) \right] \rho_f(t)$$ (13.25)

$$\times \sin \left[g\sqrt{2}(AA^\dagger)^{\frac{1}{2}}(t - t_0) \right] (AA^\dagger)^{\frac{1}{2}}(A^\dagger)^{-1}.$$ (13.26)

In the above expression, we replaced $\rho_f(t_0) \rightarrow \rho_f(t)$, assumption that is only true if the cavity time is much longer than the atomic characteristic times, and during the interaction time, ρ_f does not change appreciably.

Then we extend the lower limit to $-\infty$, because, due to the exponential damping factor in the integrand, the contribution from $t_0 < t - \gamma^{-1}$ is negligible.

Now, taking the nn' matrix elements, and replacing $(A^\dagger)^{-1} \rightarrow (AA^\dagger)^{-1}A$, when acting on the right, one gets

$$(\rho_{ac} + \rho_{bc})_{n,n'} = ir_a g(R^+_{n+1,n'} - R^-_{n+1,n'})\rho_{fn,n+1} ,$$ (13.27)

where

$$R^\pm_{n,n'} = \frac{(\sqrt{n} \pm \sqrt{n'})^2}{\gamma^2 + 2g^2(\sqrt{n} \pm \sqrt{n'})^2} .$$ (13.28)

Next, we specify the loss term in the usual way

$$L = -\frac{\nu_1}{2Q_1}(a_1^\dagger a_1 \rho_f + \rho_f a_1^\dagger a_1 - 2a_1\rho_f a_1^\dagger) - \frac{\nu_2}{2Q_2}(a_2^\dagger a_2\rho_f + \rho_f a_2^\dagger a_2 - 2a_2\rho_f a_2^\dagger) .$$ (13.29)

For convenience, we now introduce a B-mode

$$B \equiv \frac{g_2 a_1 - g_1 a_2}{(g_1^2 + g_2^2)^{\frac{1}{2}}} ,$$ (13.30)

with the properties

$$[B, B^\dagger] = 1, [A, B] = [A, B^\dagger] = 0 , \tag{13.31}$$

that is, the A and B modes are independent. One can write a_1 and a_2 in terms of A and B and use it in (13.29).

Finally, using (13.27) and it's Hermitian conjugate, (13.29) and replacing them in (13.24), we obtain the master equation for the field density operator (for simplicity, we skip the sub-index f)

$$\frac{d\rho_{n_A,n_B n'_A,n'_B}}{dt} = \left[\left(\frac{\sqrt{n_A n'_A}\mathcal{A}}{1 + \mathcal{N}_{n_A-1,n'_A-1}\frac{\mathcal{B}}{\mathcal{A}}} \right) \rho_{n_A-1,n_B n'_A-1,n'_B} \right.$$

$$\left. - \left(\frac{\mathcal{N}'_{n_A,n'_A}\mathcal{A}}{1 + \mathcal{N}_{n_A,n'_A}\frac{\mathcal{B}}{\mathcal{A}}} \right) \rho_{n_A,n_B n'_A,n'_B} \right] \tag{13.32}$$

$$-\frac{\gamma_c}{2} \left[(n_A + n'_A + n_B + n'_B)\rho_{n_A,n_B n'_A,n'_B} - 2\sqrt{(n_A+1)(n'_A+1)}\rho_{n_A+1,n_B n'_A+1,n'_B} \right.$$

$$\left. -2\sqrt{(n_B+1)(n'_B+1)}\rho_{n_A,1+n_B n'_A,1+n'_B} \right].$$

Here, for simplicity, we assumed that $\frac{\nu_1}{Q_1} = \frac{\nu_2}{Q_2} \equiv \gamma_c$, and the definitions of $\mathcal{A}, \mathcal{B}, \mathcal{N}, \mathcal{N}'$, are given in (11.27, 11.24).

As the A and B modes are independent, the solution of the master equation must be separable,

$$\rho_{n_A,n_B n'_A,n'_B}(t) = \rho^{(A)}_{n_A,n'_A}(t)\rho^{(B)}_{n_B,n'_B}(t) . \tag{13.33}$$

Photon Statistics

We take the steady-state case, that is for $\frac{d}{dt} = 0$, and the diagonal terms only, to determine the photon statistics. It is quite apparent that there is no gain term in the B-mode, which just damps away. The solution is

$$\rho^{(B)}_{n_B,n'_B} = \delta_{n_B,0} . \tag{13.34}$$

On the other hand, the A-mode satisfies the usual one mode, two-level laser difference equation

$$\left(\frac{n\mathcal{A}}{1 + \mathcal{N}_{n-1,n-1}\frac{\mathcal{B}}{\mathcal{A}}} \right) \rho^{(A)}_{n-1,n-1} - \left[\frac{(n+1)\mathcal{A}}{1 + \mathcal{N}_{n,n}\frac{\mathcal{B}}{\mathcal{A}}} \right] \rho^{(A)}_{n,n} \tag{13.35}$$

$$-\gamma_c \left[n\rho^{(A)}_{n,n} - (n+1)\rho^{(A)}_{n+1,n+1} \right] = 0.$$

As we can see, in terms of the composite mode A, the quantum beat laser exhibits the same type of behavior, as far as photon statistics, threshold or saturation properties, as the one mode laser.

Phase Diffusion

In this section, we will prove that there is, for the quantum beat laser, a complete noise quenching of the relative phase of the two quantum modes.

We start by defining a beat signal as

$$BS = \text{Re}\left\{ \exp\left[i(v_1 - v_2)t\right] Tr\left(a_1^\dagger a_2 \rho\right)\right\} \tag{13.36}$$
$$= \text{Re}\left\{ \exp iv_3 t Tr\left[(A^\dagger A - B^\dagger B + AB^\dagger - A^\dagger B)\rho\right]\right\} ,$$

where we assumed, for simplicity $g_1 = g_2$.

Now, performing the separation indicated in (13.33), we get two time-dependent master equations

$$\frac{d\rho_{n,n'}^{(A)}}{dt} = \left[\frac{\sqrt{nn'}\mathcal{A}}{1 + \mathcal{N}_{n-1,n'-1}\left(\frac{B}{A}\right)}\right] \rho_{n-1,n'-1}^{(A)} - \left[\frac{\mathcal{N}_{n,n'}'\mathcal{A}}{1 + \mathcal{N}_{n,n'}\left(\frac{B}{A}\right)}\right] \rho_{n,n'}^{(A)} \tag{13.37}$$

$$- \frac{\gamma_c}{2}\left[(n+n')\rho_{n,n'}^{(A)} - 2\sqrt{(n+1)(n'+1)}\rho_{n+1,n'+1}^{(A)}\right] ,$$

$$\frac{d\rho_{n,n'}^{(B)}}{dt} = -\frac{\gamma_c}{2}\left[(n+n')\rho_{n,n'}^{(B)} - 2\sqrt{(n+1)(n'+1)}\rho_{n+1,n'+1}^{(B)}\right] . \tag{13.38}$$

Looking at the time-dependent solutions of (13.37), one writes the general solution in the form [4]

$$\rho_{n,n+p} = \sum_{j=0}^{\infty} \phi_j(n,p) \exp(-\mu_j^{(p)}t.) \tag{13.39}$$

As we have already seen from the laser theory, the lowest eigenvalue $\mu_0^{(0)} = 0$, thus allowing a non-vanishing stationary solution for the diagonal elements. Also, $\mu_j^{(p)} > 0$ for $p \neq 0$, so that the off diagonal elements of the field density matrix goes to zero for long times.

In the case of an ordinary two-mode laser, the density matrix corresponding to the two modes factorize

$$\rho^{(1,2)} = \rho^1 \rho^2 , \tag{13.40}$$

so that the beat signal

$$\text{Re}\langle a_1^\dagger a_2\rangle_{\text{OL}} = \frac{1}{2}\left[\sum_{n,n'} \rho_{n,n+1}^{(1)}(t)\rho_{n',n'-1}^{(2)}(t)\sqrt{(n+1)n'} + cc\right] , \tag{13.41}$$

and according to (13.39), the above expression vanishes at a rate $\mu_0^{(1)}$.

This defines the phase diffusion coefficient as

$$\mu_0^{(1)} = \frac{D}{2} , \tag{13.42}$$

where D is the Schawlow–Townes linewidth.

In the case of the quantum beat laser, this does not happen, because (13.36) contains diagonal elements in $A^\dagger A$, so

$$\mathrm{Re}\langle a_1^\dagger a_2 \rangle_{\mathrm{QBL}} \sim \sum_n n \rho_{n,n}^{(A)} + \dots , \tag{13.43}$$

and because $\mu_0^{(0)} = 0$, there is always a non-zero part of the beat signal.

If we keep the definition of the diffusion coefficient as twice the lowest decay rate, then we conclude that $D = 0$, for the relative phase of the two modes.

A more formal and longer derivation can be done using Glauber's P representation to obtain a Fokker–Planck equation in α and α^*, which can be converted in a polar form, in terms of the two amplitudes ρ_1 and ρ_2 and the two phases θ_1 and θ_2 of the two modes. Then one finds that the phase difference $\theta \equiv \theta_1 - \theta_2$ locks to zero and $D(\theta) = 0$. (The details of this rather long calculation are found in the [2]).

13.1.2 Other CEL Systems

The active medium can also be prepared in a coherent superposition of the $\mid a \rangle$ and $\mid b \rangle$ states which decay to the state $\mid c \rangle$, via emission of different polarization states. This is the Hanle effect, and it can be achieved with a polarization sensitive mirror to couple the doubly resonant cavity.

Another CEL system is a ring laser whose counterpropagating modes are coupled by a spatial modulation in the gain medium. This is the holographic laser [5], where each beam is reflected in part by the thin atomic layers of the gain medium. When the reflected light interferes constructively with the light of the counterpropagating beam, noise quenching is achieved.

A last example is the two-photon CEL [6], which is the extension of the CEL principle in a system where the active medium consists in three-level atoms in the cascade configuration driving a cavity resonant with $v_1 = \frac{\omega_a - \omega_c}{2}$ as shown in the Fig. 13.3.

We are again interested in finding out the role of the atomic coherence between the most distant levels a and c, in quenching the noise.

As it turns out, this system is not only capable of quenching the quantum noise of an active system but also under certain conditions, reduces the phase noise below the shot noise level, producing a squeezed output.

We shall not go into the theoretical details of these systems, because they operate under the same basic principle.

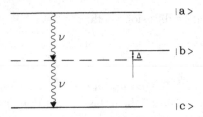

Fig. 13.3. The two-photon correlated emission laser. The active medium are three-level atoms in a cascade configuration.The cavity is tuned at ν and the intermediate level is off resonance with respect the center of the transition

Problems

13.1. Prove (13.13, 13.15, 13.16).

13.2. Show, for the quantum beat laser, that the relative phase diffusion coefficient is given by

$$D(\theta) = \frac{g^2 r}{4\gamma^2 \langle n \rangle}(1 - \cos \theta) - \frac{g^4 r}{\gamma^4}(1 - \cos 2\theta),$$

where γ is the atomic decay and g the same coupling constant for the two modes with the respective transitions.

Notice that for $\theta = 0, D(\theta) = 0$. [2]

13.3. Prove that the drift coefficient of the Fokker–Planck equation $d(\theta)$ vanishes for $\theta = 0$, thus giving the required phase locking to achieve $D(\theta) = 0$. [2]

References

1. Scully, M.O.: Phys. Rev. Lett., **55**, 2802 (1985)
2. Bergou, J.A., Orszag, M., Scully, M.O.: Phys. Rev. A, **38**, 754 (1988)
   For a general review of the subject of correlated emission laser, see also:
   Orszag, M., Bergou, J.A., Schleich, W., Scully, M.O., In: Tombesi, P., Pike, E.R. (eds) Squeezed and Non-Classical Light, (Plenum, NY (1989)
3. Scully, M.O., Zubairy, M.S.: Phys. Rev. A, **35**, 752 (1987)
4. Scully, M.O., Lamb, W.E.: Phys. Rev., **159**, 208 (1967); See also:
   SargentIII, M., Scully, M.O., Lamb, W.E.: Laser Physics. Addison Wesley, Mass.USA (1974)
5. Krause, J., Scully, M.O.: Phys., Rev. A, **36**, 1771 (1987)
6. Scully, M.O., Wodkiewicz, K., Zubairy, S., Bergou, J., Lu, N., Meyer ter Vehn, J.: Phys. Rev. Lett., **60**, 1832 (1988)

Further Reading

• Zaheer, K., Zubairy, M.S.: Phys. Rev. A, **38**, 5227 (1988)

CEL Experiments

- Ohtsu, M., Liou, K.Y.: App. Phys. Lett., **52**, 10 (1988)
- Winters, M.P., Hall, J.L., Toschek, P.E.: Phys. Rev. Lett., **65**, 3 116 (1990)
- The largest diffussion noise reduction
- Steiner, I., Toschek, P.E.: Phys. Rev. Lett., **74**, 4639 (1995)

For a Different View of CEL

- Schleich, W., Scully, M.O.: Phys. Rev. A, **37**, 1261 (1988)

A General Review of the CEL

- Dalton, B.J.: In: Barut, A.O. (ed.) New Frontiers in Quantum Electrodynamics and Quantum Optics. Plenum Press, New York (1990)

14. Quantum Noise Reduction 2

In this chapter, we will cover the area of theory and relevant experiments in the generation of squeezed states.

Before discussing the parametric oscillator, we will have a brief introduction to Non-Linear optics and the input-output theory.

Finally, we will discuss other important experiments that generated squeezed states.

14.1 Introduction to Non-Linear Optics

Light of a given frequency falling on an atomic system can give rise to different frequencies. Even stronger effects can be obtained if instead of an atomic system, one had a large number of atoms or some particular crystal such as quartz, ADP, KDP, etc.

One of the early experiments [1] was generating, with a Ruby laser falling into a quartz crystal, a blue light of $\lambda = 3472 \overset{\circ}{A}$, starting from the original red light of $\lambda = 6943 \overset{\circ}{A}$.

This experiment demonstrated for the first time the non-linear optical effect of second harmonic generation.

Since then, the field of non-linear optics has been explosive [2, 3, 4, 5].

We will be interested in some quantum non-linear effects related to the generation of squeezed states.

In order to explore the origin of the above-mentioned phenomena, we go back to the atom–radiation interaction model to study the multiple photon transitions.

14.1.1 Multiple-Photon Transitions

The atom–radiation Interaction Hamiltonian can be written as [6]

$$H = H_0 + H_1 , \tag{14.1}$$

where H_0 is the Hamiltonian of the uncoupled radiation H_r plus the bound electrons H_e

$$H_0 = H_r + H_e \,, \tag{14.2}$$

and, in the dipole approximation, H_1 is the interaction term

$$H_1 = -e\mathbf{r} \cdot \mathbf{E}(\mathbf{r}, t) \,. \tag{14.3}$$

If the initial state of the system is $\mid \phi(t_0)\rangle$, at time t it will be

$$\mid \phi(t)\rangle = \exp - \left[\frac{iH(t - t_0)}{\hbar}\right] \mid \phi(t_0)\rangle \,. \tag{14.4}$$

On the other hand, let $\mid i\rangle$ and $\mid f\rangle$ be eigenstates of the unperturbed energy

$$H_0 \mid i\rangle = \hbar\omega_i \mid i\rangle \,, \tag{14.5}$$
$$H_0 \mid f\rangle = \hbar\omega_f \mid f\rangle \,.$$

Then, if

$$\mid \phi(t_0)\rangle = \mid i\rangle \,, \tag{14.6}$$

that is, initially, the system is in an eigenstate of the unperturbed Hamiltonian H_0, then the probability for the system to be in $\mid f\rangle$ at $t = t$ is

$$P_{if} = \mid \langle f \mid \exp\left[-\frac{iH(t - t_0)}{\hbar}\right] \mid i\rangle \mid^2 \,. \tag{14.7}$$

Of course, the transition rate of the system from $\mid i\rangle \rightarrow \mid f\rangle$ is the time derivative of the above expression.

In general, there is a range of final states in an experimental observation; thus, one can define a transition rate as

$$\frac{1}{\tau} = \frac{\mathrm{d}}{\mathrm{d}t} \sum_f \mid \langle f \mid \exp\left[-\frac{iH(t - t_0)}{\hbar}\right] \mid i\rangle \mid^2 \,. \tag{14.8}$$

The above expression is not very practical, since the relevant term for transitions in the Hamiltonian is

$$H_1 \,,$$

which does not commute with

$$H_0 \,.$$

However, we make use of the following trick

$$\exp\left(\frac{iH_0 t}{\hbar}\right) H_1 \exp\left(-\frac{iHt}{\hbar}\right) = i\hbar\frac{\mathrm{d}}{\mathrm{d}t}\left[\exp\left(\frac{iH_0 t}{\hbar}\right) \exp\left(-\frac{iHt}{\hbar}\right)\right] \,. \tag{14.9}$$

The above relation can be easily verified by performing the derivative on the r.h.s. of (14.9). Next, we integrate both sides

$$\int_{t_0}^{t} \exp\left(\frac{iH_0 t_1}{\hbar}\right) H_1 \exp\left(-\frac{iH t_1}{\hbar}\right) dt_1$$

$$= i\hbar \left[\exp\left(\frac{iH_0 t}{\hbar}\right) \exp\left(-\frac{iH t}{\hbar}\right) - \exp\left(\frac{iH_0 t_0}{\hbar}\right) \exp\left(-\frac{iH t_0}{\hbar}\right) \right],$$

so

$$\exp\left(-\frac{iH t}{\hbar}\right) \qquad\qquad (14.10)$$

$$= \exp\left(-\frac{iH_0 t}{\hbar}\right) \left[\exp\left(\frac{iH_0 t_0}{\hbar}\right) \exp\left(-\frac{iH t_0}{\hbar}\right) \right.$$

$$\left. -\frac{i}{\hbar} \int_{t_0}^{t} \exp\left(\frac{iH_0 t_1}{\hbar}\right) H_1 \exp\left(-\frac{iH t_1}{\hbar}\right) dt_1 \right]$$

If one is interested in the steady-state transitions, we can assume that $t_0 \to -\infty$, and $H_1(t_0) = 0$.

Also, if one wants the interaction to be switched on in a smooth way, we introduce an $\exp(\varepsilon t_1)$, which we can conveniently eliminate at the end of the calculation by setting $\varepsilon \to 0$. Then

$$\exp\left(-\frac{iH t}{\hbar}\right) \qquad\qquad (14.11)$$

$$= \exp\left(-\frac{iH_0 t}{\hbar}\right) \left[1 - \frac{i}{\hbar} \int_{t_0}^{t} \exp\left(\frac{iH_0 t_1}{\hbar}\right) H_1 \exp(\varepsilon t_1) \exp\left(-\frac{iH t_1}{\hbar}\right) dt_1 \right].$$

The r.h.s. can be developed as power series in H_1 by iteration.

Zero-th order

$$\langle f \mid \exp\left(-\frac{iH_0 t}{\hbar}\right) \mid i \rangle = \exp(-i\omega_i t)\langle f \mid i \rangle = 0 ,$$

for $i \neq j$.

First order

$$-\frac{i}{\hbar}\langle f \mid \exp\left(-\frac{iH_0 t}{\hbar}\right) \int_{-\infty}^{t} \exp\left(\frac{iH_0 t_1}{\hbar}\right) H_1 \exp(\varepsilon t_1) \exp\left(-\frac{iH_0 t_1}{\hbar}\right) dt_1 \mid i \rangle$$

$$= -\frac{i}{\hbar}\langle f \mid H_1 \mid i \rangle \exp-(i\omega_f t) \int_{-\infty}^{t} \exp t_1(-i\omega_i + \varepsilon) dt_1$$

$$= \frac{\langle f \mid H_1 \mid i \rangle}{\hbar} \left[\frac{\exp t(i\omega_f - i\omega_i + \varepsilon)}{-\omega_f + \omega_i + i\varepsilon} \right],$$

so the transition to first order is

$$\frac{1}{\tau} = \frac{d}{dt} \sum_f \frac{|\langle f \mid H_1 \mid i \rangle|^2}{\hbar^2} \frac{\exp 2\varepsilon t}{(-\omega_f + \omega_i)^2 + \varepsilon^2} , \qquad (14.12)$$

$$\frac{1}{\tau} = \frac{2}{\hbar^2} \sum_f |\langle f \mid H_1 \mid i \rangle|^2 \frac{\varepsilon \exp 2\varepsilon t}{(-\omega_f + \omega_i)^2 + \varepsilon^2} , \qquad (14.13)$$

and when $\varepsilon \to 0$, we get

$$\frac{1}{\tau} = \frac{2\pi}{\hbar^2} \sum_f |\langle f \mid H_1 \mid i \rangle|^2 \, \delta(\omega_i - \omega_f) , \qquad (14.14)$$

which is the Fermi golden rule.

Second order

This is obtained by a first iteration of (20.62), thus getting

$$-\frac{1}{\hbar^2} \langle f \mid \exp\left(-\frac{iH_0 t}{\hbar}\right)$$

$$\int_{-\infty}^{t} dt_1 \int_{-\infty}^{t_1} dt_2 \exp\left(\frac{iH_0 t_1}{\hbar}\right) H_1 \exp(\varepsilon t_1) \exp\left(-\frac{iH_0 t_1}{\hbar}\right)$$

$$\exp\left(\frac{iH_0 t_2}{\hbar}\right) H_1 \exp(\varepsilon t_2) \exp\left(-\frac{iH_0 t_2}{\hbar}\right) \mid i \rangle,$$

and introducing a complete set of eigenstates of H_0, $1 = \sum_l |l\rangle\langle l|$, we get:

$$-\frac{1}{\hbar^2} \sum_l \exp(-i\omega_f t)\langle f \mid H_1 \mid l \rangle\langle l \mid H_1 \mid i \rangle \int_{-\infty}^{t} dt_1 \int_{-\infty}^{t_1} dt_2$$

$$\exp\left[i\omega_f t + \varepsilon t_1 + \varepsilon t_1 - i\omega_l(t_1 - t_2) + \varepsilon t_2 - i\omega_i t_2\right]$$

$$= \sum_l \frac{\langle f \mid H_1 \mid l \rangle\langle l \mid H_1 \mid i \rangle \exp(2\varepsilon t - i\omega_t t)}{\hbar^2(\omega_i - \omega_l + i\varepsilon)(\omega_i - \omega_f + 2i\varepsilon)} .$$

Adding the two contributions, we get ($\varepsilon \to \frac{\varepsilon}{2}$):

$$\frac{1}{\hbar} \left[\frac{\exp(\varepsilon t - i\omega_t t)}{\omega_i - \omega_f + i\varepsilon}\right] \qquad (14.15)$$

$$\left[\langle f \mid H_1 \mid i \rangle + \sum_l \frac{\langle f \mid H_1 \mid l \rangle\langle l \mid H_1 \mid i \rangle}{\hbar(\omega_i - \omega_l + \frac{\varepsilon}{2})}\right]$$

and the transition rate, up to second order is

$$\frac{1}{\tau} = \frac{2\pi}{\hbar^2} \sum_f |\langle f \mid H_1 \mid i \rangle + \frac{1}{\hbar} \sum_l \frac{\langle f \mid H_1 \mid l \rangle\langle l \mid H_1 \mid i \rangle}{\omega_i - \omega_l}|^2 \, \delta(\omega_i - \omega_f) . \quad (14.16)$$

n-th order

$$\frac{1}{\tau} = \frac{2\pi}{\hbar^2} \sum_f | \langle f | H_1 | i \rangle + \frac{1}{\hbar} \sum_l \frac{\langle f | H_1 | l \rangle \langle l | H_1 | i \rangle}{\omega_i - \omega_l} \delta(\omega_i - \omega_f) + ... +$$

$$\frac{1}{\hbar^{n-1}} \sum_{l_1} ... \sum_{l_{n-1}} \frac{\langle f | H_1 | l_1 \rangle \langle l_1 | H_1 | l_2 \rangle .. \langle l_{n-1} | H_1 | i \rangle}{(\omega_i - \omega_{l_1})(\omega_i - \omega_{l_2})..(\omega_i - \omega_{l_{n-1}})} |^2 \, \delta(\omega_i - \omega_f) .$$

(14.17)

The states

$$| l_1 \rangle, | l_2 \rangle ...$$

are virtual states.

The first-order term represents a direct transition

$$| i \rangle \rightarrow | f \rangle ,$$

while in the higher terms, the transitions are

$$| i \rangle \rightarrow | l_{n-1} \rangle \rightarrow | l_{n-2} \rangle ... \rightarrow | f \rangle ,$$

where there is no requirement to be met by the difference $\omega_i - \omega_l$, except that if the difference is very large, it generates a large denominator, contributing very little to the final result.

The various non-linear optical phenomena are contained in this generalized Fermi golden rule.

For example, in light scattering, where one photon is absorbed and one is emitted, the second-order term is required. In second harmonic generation, where two photons are absorbed and one photon with double frequency emitted, a third-order process is required, and so on.

14.2 Parametric Processes Without Losses

The fundamental process known as parametric amplification plays an important role in many physical effects. These include, for example, Raman and Brillouin effects.

In the case of the Raman coherent effect , a monochromatic light wave on a Raman active media gives rise to a parametric coupling between an optical vibrational mode and the mode of the radiation field, the so-called Stokes mode.

In the case of Brillouin scattering, there is a similar coupling, where the vibrations are at acoustical, rather than optical, frequencies.

In the case of parametric amplification, an intense light wave in a non-linear dielectric medium couples pairs of field modes, the **idler mode and**

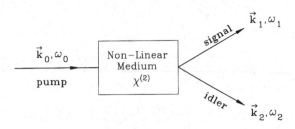

Fig. 14.1. A typical parametric amplifier, where a pump mode splits into a sigal and an idler mode, these modes obeying the conservation of energy and momentum

the signal mode, whose frequency add up to the frequency of the original strong light wave, the pump mode.

This effect is shown in Fig. 14.1.

The Hamiltonian describing the non-degenerate($\omega_1 \neq \omega_2$)parametric amplifier is described by the following effective Hamiltonian

$$H = \hbar\omega_0 a_0^\dagger a_0 + \hbar\omega_1 a_1^\dagger a_1 + \hbar\omega_2 a_2^\dagger a_2 + \qquad (14.18)$$

$$-\frac{i\chi\hbar}{2}(a_0^\dagger a_1 a_2 - a_0 a_1^\dagger a_2^\dagger) \, ,$$

where the first three terms correspond to the energies of the pump, signal and idler modes, respectively, and the last term describes the non-linear interaction, where the term $a_0 a_1^\dagger a_2^\dagger$ represents the destruction of a pump photon and the creation of an idler and a signal photon.

A particularly simple case is that of the degenerate parametric amplifier r, where a pump photon at frequency 2ω splits into two photons, each of frequency ω.

If we also assume that the pump is intense and classical [7], then we have

$$H = \hbar\omega a^\dagger a - i\hbar\frac{\chi}{2}(a^2 \exp 2i\omega t - a^{\dagger 2} \exp -2i\omega t) \, , \qquad (14.19)$$

where we have included in χ the non-linear susceptibility and the classical amplitude.

If we go to the interaction picture, we readily get

$$H = -i\hbar\frac{\chi}{2}(a^2 - a^{\dagger 2}) \, . \qquad (14.20)$$

The Heisenberg equations of motion for this system are

$$\frac{da}{dt} = \frac{1}{i\hbar}[a, H] = \chi a^\dagger \, , \qquad (14.21)$$

$$\frac{da^\dagger}{dt} = \frac{1}{i\hbar}[a^\dagger, H] = \chi a \, , \qquad (14.22)$$

and the solution can be easily calculated as

$$a(t) = a(0) \cosh \chi t + a^\dagger(0) \sinh \chi t \ . \tag{14.23}$$

Also, if we combine the two differential equations (14.21, 14.22), we get

$$\frac{dX}{dt} = \chi X \ , \tag{14.24}$$

$$\frac{dY}{dt} = -\chi Y \ , \tag{14.25}$$

getting, as solutions

$$X(t) = X(0) \exp(\chi) \ , \tag{14.26}$$

$$Y(t) = Y(0) \exp(-\chi t) \ , \tag{14.27}$$

$$\langle (\Delta X)^2 \rangle(t) = \exp(2\chi t) \langle (\Delta X)^2 \rangle(0) \ , \tag{14.28}$$

$$\langle (\Delta Y)^2 \rangle(t) = \exp(-2\chi t) \langle (\Delta Y)^2 \rangle(0) \ . \tag{14.29}$$

If the field is initially in the vacuum state, that is

$$\langle (\Delta X)^2 \rangle(0) = \langle (\Delta Y)^2 \rangle(0) = \frac{1}{4} \ , \tag{14.30}$$

then $\langle (\Delta X)^2 \rangle(t) = \frac{1}{4} \exp(2\chi t)$, $\langle (\Delta Y)^2 \rangle(t) = \frac{1}{4} \exp(-2\chi t)$, and the deamplified quadrature (Y) is squeezed at the expense of the other one, the product satisfying the minimum uncertainty relation

$$\langle (\Delta X)^2 \rangle(t) \langle (\Delta Y)^2 \rangle(t) = \frac{1}{16} \ . \tag{14.31}$$

The amount of squeezing, from the above results, is proportional to the interaction time, the non-linear parameter and the pump amplitude.

Actually, the theory presented here is a bit naive, since no loss mechanism is present, and one has always fluctuations in the pump intensity, in the case of a parametric amplifier, or if we place the non-linear crystal in a cavity, then in the parametric oscillator we would have cavity losses.

Since, for the case of the parametric oscillator, only the outside field is available for detection, we have to connect the field inside the cavity, with the field outside. The input–output theory is quite suitable for this type of problem. This is the subject of the next section.

14.3 The Input–Output Theory

In Quantum Mechanics, the S-matrix theory relates input and output fields, having in mind situations such as scattering experiments.

The input–output theory is a particular model [8, 9], that assumes a heat bath coupled to a system, with the following assumptions:

a) We consider a particular class of system–bath interaction that is linear in the bath operators. The vast majority of the models in Quantum Optics satisfy the above requirement.

b) We make the rotating wave approximation.

c) The spectrum of the bath is flat, that is independent of frequency.

These assumptions are quite common, and we already made them when dealing with the damped harmonic oscillator.

Next, we will define the "input" and "output" operators in terms of the bath operators, evaluated at the remote past and future. Then, we can derive quantum Langevin equations for the system and bath operators.

We start by considering the system-bath Hamiltonian

$$H = H_{sys} + H_B + H_{Int} , \tag{14.32}$$

$$H_B = \hbar \int_{-\infty}^{\infty} d\omega\, \omega b^{\dagger}(\omega) b(\omega) , \tag{14.33}$$

$$H_{\text{Int}} = i\hbar \int_{-\infty}^{\infty} d\omega\, K(\omega) \left[b^{\dagger}(\omega)c - b(\omega)c^{\dagger} \right] , \tag{14.34}$$

where $b(\omega)$ are the boson annihilation operators for the bath, satisfying

$$[b(\omega), b^{\dagger}(\omega')] = \delta(\omega - \omega') , \tag{14.35}$$

and c is any system operator.

Of course, the real frequency range is $(0, \infty)$, but for convenience, we have extended it to $(-\Omega, \infty)$, which is acceptable if one goes in a rotating frame with angular frequency Ω, and then take Ω larger than a typical bandwidth.

We follow now the same procedure as in the quantum theory of damping.

We write the Heisenberg equation for an arbitrary system operator a and $b(\omega)$

$$\dot{b}(\omega, t) = -i\omega b(\omega, t) + K(\omega)c, \tag{14.36}$$

$$\dot{a} = -\frac{i}{\hbar}[a, H_{\text{sys}}] + \int_{-\infty}^{\infty} d\omega\, K(\omega) \left\{ b^{\dagger}(\omega, t)[a, c] - [a, c^{\dagger}] b(\omega, t) \right\} , \tag{14.37}$$

and integrating (14.36), we get:

$$b(\omega, t) = \exp\left[-i\omega(t - t_0)\right] b(\omega, t_0) + \int_{t_0}^{t} K(\omega) \exp\left[-i\omega(t - t')\right] c(t') dt' . \tag{14.38}$$

Now, we substitute $b(\omega)$ in (14.37), obtaining

$$\dot{a} = -\frac{i}{\hbar}[a, H_{\text{sys}}]$$

$$+ \int_{-\infty}^{\infty} d\omega K(\omega) \left\{ \exp\left[i\omega(t - t_0)\right] b^{\dagger}(\omega, t_0) [a, c] - \left[a, c^{\dagger}\right] \exp\left[-i\omega(t - t_0)\right] b(\omega, t_0) \right\}$$

$$+ \int_{-\infty}^{\infty} d\omega K(\omega)^2 \int_{t_0}^{t} dt' \left\{ \exp\left[i\omega(t - t')\right] c^{\dagger}(t') [a, c] - \left[a, c^{\dagger}\right] \exp\left[-i\omega(t - t')\right] c(t') \right\} .$$

$$(14.39)$$

So far, our Heisenberg equation of motion is exact. But not for too long. We introduce now the first Markov approximation:

$$K(\omega) = \sqrt{\gamma} , \tag{14.40}$$

which is the broadband assumption mentioned at the beginning.

Also, we make use of the properties of the δ function

$$\int_{-\infty}^{\infty} d\omega \exp\left[-i\omega(t - t')\right] = 2\pi\delta(t - t') , \tag{14.41}$$

$$\int_{t_0}^{t} c(t')\delta(t - t')dt' = \frac{1}{2}c(t) ,$$

to get

$$\dot{a} = -\frac{i}{\hbar} [a, H_{\text{sys}}] - [a, c^{\dagger}] \left[\frac{\gamma}{2} c + \sqrt{\gamma}a_{\text{IN}}(t)\right] \tag{14.42}$$

$$- \left[\frac{\gamma}{2} c^{\dagger} + \sqrt{\gamma}a_{\text{IN}}^{\dagger}(t)\right] [a, c] ,$$

where we defined the input field as

$$a_{\text{IN}}(t) = -\int_{-\infty}^{\infty} d\omega \exp\left[-i\omega(t - t_0)\right] b(\omega, t_0) . \tag{14.43}$$

We notice that the $\frac{\gamma}{2}c$ and $-\frac{\gamma}{2}c^{\dagger}$ are the damping terms.

Also, the $a_{\text{IN}}(t)$ and $a_{\text{IN}}^{\dagger}(t)$ terms represent the noise, since they depend on the bath operators at the initial time t_0.

We may assume that at this initial time, the system and bath density operators factorize and a typical bath state correspond to a thermal state.

In the case that the bath state corresponds to a coherent or squeezed state, we no longer can interpret these terms as noise.

Foe the particular case c=a, we get

$$\dot{a} = -\frac{i}{\hbar} [a, H_{\text{sys}}] - \frac{\gamma}{2} a + \sqrt{\gamma}a_{\text{IN}}(t) . \tag{14.44}$$

Now, if we take $t_1 > t$, we can integrate (14.36) again, getting

$$b(\omega, t) = \exp\left[-i\omega(t - t_0)\right] b(\omega, t_1) - \int_{t_0}^{t} K(\omega) \exp\left[-i\omega(t - t')\right] c(t')dt' , \tag{14.45}$$

and we define an "output field"

$$a_{\text{OUT}}(t) = \int_{-\infty}^{\infty} d\omega \exp\left[-i\omega(t - t_1)\right] b(\omega, t_1) . \qquad (14.46)$$

If we follow the same procedure as with the input field, we readily get

$$\dot{a} = -\frac{i}{\hbar}[a, H_{sys}] - [a, c^\dagger]\left[-\frac{\gamma}{2}c + \sqrt{\gamma}a_{\text{OUT}}(t)\right]$$

$$-\left[-\frac{\gamma}{2}c^\dagger + \sqrt{\gamma}a_{\text{OUT}}^\dagger(t)\right][a, c] , \qquad (14.47)$$

and for the a=c case

$$\dot{a} = -\frac{i}{\hbar}[a, H_{\text{sys}}] + \frac{\gamma}{2}a - \sqrt{\gamma}a_{\text{OUT}}(t) . \qquad (14.48)$$

By comparing (14.48) with (14.44), we get

$$\sqrt{\gamma}a(t) = a_{\text{IN}} + a_{\text{OUT}} . \qquad (14.49)$$

Equation (14.48) can be interpreted as a boundary condition relating the input, output and internal field, at the mirrors of the cavity.

Although this analysis refers to a system driven by a bath, this is not necessarily a theory about noise, since no assumption was made about the bath, except for the broad spectrum.

For a linear system, (14.44) and (14.48) can be cast in a convenient matrix form

$$\frac{d\mathbf{a}}{dt} = \left(\mathbf{A} - \frac{\gamma}{2}\mathbf{1}\right)\mathbf{a} + \sqrt{\gamma}\mathbf{a}_{\text{IN}}(t) , \qquad (14.50)$$

$$= \left(\left[\mathbf{A} + \frac{\gamma}{2}\mathbf{1}\right]\right) - \sqrt{\gamma}\mathbf{a}_{\text{OUT}}(t) ,$$

with

$$\mathbf{a} = \begin{pmatrix} a \\ a^\dagger \end{pmatrix} , \qquad (14.51)$$

and $\mathbf{A}$ is a matrix.

It is convenient to define the Fourier transform:

$$\tilde{a}(\omega) = \frac{1}{2\pi}\int_{-\infty}^{\infty} \exp(i\omega t)a(t)dt . \qquad (14.52)$$

Now, (14.50) can be written as

$$-i\omega\,\tilde{\mathbf{a}}(\omega) = \left(\mathbf{A} - \frac{\gamma}{2}\mathbf{1}\right)\tilde{\mathbf{a}}(\omega) + \sqrt{\gamma}\,\tilde{\mathbf{a}}(\omega)_{\text{IN}} ,$$

$$= \left(\mathbf{A} + \frac{\gamma}{2}\mathbf{1}\right)\tilde{\mathbf{a}}(\omega) - \sqrt{\gamma}\,\tilde{\mathbf{a}}(\omega)_{\text{OUT}} , \qquad (14.53)$$

where

$$\tilde{\mathbf{a}}(\omega) = \begin{pmatrix} \tilde{a}(\omega) \\ \tilde{a}^\dagger(-\omega) \end{pmatrix} . \qquad (14.54)$$

From the above equations, we can eliminate the internal field and relate the input and output fields.

Since

$$-\left[\mathbf{A} + (i\omega - \frac{\gamma}{2})\mathbf{1}\right] \tilde{\mathbf{a}}(\omega) = \sqrt{\gamma}\,\tilde{\mathbf{a}}(\omega)_{\text{IN}} \qquad (14.55)$$

$$-\left[\mathbf{A} + (i\omega + \frac{\gamma}{2})\mathbf{1}\right] \tilde{\mathbf{a}}(\omega) = -\sqrt{\gamma}\,\tilde{\mathbf{a}}(\omega)_{\text{OUT}}, \qquad (14.56)$$

we get

$$\left[\mathbf{A} + (i\omega - \frac{\gamma}{2})\mathbf{1}\right]^{-1} \tilde{\mathbf{a}}(\omega)_{\text{IN}} = -\left[\mathbf{A} + (i\omega + \frac{\gamma}{2})\mathbf{1}\right]^{-1} \tilde{\mathbf{a}}(\omega)_{\text{OUT}},$$

or, finally

$$\tilde{\mathbf{a}}(\omega)_{\text{OUT}} = -\left[\mathbf{A} + (i\omega + \frac{\gamma}{2})\mathbf{1}\right]\left[\mathbf{A} + (i\omega - \frac{\gamma}{2})\mathbf{1}\right]^{-1} \tilde{\mathbf{a}}(\omega)_{\text{IN}} . \qquad (14.57)$$

14.4 The Degenerate Parametric Oscillator

The physical system is described in the Fig. 14.2.

A non-linear crystal acts as the amplifying medium inside an optical cavity, with two mirrors chosen with the following properties

a) Both mirrors are almost 100% transmitting at the pump frequency $\omega_p = 2\omega_0$.
b) One of the end mirrors is 100% reflecting at ω_0 and the other one (right mirror) is partially transmitting at ω_0.

The input–output theory developed in the last section is quite suitable to be used in this example.

The Hamiltonian of the degenerate parametric oscillator can be written as

$$H_{\text{sys}} = \hbar\omega_0 a^\dagger a + \frac{i\hbar\left[\varepsilon\exp(-i\omega_p t)a^{\dagger 2} - \varepsilon^*\exp(i\omega_p t)a^2\right]}{2}, \qquad (14.58)$$

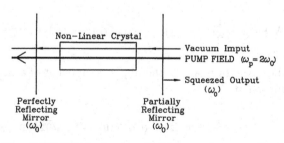

Fig. 14.2. Physical setup of the parametric oscillator

where ε is the classical pump amplitude (including the non-linear coefficient) and $\omega_p = 2\omega_0$ is the pump frequency.

The equation of motion for the internal field is

$$\frac{da}{dt} = -i\omega_0 a + \varepsilon \exp(-i\omega_p t)a^\dagger - \frac{1}{2}(\gamma_1 + \gamma_2)a \qquad (14.59)$$
$$+\sqrt{\gamma_1}a_{1\mathrm{IN}}(t) + \sqrt{\gamma_2}a_{2\mathrm{IN}}(t) \, ,$$

where γ_1 and γ_2 are the two damping coefficients of the two mirrors and $a_{1\mathrm{IN}}(t), a_{2\mathrm{IN}}(t)$ are the two respective input fields.

Now, we go to a rotating frame

$$a \to a \exp\left(i\frac{\omega_p t}{2}\right) \, ,$$

$$a_{1\mathrm{IN}}(t) \to \exp\left(i\frac{\omega_p t}{2}\right) a_{1\mathrm{IN}}(t),$$

$$a_{2\mathrm{IN}}(t) \to \exp\left(i\frac{\omega_p t}{2}\right) a_{2\mathrm{IN}}(t) \, .$$

If we call the new fields with the same symbols as the old ones, we write

$$\frac{d\mathbf{a}}{dt} = \left[\mathbf{A} - \frac{1}{2}(\gamma_1 + \gamma_2)\mathbf{1}\right]\mathbf{a} + \sqrt{\gamma_1}\mathbf{a}_{1\mathrm{IN}}(t) + \sqrt{\gamma_2}\mathbf{a}_{2\mathrm{IN}}(t) \, , \qquad (14.60)$$

with

$$\mathbf{A} = \begin{bmatrix} 0 & |\varepsilon| \exp(i\theta) \\ |\varepsilon| \exp(-i\theta) & 0 \end{bmatrix}, \qquad (14.61)$$

$$\mathbf{a}(t) = \begin{bmatrix} a \\ a^\dagger \end{bmatrix},$$

$$\mathbf{a}_{i\mathrm{IN}}(t) = \begin{bmatrix} a_{i\mathrm{IN}} \\ a_{i\mathrm{IN}}^\dagger \end{bmatrix}, i = 1, 2.$$

Now, performing the Fourier transform and using the property $a^\dagger(\omega) = [a(-\omega)]^\dagger$, we get

$$\tilde{\mathbf{a}}(\omega) = -\left[\mathbf{A} + (i\omega - \frac{1}{2}(\gamma_1 + \gamma_2))\mathbf{1}\right]^{-1}\left[\sqrt{\gamma_1}\,\tilde{\mathbf{a}}_{1\mathrm{IN}}(\omega) + \sqrt{\gamma_2}\,\tilde{\mathbf{a}}_{2\mathrm{IN}}(\omega)\right] \, . \qquad (14.62)$$

From the above vector relation, the upper component of $\tilde{\mathbf{a}}(\omega)$ is given by (after matrix inversion)

$$\tilde{a}(\omega) = \frac{(-i\omega + \frac{1}{2}(\gamma_1 + \gamma_2))\left[\sqrt{\gamma_1}\,\tilde{a}_{1\mathrm{IN}}(\omega) + \sqrt{\gamma_2}\,\tilde{a}_{2\mathrm{IN}}(\omega)\right]}{(i\omega - \frac{1}{2}(\gamma_1 + \gamma_2))^2 - |\varepsilon|^2}$$

$$+\frac{\varepsilon\left[\sqrt{\gamma_1}\,\tilde{a}^\dagger_{1\mathrm{IN}}\,(-\omega)+\sqrt{\gamma_2}\,\tilde{a}^\dagger_{2\mathrm{IN}}\,(-\omega)\right]}{(i\omega-\frac{1}{2}(\gamma_1+\gamma_2))^2-\mid\varepsilon\mid^2}\ . \tag{14.63}$$

From (14.49), we get the output field

$$\tilde{a_{\mathrm{OUT}}}\,(\omega)=\frac{\left[(\frac{\gamma_1}{2})^2-(\frac{\gamma_2}{2}-i\omega)^2+\mid\varepsilon\mid^2\right]\tilde{a}_{1\mathrm{IN}}\,(\omega)+\varepsilon\gamma_1\,\tilde{a}^\dagger_{1\mathrm{IN}}\,(-\omega)}{(i\omega-\frac{1}{2}(\gamma_1+\gamma_2))^2-\mid\varepsilon\mid^2} \tag{14.64}$$

$$+\frac{\sqrt{\gamma_2\gamma_1}\left[(\frac{\gamma_1+\gamma_2}{2})-i\omega\right]\tilde{a}_{2\mathrm{IN}}\,(\omega)+\varepsilon\sqrt{\gamma_2\gamma_1}\,\tilde{a}^\dagger_{2\mathrm{IN}}\,(-\omega)}{(i\omega-\frac{1}{2}(\gamma_1+\gamma_2))^2-\mid\varepsilon\mid^2}\ .$$

We are mostly interested in the quadrature fluctuations.
It takes a little algebra to show the following results

$$\langle:X_{1\mathrm{OUT}}(\omega),X(\omega)_{1\mathrm{OUT}}:\rangle \tag{14.65}$$

$$=\frac{\mid\varepsilon\mid\frac{\gamma_1}{2}}{\left[\frac{1}{2}(\gamma_1+\gamma_2)-\mid\varepsilon\mid\right]^2+\omega^2}\delta(\omega+\omega)\ ,$$

$$\langle:Y_{1\mathrm{OUT}}(\omega),Y(\omega)_{1\mathrm{OUT}}:\rangle \tag{14.66}$$

$$=-\frac{\mid\varepsilon\mid\frac{\gamma_1}{2}}{\left[\frac{1}{2}(\gamma_1+\gamma_2)+\mid\varepsilon\mid\right]^2+\omega^2}\delta(\omega+\omega)\ ,$$

where

$$\langle a,b\rangle\equiv\langle ab\rangle-\langle a\rangle\langle b\rangle\ .$$

The maximum squeezing is obtained for $\frac{1}{2}(\gamma_1+\gamma_2)=\mid\varepsilon\mid$, getting

$$\langle:Y_{1\mathrm{OUT}}(\omega),Y(\omega)_{1\mathrm{OUT}}:\rangle \tag{14.67}$$

$$=-\frac{\gamma_1}{4}\frac{(\gamma_1+\gamma_2)}{(\gamma_1+\gamma_2)^2+\omega^2}\delta(\omega+\omega)\ .$$

The normally ordered spectrum of $Y_{1\mathrm{OUT}}(\omega)$ is the coefficient of the above
formula

$$:S_{Y1\mathrm{OUT}}(\omega):=-\frac{\gamma_1}{4}\frac{(\gamma_1+\gamma_2)}{(\gamma_1+\gamma_2)^2+\omega^2}\ . \tag{14.68}$$

Two important cases are

a) $\gamma=\gamma_1=\gamma_2$, the double ended cavity, for which, at resonance

$$:S_{Y1\mathrm{OUT}}(0):=-\frac{1}{8}\ . \tag{14.69}$$

b) $\gamma_2=0$, single ended cavity

$$:S_{Y1\mathrm{OUT}}(0):=-\frac{1}{4}\ . \tag{14.70}$$

We notice that the single ended cavity has a perfect squeezing in the
output signal.

14.5 Experimental Results

Quadrature squeezing has been observed experimentally in parametric oscillators, and also other non-linear effects such as second harmonic generation [10, 11], optical bistability [12], four wave mixing [13].

Squeezing greater than 50% relative to the vacuum noise level was obtained by Wu et al. [14], using degenerate parametric down conversion in an optical cavity.

The diagram of the experimental setup is shown in the Fig. 14.3.

The downconversion occurs in the cavity M'M'' that contains the non-linear crystal MgO:LiNbO$_3$, phase matched at 98°C. The pump for the parametric oscillator is an Nd:YAG laser, whose frequency was doubled from 1.06 µm to 0.53 µm , using a $Ba_2NaNb_5O_{15}$ crystal inside the laser cavity.

The pump field enters the OPO cavity through the M' mirror with a transmission coefficient of 3.5% at 0.53 µm, and 0.06% at 1.06 µm.

On the other hand, M'' is coated for low transmission at 0.53 µm and either 4.3% or 7.3% transmission at 1.06 µm.

A fraction of the downconverted light from the OPO exits through M'' and combines with the original Nd:YAG laser that acts as a strong local oscillator at one of the ports of a balanced homodyne detector and the squeezed light enters through the other port.

The squeezed signal is observed from the spectra of the intensity difference of the photocurrents coming from the detectors A and B.

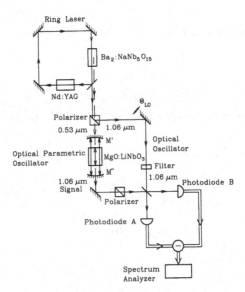

Fig. 14.3. Diagram of the experimental setup of the Parametric Oscillator that generates squeezed states

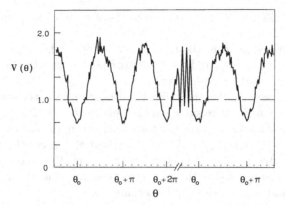

Fig. 14.4. Phase dependence of the rms noise voltage $V(\theta)$ from a balanced homodyne detector as a function of the local oscillator phase θ at a fixed analysis frequency 1.8 MHz. The dashed line corresponds to the vacuum noise voltage, with the OPO blocked and no phase dependence (After [14])

In the Fig. 14.4, we display the noise voltage $V(\theta)$ from the detector, as a function of the phase θ, of the local oscillator.

With the OPO input blocked, the vacuum field entering the signal port of the detector produces the noise drawn in dotted line, with no θ dependence.

With the OPO unblocked, there are several dips below the vacuum level, some of which correspond to a reduction of 50%.

Problems

14.1. Show that after matrix inversion, $\tilde{a}\,(\omega)$ is given by (14.63).

14.2. Show that the optimum normally ordered squeezing spectrum for Y is given by (14.68).

14.3. Using the input–output theory, show that

$$[a(t), a_{\mathrm{IN}}(t')] = u(t-t')\sqrt{\gamma}\,[a(t), c(t)]\,,$$

$$[a(t), a_{\mathrm{OUT}}(t')] = u(t'-t)\sqrt{\gamma}\,[a(t), c(t)]\,,$$

where $u(t)$ is the step function, defined as

$$u(t) = 1, t > 0$$
$$= \frac{1}{2}, t = 0$$
$$= 0, t < 0\,.$$

14.4. Light scattering by a two-level atom.

For a weak incident beam on a two-level atom, the usual interaction Hamiltonian is

$$H_1 = e\mathbf{d} \cdot \mathbf{E}(0).$$

The initial state of the system consists of photons at frequency ω and wavevector $\mathbf{k}$.

The scattered photons (by the two-level atom) will have a frequency ω_s and a wavevector $\mathbf{k}_s$, and the fields are characterized by the annihilation operators a and a_s, respectively.

Also, we assume that the atom is initially in its lower state $\mid b\rangle$, with zero energy and left in some final atomic state $\mid f\rangle$ with an energy $\hbar\omega_f$. If the incident field has n photons, then we make use of (20.67) with $\mid f\rangle_{at-f} = \mid n-1, 1, f\rangle$ and $\mid i\rangle_{at-f} = \mid n, 0, b\rangle$, where the first number refers to the number of photons in the beam, the second number the number of photons in the scattered field and the third index is the atomic state, and write (up to second order)

$$\frac{1}{\tau} = \frac{2\pi}{\hbar^2} \sum_f \mid \langle n-1, 1, f \mid H_1 \mid n, 0, b\rangle$$

$$+ \frac{1}{\hbar} \sum_l \frac{\langle n-1, 1, f \mid H_1 \mid l\rangle\langle l \mid H_1 \mid n, 0, b\rangle}{\omega_i - \omega_l} \mid^2 \delta(\omega - \omega_s - \omega_f).$$

a) Prove that the linear term does not contribute and that the contributions of the quadratic terms come from
   $aa_s^\dagger$ term with the relevant $\mid l\rangle = \mid n, 1, j\rangle, \omega_l = (n)\omega + \omega_s + \omega_j$ and
   $a_s^\dagger a$ term with the relevant $\mid l\rangle = \mid n-1, 0, j\rangle, \omega_l = (n-1)\omega + \omega_j$.

b) Show that

$$\frac{1}{\tau} = \sum_f \sum_{\mathbf{k}_s} \frac{\pi e^4 \omega_s \omega n}{2\hbar\varepsilon_0^2 v^2} \mid \sum_j \frac{(\varepsilon_s \cdot \mathbf{d}_{fj})(\varepsilon \cdot \mathbf{d}_{jf})}{\omega_j - \omega} + \frac{(\varepsilon \cdot \mathbf{d}_{fj})(\varepsilon_s \cdot \mathbf{d}_{jf})}{\omega_j + \omega_s} \mid^2$$
$$\delta(\omega - \omega_s - \omega_f).$$

c) Converting the sum over $\mathbf{k}_s$ into an integral, and using the notion of cross section, usually defined as the ratio of energy removed from the beam\energy rate crossing a unit area perpendicular to the beam, or

$$\sigma = \frac{\frac{\hbar\omega}{\tau}}{\frac{c\hbar n\omega}{v}} = \frac{v}{\tau n c},$$

show that

$$\frac{d\sigma}{d\Omega} = \sum_f^{\omega_f < \omega} \frac{e^4}{16\pi^2\hbar^2\varepsilon_0^2 c^4} \mid \sum_j \frac{(\varepsilon_s \cdot \mathbf{d}_{fj})(\varepsilon \cdot \mathbf{d}_{jf})}{\omega_j - \omega} + \frac{(\varepsilon \cdot \mathbf{d}_{fj})(\varepsilon_s \cdot \mathbf{d}_{jf})}{\omega_j - \omega_f + \omega} \mid^2$$

This is the Kramer–Heisenberg formula. The cross-section includes the elastic Rayleigh scattering, corresponding to the $f = b$ term, $\omega_f = 0$, and the inelastic Raman scattering, corresponding to the rest. [6].

14.5. Prove that for the elastic Raman scattering (refer to problem 14.4) one has

$$\frac{d\sigma}{d\Omega} = \sum_{f}^{\omega_f < \omega} \frac{e^4(\omega - \omega_f)^3 \omega}{16\pi^2 \hbar^2 \varepsilon_0^2 c^4} \mid \sum_{j} \omega_j \{(\varepsilon_s \cdot \mathbf{d}_{bj})(\varepsilon \cdot \mathbf{d}_{jb}) + (\varepsilon \cdot \mathbf{d}_{bj})(\varepsilon_s \cdot \mathbf{d}_{jb})\} \mid^2,$$

in the limit $\omega \gg \omega_j$, when ω is much larger than the atomic excitation frequencies. [6]

References

1. Franken, P.A., Hill, A.E., Peters, C., Weinreich, G.: Phys. Rev. Lett., **7**, 118 (1961)
2. Blombergen, N.: Nonlinear Optics., Benjamin, New York (1965)
3. Hopf, F.A., Stegeman, G.: Applied Classical Electrodynamics. Volume 2, J.Wiley, New York (1986)
4. Shen, Y.R.: Principles of Nonlinear Optics. John Wiley and Sons, New York (1984)
5. Kielich, S.: Nonlinear Molecular Optics. Nauk, Moscow (1981)
6. Loudon, R.: The Quantum Theory of Light. Clarendon Press, Oxford (1983)
7. Walls, D.F., Milburn, G.: Quantum Optics. Springer Verlag, Berlin (1994)
8. Collett, M.J., Gardiner, C.W.: Phys. Rev. A, **30**, 1386 (1984)
9. Gardiner, C.W., Collett, M.J.: Phys. Rev. A, **31**, 3761 (1985)
10. Pereira, S.F., Xiao, M., Kimble, H.J.: Phys. Rev. A, **38**, 4931 (1989)
11. Sizman, A., Horowicz, R., Wagner, G., Leuchs, G.: Opt. Comm., **80**, 138 (1990)
12. The theoretical approach of the generation of squeezed states in dispersive optical bistability can be found in: Drummond, P.D., Walls, D.F., J. Phys. A, **13**, 725 (1980)
13. Slusher, R.E., Hollberg, L.W., Yurke, B., Mertz, H.J.C., Valley, J.F.: Phys. Rev. Lett., **55**, 2409 (1985)
14. Wu, L.A., Kimble, H.J., Hall, J.L., Wu, H.: Phys. Rev. Lett., **57**, 2520 (1986)

15. Quantum Phase

We study, in this chapter, the various approaches to the problem of the Quantum Phase.

Dirac was the first one to postulate the existence of a Hermitian phase variable in the early days of Quantum Electrodynamics [1].

15.1 The Dirac Phase

Comparison with Classical Mechanics led Dirac to assume a commutation relation of the number and phase operator

$$[\Phi_D, \mathbf{n}] = -i \, , \tag{15.1}$$

which immediately leads to the uncertainty relation

$$\Delta \mathbf{n} \Delta \Phi_D \geq \frac{1}{2} \, . \tag{15.2}$$

There are some obvious difficulties with the above commutation rule.

If we take matrix elements between n and n', we get

$$(n - n')\langle n' \mid \Phi_D \mid n \rangle = -i\delta_{nn'} \, , \tag{15.3}$$

which is inconsistent when $n = n'$, giving $0 = -i$.

Also, when trying the polar decomposition of the annihilation operator:

$$a = \exp i\Phi_D \sqrt{\mathbf{n}} = \sqrt{\mathbf{n} + 1} \exp(i\Phi_D) \, , \tag{15.4}$$

$$a^\dagger = \sqrt{\mathbf{n}} \exp(-i\Phi_D) = \exp(-i\Phi_D \sqrt{\mathbf{n} + 1}) \, ,$$

leads to difficulties, because if one assumes that Φ_D is a Hermitian operator, as any respectable observable should, then $\exp i\Phi_D$ is not unitary, and therefore, $\exp(-i\Phi_D) \neq [\exp(i\Phi_D)]^\dagger$.

These difficulties were pointed out by **Dirac** himself and also by **Susskind and Glogower**. [2]

Another difficulty is that the uncertainty relation (15.2), implies, for small Δn that $\Delta \Phi_D$ can have values larger than 2π, which makes no physical sense, and basically does not take into account the periodic nature of the phase.

If one inverts the relations (15.4), we define

$$\exp(i\Phi_D) = (\mathbf{n}+1)^{-\frac{1}{2}}a \,, \tag{15.5}$$
$$\exp(-i\Phi_D) = a^\dagger(\mathbf{n}+1)^{-\frac{1}{2}} \,,$$

and from (15.5), it follows that

$$[\exp(\pm i\Phi_D), \mathbf{n}] = \pm \exp(\pm i\Phi_D) \,. \tag{15.6}$$

15.2 The Louisell Phase

Louisell [3] tried to solve the problem of periodicity, by defining trigonometric functions of the **Dirac** phase

$$\cos\Phi_D = \frac{1}{2} [\exp(i\Phi_D) + \exp(-i\Phi_D)] \,, \tag{15.7}$$
$$\sin\Phi_D = \frac{1}{2i} [\exp(i\Phi_D) - \exp(-i\Phi_D)] \,,$$

leading to the commutation relation

$$[\cos\Phi_D, \mathbf{n}] = i\sin\Phi_D \tag{15.8}$$

$$[\sin\Phi_D, \mathbf{n}] = -i\cos\Phi_D$$

thus $\mathbf{n}$ and $\sin\Phi_D$ or $\cos\Phi_D$ obey uncertainty relations

$$\Delta \mathbf{n}\Delta\cos\Phi_D \geq \frac{1}{2} |\langle\sin\Phi_D\rangle| \,, \tag{15.9}$$
$$\Delta \mathbf{n}\Delta\sin\Phi_D \geq \frac{1}{2} |\langle\cos\Phi_D\rangle| \,.$$

Both approaches of Dirac and Louisell have a common difficulty that $\exp(i\Phi_D)$ is not unitary and that (15.4) do not define a Hermitian phase operator. [4]

15.3 The Susskind–Glogower Phase

Susskind and Glogower [2] have a phase definition that is similar to **Dirac**'s

$$a = \mathbf{A}\sqrt{\mathbf{n}}, \mathbf{A} = \exp(i\Phi_S) \,, \tag{15.10}$$
$$a^\dagger = \sqrt{\mathbf{n}}\mathbf{A}^\dagger, \mathbf{A}^\dagger = [\exp(i\Phi_S)]^\dagger \,,$$

where $\mathbf{A}$ and $\mathbf{A}^\dagger$ do not commute.

As a matter of fact

$$\mathbf{A} = a(\mathbf{n})^{-\frac{1}{2}} = (\mathbf{n}+1)^{-\frac{1}{2}}a, \tag{15.11}$$
$$\mathbf{A}^{\dagger} = (\mathbf{n})^{-\frac{1}{2}}a^{\dagger} = a^{\dagger}(\mathbf{n}+1)^{-\frac{1}{2}}.$$

We see that

$$\mathbf{AA}^{\dagger} = (\mathbf{n}+1)^{-\frac{1}{2}}aa^{\dagger}(\mathbf{n}+1)^{-\frac{1}{2}} = 1, \tag{15.12}$$

but on the other hand

$$
\begin{aligned}
\mathbf{A}^{\dagger}\mathbf{A} &= a^{\dagger}(\mathbf{n}+1)^{-\frac{1}{2}}(\mathbf{n}+1)^{-\frac{1}{2}}a \\
&= \sum_{m=0}^{\infty} a^{\dagger} \mid m\rangle\langle m \mid (\mathbf{n}+1)^{-1}a \\
&= \sum_{m=0}^{\infty} \sqrt{m+1} \mid m+1\rangle\langle m+1 \mid (m+1)^{-1}\sqrt{m+1} \\
&= \sum_{m=0}^{\infty} \mid m+1\rangle\langle m+1 \mid = 1 - \mid 0\rangle\langle 0 \mid .
\end{aligned}
$$

Thus

$$\left[\mathbf{A}, \mathbf{A}^{\dagger}\right] = \mid 0\rangle\langle 0 \mid . \tag{15.13}$$

The two operators $\mathbf{A}$ and $\mathbf{A}^{\dagger}$ do not commute, and they are not unitary. However, these two operators act like raising and lowering operators

$$a \mid n\rangle = \sqrt{n} \mid n-1\rangle = A\sqrt{\mathbf{n}} \mid n\rangle = A\sqrt{n} \mid n\rangle , \tag{15.14}$$

and therefore

$$\mathbf{A} \mid n\rangle = \mid n-1\rangle , \tag{15.15}$$

and similarly

$$\mathbf{A}^{\dagger} \mid n\rangle = \mid n+1\rangle . \tag{15.16}$$

Equation (15.15) has one exception. In order not to create a Fock state with a negative photon number, we must have

$$\mathbf{A} \mid 0\rangle = 0 , \tag{15.17}$$

which is the mathematical origin of the non-unitary character of $\mathbf{A}$ and $\mathbf{A}^{\dagger}$.
Thus

$$\langle n-1 \mid \mathbf{A} \mid n\rangle = 1 , \tag{15.18}$$
$$\langle n+1 \mid \mathbf{A}^{\dagger} \mid n\rangle = 1 ,$$

and all the other matrix elements are zero.
One can also write

$$\mathbf{A} = |\,0\rangle\langle 1\,| + |\,1\rangle\langle 2\,| + |\,2\rangle\langle 3\,| + \dots . \qquad (15.19)$$

One can also show, similarly to the Dirac phase, that

$$[\mathbf{A}, \mathbf{n}] = \mathbf{A}, \qquad (15.20)$$
$$[\mathbf{A}^\dagger, \mathbf{n}] = -\mathbf{A}^\dagger .$$

We can define the trigonometric functions

$$\cos\Phi_S = \frac{1}{2}(\mathbf{A} + \mathbf{A}^\dagger), \qquad (15.21)$$

$$\sin\Phi_S = \frac{1}{2i}(\mathbf{A} - \mathbf{A}^\dagger),$$

where the non-vanishing elements are

$$\langle n-1\,|\,\cos\Phi_S\,|\,n\rangle = \langle n\,|\,\cos\Phi_S\,|\,n-1\rangle = \frac{1}{2}, \qquad (15.22)$$

$$\langle n-1\,|\,\sin\Phi_S\,|\,n\rangle = -\langle n\,|\,\sin\Phi_S\,|\,n-1\rangle = \frac{1}{2i}.$$

The reader can readily verify that the condition for a Hermitian operator $\langle n\,|\,O\,|\,n'\rangle = \langle n'\,|\,O\,|\,n\rangle^*$ is satisfied by both $\cos\Phi_S$ and $\sin\Phi_S$.

From the commutation relations (15.20), one can verify that

$$[\mathbf{n}, \cos\Phi_S] = -i\sin\Phi_S, \qquad (15.23)$$
$$[\mathbf{n}, \sin\Phi_S] = i\cos\Phi_S,$$

which implies that both n and $\cos\Phi_S$ or $\sin\Phi_S$ cannot be precisely specified. The results of measurements of amplitude and phase are governed by the uncertainty relations

$$\Delta n \Delta \cos\Phi_S \geq \frac{1}{2}\,|\,\langle \sin\Phi_S\rangle\,|, \qquad (15.24)$$

$$\Delta n \Delta \sin\Phi_S \geq \frac{1}{2}\,|\,\langle \cos\Phi_S\rangle\,|.$$

One can also prove that the operators $\cos\Phi_S$ and $\sin\Phi_S$ do not commute

$$[\cos\Phi_S, \sin\Phi_S] = \frac{1}{4i}\left[\mathbf{A} + \mathbf{A}^\dagger, \mathbf{A} - \mathbf{A}^\dagger\right] \qquad (15.25)$$

$$= -\frac{1}{2i}\,|\,0\rangle\langle 0\,|.$$

The last property, that the $\cos\Phi_S$ and $\sin\Phi_S$ do not commute is rather strange. Normally, in Classical Mechanics or Electromagnetism, the phase is a simple quantity and it is not necessary to define separately both the cosine and the sine of that phase.

Of course, the above property as well as the failure of A and $A^\dagger$ to commute is directly related to the vacuum state.

If one has to take the average of these commutation relations, with classical strong fields, for which, the probability of being in the vacuum is very small, then all the difficulties vanish, which shows that the real problems arise only when one is dealing with highly quantum mechanical field states with low photon numbers.

We can calculate the expectation values of these trigonometric functions, taking a coherent state [5] $| \alpha \rangle$ with $\alpha = | \alpha | \exp(i\theta)$

$$\langle \alpha \mid \cos \Phi_S \mid \alpha \rangle = \frac{1}{2} \left[\langle \alpha \mid A \mid \alpha \rangle + \langle \alpha \mid A^\dagger \mid \alpha \rangle \right] , \tag{15.26}$$

$$\langle \alpha \mid \cos^2 \Phi_S \mid \alpha \rangle = \frac{1}{4} \left[\langle \alpha \mid A^2 \mid \alpha \rangle + \langle \alpha \mid AA^\dagger + A^\dagger A \mid \alpha \rangle + \langle \alpha \mid A^{\dagger 2} \mid \alpha \rangle \right] ,$$

$$\langle \alpha \mid \sin \Phi_S \mid \alpha \rangle = \frac{1}{2i} \left[\langle \alpha \mid A \mid \alpha \rangle - \langle \alpha \mid A^\dagger \mid \alpha \rangle \right] ,$$

$$\langle \alpha \mid \sin^2 \Phi_S \mid \alpha \rangle = -\frac{1}{4} \left[\langle \alpha \mid A^2 \mid \alpha \rangle - \langle \alpha \mid AA^\dagger + A^\dagger A \mid \alpha \rangle + \langle \alpha \mid A^{\dagger 2} \mid \alpha \rangle \right] .$$

After some algebraic work, one can show the following properties:

$$\langle \alpha \mid \cos \Phi_S \mid \alpha \rangle = \cos \theta (1 - \frac{1}{8 \mid \alpha \mid^2} ...) , \tag{15.27}$$

$$\langle \alpha \mid \cos^2 \Phi_S \mid \alpha \rangle = \cos^2 \theta - \frac{\cos^2 \theta - \frac{1}{2}}{2 \mid \alpha \mid^2} + ... ,$$

valid only for $\mid \alpha \mid^2 \gg 1$.

Also

$$\langle \alpha \mid \cos^2 \Phi_S + \sin^2 \Phi_S \mid \alpha \rangle = 1 - \frac{\exp(- \mid \alpha \mid^2)}{2} \tag{15.28}$$

valid for all α.

It is interesting to notice that when $\mid \alpha \mid \to 0$, $\langle \alpha \mid (\cos^2 \Phi_S + \sin^2 \Phi_S) \mid \alpha \rangle \to \frac{1}{2}$, which again is a strange property.

As a conclusion of this section, the Susskind–Glogower is formalism of the quantum phase, is a fairly consistent one, where the main difficulties are not mathematical but with the physical interpretation of $\cos \Phi_S$ and $\sin \Phi_S$ and their relationship with the actual experiments.

Finally, a long time ago, **F. London** defined a phase state:

$$\mid \phi \rangle = \frac{1}{\sqrt{2\pi}} \sum_{n=0}^{\infty} \exp(in\phi) \mid n \rangle \tag{15.29}$$

which are neither orthogonal nor be normalized. However, as we shall see, it can be useful to define probability distributions that can be normalized [6, 7].

The shortcomings of the above phase operators have led a number of investigators to explore different possibilities [8, 9, 10, 11, 12, 13, 14, 15, 16, 17, 18, 19, 20, 21, 22, 23, 24, 25].

We turn now to the **Pegg and Barnett** description.

15.4 The Pegg–Barnett Phase

A phase state $|\theta\rangle$ is defined in a finite $(s+1)$ dimensional space, in the limit $s \to \infty$, as follows [13, 14, 15, 16]

$$|\theta\rangle = \mathrm{Lim}_{s\to\infty}(s+1)^{-\frac{1}{2}} \sum_{n=0}^{s} \exp(in\theta) \, |\, n\rangle \,. \tag{15.30}$$

The way to operate the limit is to perform the calculation in a finite space, and after the physical averages are calculated, one is to take the limit $s \to \infty$.

The parameter θ can take any values between 0 and 2π; thus, there are an infinite number of these states, which are overcomplete and non-orthogonal.

However, one can construct a set of orthogonal states if one picks only specified values of $\theta = \theta_m$

$$\theta_m = \theta_0 + \frac{2\pi m}{s+1} \,. \tag{15.31}$$

Thus, if one starts from a reference state $|\,\theta_0\rangle$, one can find a complete set of $s+1$ orthonormal states

$$|\,\theta_m\rangle = \exp i \left[\mathbf{n} \left(\frac{m 2\pi}{s+1} \right) \right] |\,\theta_0\rangle \,, \tag{15.32}$$

$$m = 0, 1, 2 \, \dots \, s \,.$$

The above equation is simple to verify, because

$$\exp(in\gamma) \, |\, n\rangle = |\, n+\gamma\rangle \,.$$

The orthonormal condition can be seen as follows:

$$\langle\theta_p \,|\, \theta_m\rangle = \langle\theta_0 \,|\, \exp\left[-in \left(\frac{p 2\pi}{s+1} \right) \right] \exp\left[in \left(\frac{m 2\pi}{s+1} \right) \right] |\,\theta_0\rangle$$

$$= \mathrm{Lim}_{s\to\infty} \frac{1}{s+1} \sum_{q,r=0}^{s} \langle q \,|\, \exp(-iq\theta_0) \exp\left[-in \left(\frac{p 2\pi}{s+1} \right) \right]$$

$$\exp i \left[\mathbf{n} \left(\frac{m 2\pi}{s+1} \right) \right] \exp(ir\theta_0) \, |\, r\rangle \,,$$

where $|\, q\rangle$ and $|\, r\rangle$ are Fock states.

Thus

$$\langle\theta_p \,|\, \theta_m\rangle = \mathrm{Lim}_{s\to\infty} \frac{1}{s+1} \sum_{q=0}^{s} \langle q \,|\, q\rangle \exp\left[i\frac{2\pi}{s+1} q(m-p) \right] = \delta_{mp} \,. \tag{15.33}$$

The Hermitian phase operator is defined as

$$\Phi_\theta = \sum_{m=0}^{s} \theta_m \mid \theta_m \rangle \langle \theta_m \mid , \tag{15.34}$$

or

$$\Phi_\theta = \theta_0 + \frac{2\pi}{s+1} \sum_{m=0}^{s} m \mid \theta_m \rangle \langle \theta_m \mid . \tag{15.35}$$

We notice that Φ_θ depends on an arbitrary reference phase θ_0, which also happens in its classical counterpart.

Clearly, the phase operator defined by (15.34) is Hermitian and satisfies the eigenvalue equation

$$\Phi_\theta \mid \theta_m \rangle = \theta_m \mid \theta_m \rangle . \tag{15.36}$$

One of the problems with Dirac's phase was that the matrix element $\langle n' \mid \Phi_D \mid n \rangle$ was undefined. In the Pegg–Barnett formalism, that problem is not present.

We start from

$$\mid \theta_m \rangle \langle \theta_m \mid = (s+1)^{-1} \sum_{n,n'=0}^{s} \exp \left[i(n'-n)\theta_m \right] \mid n \rangle \langle n \mid , \tag{15.37}$$

thus the phase operator can be written as

$$\Phi_\theta = \theta_0 + \frac{2\pi}{s+1} \sum_{m=0}^{s} m(s+1)^{-1} \sum_{n,n'=0}^{s} \exp \left[i(n'-n)\theta_m \right] \mid n \rangle \langle n \mid \tag{15.38}$$

$$= \theta_0 + \frac{\pi s}{s+1} + \frac{2\pi}{s+1} \sum_{n \neq n'}^{s} \frac{\exp \left[i(n'-n)\theta_0 \right]}{\exp \left[\frac{i(n'-n)2\pi}{s+1} \right] - 1} \mid n \rangle \langle n \mid .$$

Now, taking matrix elements of Φ_θ

$$\langle n \mid \Phi_\theta \mid n \rangle = \theta_0 + \frac{\pi s}{s+1}, \tag{15.39}$$

$$\langle n' \mid \Phi_\theta \mid n \rangle = \frac{2\pi}{s+1} \frac{\exp \left[i(n'-n)\theta_0 \right]}{\exp \left[\frac{i(n'-n)2\pi}{s+1} \right] - 1},$$

$$n \neq n'.$$

From the above formula, we observe that the matrix elements of Φ_θ are well defined, implying that the commutator $[\Phi, \mathbf{n}] = -i$ must be incorrect.

From (15.39), we immediately see that

$$\langle n \mid [\mathbf{n}, \Phi_\theta] \mid n \rangle = 0, \tag{15.40}$$

$$\langle n' \mid [\mathbf{n}, \Phi_\theta] \mid n \rangle = \frac{2\pi(n'-n)}{s+1} \frac{\exp\left[i(n'-n)\theta_0\right]}{\exp\left[\frac{i(n'-n)2\pi}{s+1}\right] - 1},$$

$$n \neq n'.$$

If we take 'finite' or 'physical' states with n, $n' \ll s$, then we get approximately

$$\langle n' \mid \Phi_\theta \mid n \rangle \approx \frac{i}{n-n'} \exp\left[i(n'-n)\theta_0\right] , \tag{15.41}$$

$$\langle n' \mid [\mathbf{n}, \Phi_\theta] \mid n \rangle \approx -i(1-\delta_{nn'})\exp\left[i(n'-n)\theta_0\right] .$$

If $\theta_0 = 0$, and for $n \neq n'$, we get

$$[\mathbf{n}, \Phi_\theta]_{n'n} = -i , \tag{15.42}$$

which resembles the Dirac commutator.

Now, we concentrate in the exponential operators $\exp(\pm i\Phi_\theta)$. They are both eigenstates of $\mid \theta_m \rangle$

$$\exp(\pm i\Phi_\theta) \mid \theta_m \rangle = \exp(\pm i\theta_m) \mid \theta_m \rangle . \tag{15.43}$$

The action of $\exp(\pm i\Phi_\theta)$ on the Fock states is

$$\exp(i\Phi_\theta) \mid n \rangle = \exp(i \sum_{m=0}^{s} \theta_m \mid \theta_m \rangle\langle \theta_m \mid) \mid n \rangle , \tag{15.44}$$

but because

$$\mid n \rangle = \sum_{p=0}^{s} \mid \theta_p \rangle\langle \theta_p \mid\mid n \rangle = (s+1)^{-\frac{1}{2}} \sum_{p=0}^{s} \exp(-in\theta_p) \mid \theta_p \rangle, \tag{15.45}$$

then replacing (15.45) in (15.44), we get

$$\exp(i\Phi_\theta) \mid n \rangle = (s+1)^{-1} \sum_{m=0}^{s} \exp\left[-i(n-1)\theta_m\right] \mid \theta_m \rangle = \mid n-1 \rangle, \tag{15.46}$$

for $n > 0$.

For $n = 0$, we get the unphysical state $\mid -1 \rangle$, but we can call that state $\mid s \rangle$, thus

$$(s+1)^{-1} \sum_{m=0}^{s} \exp(i\theta_m) \mid \theta_m \rangle = (s+1)^{-1} \exp\left[i\theta_0(s+1)\right] \mid s \rangle$$

$$\sum_{m=0}^{s} \exp(-is\theta_m) \mid \theta_m\rangle = \exp\left[i\theta_0(s+1)\right] \mid s\rangle .$$

Thus the number state representation of $\exp(i\Phi_\theta)$ is

$$\exp(i\Phi_\theta) = \mid 0\rangle\langle 1 \mid + \mid 1\rangle\langle 2 \mid +...+ \mid s-1\rangle\langle s \mid + \exp\left[i\theta_0(s+1)\right] \mid s\rangle\langle 0 \mid .$$
$$(15.47)$$

The above expansion is similar to **Susskind–Glogower's**, except for the last term, which makes the exponential operator unitary.

Finally, in the **Pegg–Barnett** phase, the trigonometric functions behave in a more normal way.

The reader may verify the following properties:

$$\cos^2 \Phi_\theta + \sin^2 \Phi_\theta = 1 , \qquad (15.48)$$

$$[\cos \Phi_\theta, \sin \Phi_\theta] = 0 , \qquad (15.49)$$

$$\langle n \mid \cos^2 \Phi_\theta \mid n\rangle + \langle n \mid \sin^2 \Phi_\theta \mid n\rangle = 1 . \qquad (15.50)$$

A drawback in the **Pegg–Barnett** formalism is that the state space is finite, thus

$$[a, a^\dagger] \neq 1 . \qquad (15.51)$$

This can be easily seen as follows:

$$a = \exp(i\Phi_\theta)\sqrt{\mathbf{n}} = \mid 0\rangle\langle 1 \mid +\sqrt{2} \mid 1\rangle\langle 2 \mid +... + \sqrt{s} \mid s-1\rangle\langle s \mid , \quad (15.52)$$
$$a^\dagger = \sqrt{\mathbf{n}}\exp(-i\Phi_\theta) = \mid 1\rangle\langle 0 \mid +\sqrt{2} \mid 2\rangle\langle 1 \mid +... + \sqrt{s} \mid s\rangle\langle s-1 \mid ,$$

so

$$[a, a^\dagger] = 1 - (s+1) \mid s\rangle\langle s \mid . \qquad (15.53)$$

15.4.1 Applications

Fock states

The Fock states should be states with random phase. This is actually also true for any mixed state of the field with only diagonal elements in the density matrix.

The expectation value of Φ_θ is

$$\langle n \mid \Phi_\theta \mid n\rangle = \sum_{m=0}^{s} \theta_m \mid \langle \theta_m \mid n\rangle \mid^2 \qquad (15.54)$$

$$= \frac{1}{s+1} \sum_{m=0}^{s} \theta_m = \theta_0 + \frac{\pi s}{s+1} .$$

We notice that when $s \to \infty$

$$\langle n \mid \Phi_\theta \mid n \rangle = \theta_0 + \pi .$$ (15.55)

On the other hand

$$\langle n \mid \Phi_\theta^2 \mid n \rangle = \sum_{m=0}^{s} \theta_m^2 \mid \langle \theta_m \mid n \rangle \mid^2$$

$$= \frac{1}{s+1} \sum_{m=0}^{s} \theta_m^2 = \theta_0^2 + \frac{2\pi\theta_0 s}{s+1} + \frac{4\pi^2 s(s+\frac{1}{2})}{3(s+1)^2} ,$$

and when $s \to \infty$

$$\langle n \mid \Phi_\theta^2 \mid n \rangle = \theta_0^2 + 2\pi\theta_0 + \frac{4\pi^2}{3} ,$$ (15.56)

and

$$(\Delta\Phi_\theta^2)_n = \frac{\pi^2}{3} .$$ (15.57)

This corresponds exactly to the classical result. If one has a uniform phase distribution, corresponding to a random phase, then

$$\langle \phi \rangle = \frac{1}{2\pi} \int_{\theta_0}^{\theta_0+2\pi} \phi d\phi = \theta_0 + \pi ,$$ (15.58)

$$\langle \phi^2 \rangle = \frac{1}{2\pi} \int_{\theta_0}^{\theta_0+2\pi} \phi^2 d\phi = \theta_0^2 + 2\pi\theta_0 + \frac{4\pi^2}{3} ,$$

$$(\Delta\phi^2)_{\text{class}} = \frac{\pi^2}{3} .$$

Coherent states

$$\langle \alpha \mid \Phi_\theta \mid \alpha \rangle = \sum_{m=0}^{s} \theta_m \mid \langle \theta_m \mid \alpha \rangle \mid^2 ,$$ (15.59)

$$\langle \alpha \mid \Phi_\theta^2 \mid \alpha \rangle = \sum_{m=0}^{s} \theta_m^2 \mid \langle \theta_m \mid \alpha \rangle \mid^2 .$$ (15.60)

We have to calculate $\langle \theta_m \mid \alpha \rangle$.

$$\langle \theta_m \mid \alpha \rangle = \exp\left(-\frac{r^2}{2}\right)(s+1)^{-\frac{1}{2}} \sum_{n=0}^{\infty} \left(\frac{r^n}{\sqrt{n!}}\right) \exp\left[in(\phi - \theta_m)\right],$$ (15.61)

$$\alpha = r\exp(i\phi) ,$$

and

$$\mid \langle \theta_m \mid \alpha \rangle \mid^2 = \frac{1}{s+1} + \left[\frac{2\exp(-r^2)}{s+1}\right] \sum_{n>n'} \frac{r^n r^{n'}}{\sqrt{n!n'!}} \cos\left[(n-n')(\phi - \theta_m)\right],$$

$$(15.62)$$

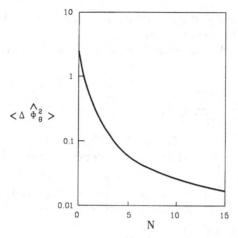

Fig. 15.1. Phase fluctuations of a coherent state versus number of photons

and choosing, for convenience $(\phi - \theta_0) = \frac{\pi s}{s+1}$, and defining $\mu \equiv m - \frac{s}{2}$, so $-\frac{s}{2} \leq \mu \leq \frac{s}{2}$, then when $s \to \infty$, the above summations can be converted into integrals, getting [17]

$$\langle \Phi_\theta \rangle_\alpha = \phi \,, \tag{15.63}$$

$$\langle \Delta \Phi_\theta^2 \rangle_\alpha = \frac{\pi^2}{3} + 4 \exp(-r^2) \left[\sum_{n>n'} \frac{(-1)^{n+n'} r^n r^{n'}}{\sqrt{n! n'!} (n-n')^2} \right] \,,$$

which can be evaluated numerically, giving the result shown in the Fig. 15.1.

15.5 Phase Fluctuations in a Laser

From the quantum Theory of the laser, in the master equation approach, the off diagonal matrix elements of the field density operator can be written as [26]

$$\rho_{n,n+k} = \sqrt{\rho_{n,n} \rho_{n+k,n+k}} \exp -\mu_k t \,, \tag{15.64}$$

where, according to the Scully–Lamb theory

$$\mu_k = \frac{1}{2} k^2 D \,, \tag{15.65}$$

D being the phase diffusion coefficient

$$D = \frac{C}{2\langle n \rangle} \,, \tag{15.66}$$

and we have assumed that $\rho_{n,n}$ and $\rho_{n+k,n+k}$ are the steady- state diagonal elements of the field density operator.

Now, assuming that we start from a coherent state $\mid \alpha \rangle$, one can write [26]

$$\rho(t) = \exp(-r^2) \sum_{n,l=0}^{s} \left\{ \frac{r^n r^l}{\sqrt{n! l!}} \exp\left[i\xi_0(n-l) - \frac{Dt(n-l)^2}{2} \right] \right. \tag{15.67}$$

$$\left. \times \mid n\rangle\langle l \mid \right\}, \alpha = r \exp(i\xi_0) .$$

Now we calculate the variance of the **Pegg–Barnett** phase operator

$$\langle \Delta\Phi_\theta^2 \rangle_{\text{Laser}} = \sum_m \theta_m^2 \langle \theta_m \mid \rho \mid \theta_m \rangle - \left[\sum_m \theta_m \langle \theta_m \mid \rho \mid \theta_m \rangle \right]^2 . \tag{15.68}$$

From (15.67), we calculate

$$\langle \theta_m \mid \rho \mid \theta_m \rangle \tag{15.69}$$

$$= \exp\left(-r^2\right) \sum_{n,l=0}^{s} \frac{r^n r^l}{\sqrt{n! l!}} \exp\left[i\xi_0(n-l) - \frac{Dt(n-l)^2}{2} \right] \langle \theta_m \mid n\rangle\langle l \mid \theta_m \rangle$$

$$= \frac{\exp(-r^2)}{s+1} \sum_{n,l=0}^{s} \frac{r^n r^l}{\sqrt{n! l!}} \exp\left[i(\xi_0 - \theta_m)(n-l) - \frac{Dt(n-l)^2}{2} \right]$$

$$= \frac{1}{s+1} \left[1 + 2\exp(-r^2) \sum_{n=1}^{s} \right.$$

$$\left. \times \sum_{l=0}^{n-1} \frac{r^n r^l}{\sqrt{n! l!}} \cos\left[(\xi_0 - \theta_m)(n-l) \right] \exp\left(-\frac{Dt(n-l)^2}{2} \right) \right] .$$

Now, we take the continuous limit $s \to \infty, \theta_m \to \theta$, and get the probability density distribution

$$P(\theta)^{(\xi_0)} = \frac{s+1}{2\pi} \langle \theta \mid \rho \mid \theta \rangle \tag{15.70}$$

$$= \frac{1}{2\pi} \left\{ 1 + 2\exp(-r^2) \sum_{n=1}^{\infty} \right.$$

$$\left. \times \sum_{l=0}^{n-1} \frac{r^n r^l}{\sqrt{n! l!}} \cos\left[(\xi_0 - \theta)(n-l) \right] \exp\left(-\frac{Dt(n-l)^2}{2} \right) \right\} .$$

The above distribution is normalized in the range $[0, 2\pi]$, which corresponds to the entire range of θ_m, when $s \to \infty$.

Furthermore, this probability density is invariant under

$$\xi_0 \to \xi_0 + 2\pi k , \tag{15.71}$$

$$k = 0, 1, 2, \ldots .$$

For computational convenience, we use the following Gaussian approximation

$$P(k) = \exp(-r^2)\frac{r^{2k}}{k!} \approx \frac{1}{(2\pi r^2)^{\frac{1}{2}}} \exp\left[-\frac{(r^2-k)^2}{2r^2}\right], \tag{15.72}$$

and substituting $\sqrt{P(k)}$ in (15.70), and replacing summations by integrals, we write

$$P(\theta)^{(\xi_0)} = \frac{1}{2\pi}\left[\frac{1}{(2\pi r^2)^{\frac{1}{2}}}\int_0^\infty dn \int_0^\infty dl\right]$$

$$\exp\left\{\frac{-1}{4r^2}\left[(r^2-n)^2 + (r^2-l)^2\right] + i(\xi_0-\theta)(n-l) - \frac{Dt(n-l)^2}{2}\right\}$$

or

$$P(\theta)^{(\xi_0)} = \frac{1}{\left[2\pi(\frac{1}{4r^2}+Dt)\right]^{\frac{1}{2}}}\exp\left[-\frac{(\xi_0-\theta)^2}{2(\frac{1}{4r^2}+Dt)}\right]. \tag{15.73}$$

The above probability density is normalized in the θ range $[-\infty,\infty]$, and it is a Gaussian with a variance increasing in time.

We would like to have, instead, a probability distribution in the $[\theta_0,\theta_0+2\pi]$ range.

Adding an infinite number of Gaussians with the same width, with their center displaced in $2\pi k$, k integer, we obtain a probability density that is normalized in the desired range and is periodic

$$P(\theta,\xi_0) = \sum_{k=-\infty}^\infty P(\theta)^{(\xi_0+2\pi k)} = \frac{1}{[\pi a^2]^{\frac{1}{2}}}\sum_{k=-\infty}^\infty \exp\left[-\frac{(\xi_0+2\pi k-\theta)^2}{2\left(\frac{1}{4r^2}+Dt\right)}\right], \tag{15.74}$$

with

$$a^2 = 2\left(\frac{1}{4r^2}+Dt\right). \tag{15.75}$$

One can show [26] that the phase variance can be written in terms of error functions

$$\langle(\Delta\Phi_\theta)^2\rangle = \left(\frac{1}{4r^2}+Dt\right) + \sum_{k=1}^\infty k^2\left\{\Phi\left[\frac{-\pi+2\pi(k+1)}{a^2}\right]\right.$$

$$\left. - \Phi\left[\frac{-\pi+2\pi k}{a^2}\right]\right\} \tag{15.76}$$

$$- 4(\pi a^2)^{\frac{1}{2}}\sum_{k=1}^\infty\left\{\exp\left[-\frac{(-\pi+2\pi k)^2}{a^2}\right]\right.$$

$$\left. - \exp\left[-\frac{(-\pi+2\pi(k+1))^2}{a^2}\right]\right\},$$

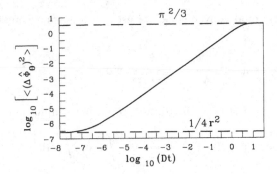

Fig. 15.2. Phase fluctuations of a laser versus time

where Φ is the error function.

The results are shown in the Fig. 15.2

We notice, from the Fig. 15.2 that in the lower end, for $Dt \ll 1$, the phase fluctuations correspond to the shot noise $1/4r^2$, and in the upper end, $Dt \gg 1$, it corresponds to a random phase $\pi^2/3$.

There is an extensive region of Dt where there is a linear dependence. In this last region, the phase diffusion model is valid, and the slope of the curve is the well-known Schawlow–Townes phase diffusion coefficient.

Problems

15.1. Show that
$$\langle n \mid \cos \phi_D \mid n \rangle = \langle n \mid \sin \phi_D \mid n \rangle = 0$$
and
$$\langle n \mid \cos^2 \phi_D \mid n \rangle = \langle n \mid \sin^2 \phi_D \mid n \rangle$$
$$= \frac{1}{2}, n \neq 0$$
$$= \frac{1}{4}, n = 0.$$

15.2. Prove that
$$[\cos \phi_D, \sin \phi_D] = \frac{[a^\dagger (n+1)^{-1} a - 1]}{2i},$$
and hence, all the matrix elements are zero, except for
$$\langle 0 \mid [\cos \phi_D, \sin \phi_D] \mid 0 \rangle = -\frac{1}{2i}.$$

15.3. Prove (15.6).

15.4. Prove (15.27).

15.5. Verify the properties given by (15.48–15.50).

15.6. Prove (15.63).

References

1. Dirac, P.A.M.: Proc. Roy. Soc. Lond. A, **114**, 243 (1927)
2. Susskind, L., Glogower, I.: Physics 1, **49** (1964)
3. Louisell, W.H.: Phys. Lett., **7**, 60 (1963)
4. Vogel, W., Welsch, D.G.: Lectures on Quantum Optics. Akademic Verlag, Berlin (1994)
5. Carruthers, P., Nieto, M.M.: Phys. Rev. Lett., **14**, 387 (1965)
6. London, F.: Z. Phys., **37**, 915 (1926)
7. London, F.: Z. Phys., **40**, 193 (1926)
8. Lerner, E.C.: Nuovo Cimento B, **56**, 183 (1968)
9. Zak, J.: Phys. Rev., **187**, 1803 (1969)
10. Turski, L.A.: Physics, **57**, 432 (1972)
11. Paul, H.: Forts. d, Physik, **22**, 657 (1974)
12. Schubert, M., Vogel, W.: Phys. Lett., **68A**, 321 (1978)
13. Barnett, S.M., Pegg, D.T.: J. Phys. A, **19**, 3849 (1986)
14. Barnett, S.M., Pegg, D.T.: J. Mod. Opt, **36**, 7 (1989)
15. Pegg, D.T., Barnett, S.M.: Europhys. Lett., **6**, 483 (1988)
16. Pegg, D.T., Barnett, S.M.: Phys. Rev. A, **39**, 1665 (1986)
17. Lynch, R.: Phys. Rev. A, **41**, 2841 (1990)
18. Shapiro, J.H., Shepard, S.R., Wong, N.C.: Phys. Rev. Lett., **62**, 2377 (1989)
19. Schleich, W., Horowitz, R.J., Varro, S.: Phys. Rev. A, **40**, 7405 (1989)
20. Bandilla, A., Paul, H., Ritze, H.H.: Quant. Opt., **3**, 267 (1991)
21. Vogel, W., Schleich, W.: Phys. Rev. A, **44**, 7642 (1991)
22. Noh, J.W., Fougieres, A., Mandel, L.: Phys. Rev. Lett., **67**, 1426 (1991)
23. Noh, J.W., Fougieres, A., Mandel, L.: Phys. Rev. A., **45**, 424 (1992a)
24. Noh, J.W., Fougieres, A., Mandel, L.: Phys. Rev. A, **46**, 2840 (1992b)
25. Noh, J.W., Fougieres, A., Mandel, L.: Phys. Rev. Lett., **71**, 2579 (1993)
26. Orszag, M., Saavedra, C.: Phys. Rev. A, **43**, 554 (1991)

Further reading

• Loudon, R. The Quantum Theory of Light. Clarendon Press, Oxford (1973)

16. Quantum Trajectories

Einstein, in his classical paper on the A and B coefficients for spontaneous and stimulated emission, assumed the existence of quantum jumps, which greatly stimulated quantum mechanics. However, until very recently, quantum jumps played practically no role in various theories coupling radiation and matter, and this type of interactions are well described by the Schrödinger wavefunction describing the properties of an ensemble, rather than individual systems.

Now, in all the examples we have seen of systems coupled to reservoirs, we have followed the procedure of starting from Liouville Equation for the system coupled to the reservoir; next, we trace or average over the reservoir variables, after a Markovian approximation, to end up in a master equation of the form

$$\frac{\mathrm{d}\rho_S}{\mathrm{d}t} = \frac{i}{\hbar}[\rho_S, H_s] + L_{\text{relax}}(\rho_S), \tag{16.1}$$

where

$$L_{\text{relax}}(\rho_S) = -\frac{1}{2}\sum_m (C_m^\dagger C_m \rho_S + \rho_S C_m^\dagger C_m - 2C_m \rho_S C_m^\dagger). \tag{16.2}$$

The above form, normally called Lindblad form, describes many systems coupled to reservoirs .

For example, if we want to describe the spontaneous Emission in a two-level atomic system, then $C_1 = \sqrt{\gamma}\sigma^-$, and $L_{\text{relax}}(\rho_S) = -\frac{1}{2}\gamma(\sigma^\dagger \sigma \rho_S + \rho_S \sigma^\dagger \sigma - 2\sigma\rho_S\sigma^\dagger)$, γ being the inverse lifetime of the atomic transition. Similarly, if we are describing a damped harmonic oscillator, then the same applies after replacing $\sigma^\dagger \to a^\dagger, \sigma \to a$.

Now, we present a method based on the wavefunction to describe such a system. **In general, we are not allowed to use Schrödinger Equation to describe System–Reservoir-type interactions, since even if we start initially with a pure state, the coupling to the bath will produce statistical mixtures, and we traditionally are forced to go to the Liouville Equation.**

However, two alternatives have been recently proposed

1) The system evolves with a non-Hermitian Hamiltonian, interrupted, once in a while by instantaneous quantum jumps. This process was baptized by

Carmichael as "Quantum Trajectories" or also called Montecarlo wavefunction method. [1, 2, 3].

2) Schrödinger's Equation is reinterpreted as representing an individual system following a stochastic dynamics of the diffusive type. We will write a Stochastic Schrödinger Equation.

16.1 Montecarlo Wavefunction Method

We calculate the change of the wavefunction $\mid \phi(t)\rangle \rightarrow \mid \phi(t+\delta t)\rangle$ in two steps:

a) Calculate $\mid \phi^1(t+\delta t)\rangle$ obtained from the evolution of $\mid \phi(t)\rangle$ with a non-Hermitian Hamiltonian given by

$$H = H_s - \frac{i\hbar}{2}\sum_m C_m^\dagger C_m \ . \tag{16.3}$$

For δt small, we get

$$\mid \phi^1(t+\delta t)\rangle = (1 - \frac{iH\delta t}{\hbar})\mid \phi(t)\rangle \ . \tag{16.4}$$

Since H is non-Hermitian $\mid \phi^1(t+\delta t)\rangle$ is not normalized. Thus,

$$\langle \phi^1(t+\delta t)\mid \phi^1(t+\delta t)\rangle = \langle \phi(t)\mid (1+\frac{iH^\dagger\delta t}{\hbar})(1-\frac{iH\delta t}{\hbar})\mid \phi(t)\rangle \tag{16.5}$$
$$\equiv 1 - \delta p \ ,$$

$$\delta p = \delta t\frac{i}{\hbar}\langle \phi(t)\mid H - H^\dagger \mid \phi(t)\rangle \equiv \sum_m \delta p_m \ , \tag{16.6}$$

$$\delta p_m \equiv \delta t\langle \phi(t)\mid C_m^\dagger C_m \mid \phi(t)\rangle \geq 0 \ .$$

We can always adjust δt such that $\delta p \ll 1$.

b) The second step corresponds to a **gedanken experiment** of a measurement process. We consider the possibility of a quantum jump. In order to decide whether a quantum jump has occurred, we define a random number ε uniformly distributed between zero and one and compare it to δp. Two cases may arise:

I) $\varepsilon \geq \delta p$

This will be the large majority of the cases, since $\delta p \ll 1$. In this case, there is no quantum jump and $\mid \phi(t+\delta t)\rangle = \frac{\mid \phi^1(t+\delta t)\rangle}{\sqrt{1-\delta p}}$.

II) $\varepsilon < \delta p$

A quantum jump occurs to one of the states $C_m \mid \phi(t)\rangle$ according to the relative probability among the various possible types of jumps, $\Pi_m = \frac{\delta p_m}{\delta p}$ (notice that $\sum_m \Pi_m = 1$). So

$$| \phi(t + \delta t)\rangle = \frac{C_m \mid \phi(t)\rangle}{\sqrt{\frac{\delta p_m}{\delta t}}} . \tag{16.7}$$

Milburn et al. [10] showed that a mode of the electromagnetic field in a cavity at $T = 0$ is described by a quantum jump equation, if the outgoing light is detected directly by a photodetector. We will generalize these arguments in the next sections.

16.1.1 The Montecarlo Method is Equivalent, on the Average, to the Master Equation

We define $\overline{\sigma}(t) = Av\,[\sigma(t) =\mid \phi(t)\rangle\langle \phi(t)\mid]$, where Av means the average of many Montecarlo results at time t, all of them starting from $\mid \phi(0)\rangle$.

We will now show that $\overline{\sigma}(t)$ coincides with ρ_S.

We calculate $\overline{\sigma}(t + \delta t)$

$$\overline{\sigma}(t + \delta t) = (1 - \delta p)\frac{\mid \phi^1(t + \delta t)\rangle\langle \phi^1(t + \delta t)\mid}{\sqrt{1 - \delta p}\sqrt{1 - \delta p}} \tag{16.8}$$

$$+\delta p \sum_m \pi_m \frac{C_m \mid \phi(t)\rangle\langle \phi(t)\mid C_m^\dagger}{\sqrt{\frac{\delta p_m}{\delta t}}\sqrt{\frac{\delta p_m}{\delta t}}}$$

or

$$\overline{\sigma}(t + \delta t) = (1 - \frac{i\delta t}{\hbar}(H_s - \frac{i\hbar}{2}\sum_m C_m^\dagger C_m)) \mid \phi(t)\rangle\langle \phi(t)\mid$$

$$\left[1 + \frac{i\delta t}{\hbar}(H_s + \frac{i\hbar}{2}\sum_m C_m^\dagger C_m)\right]$$

$$+\delta t \sum_m C_m \mid \phi(t)\rangle\langle \phi(t)\mid C_m\dagger ,$$

which can be put as

$$\overline{\sigma}(t + \delta t) = \sigma(t) + \frac{i\delta t}{\hbar}\left\{\sigma(H_s + \frac{i\hbar}{2}\sum_m C_m^\dagger C_m) - (H_s - \frac{i\hbar}{2}\sum_m C_m^\dagger C_m)\sigma\right\}$$

$$+\delta t \sum_m C_m \sigma C_m\dagger = \sigma + \frac{i\delta t}{\hbar}\,[\sigma, H_s] + \delta t L_{\text{relax}}(\sigma) . \tag{16.9}$$

Finally, if we average over a large number of trajectories, we recover the master equation (16.1).

Similarly to the master equation methods, one is interested in computing averages of interesting observables. Here, for each trajectory, we get $\langle \phi^i(t)\mid$

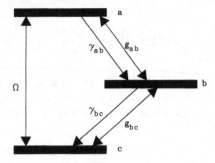

Fig. 16.1. Three-level atom inside a double cavity

$A \mid \phi^i(t)\rangle$ for many solutions $\mid \phi^i(t)\rangle$, thus $\langle A \rangle_n = \frac{1}{n} \sum_n \langle \phi^i(t) \mid A \mid \phi^i(t) \rangle$ and $\langle A \rangle_n \to \langle A \rangle$ as $n \to \infty$.

The equivalency between the master equation and the Montecarlo Method is valid as long as $\eta_i \delta t \ll 1$, where $\eta_i \hbar$ is a typical energy eigenvalue of the system.

An example to illustrate the procedure is the one atom Raman Laser [4, 5]. It consists in a three-level atom interacting with two quantum fields and a classical coherent pump, as shown in Fig. 16.1.

The Hamiltonian of the system is

$$H = i\hbar g_{ab}(a\sigma_{ab} - a^\dagger \sigma_{ba}) + i\hbar g_{bc}(b\sigma_{bc} - b^\dagger \sigma_{cb}) + i\hbar \Omega(\sigma_{ac} - \sigma_{ca}) , \quad (16.10)$$

and the damping terms are

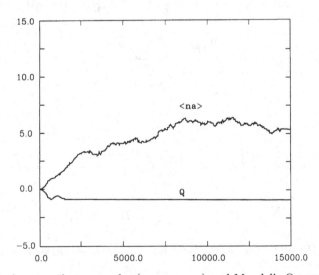

Fig. 16.2. Average photon number(*upper curve*) and Mandel's Q parameter as a function of time. The parameters taken are $g_{ab} = g_{bc} = 1 \ \Omega = 1$, $\gamma_a = 0.5$, $\gamma_b = 3$, $\gamma_{ab} = 0$, $\gamma_{bc} = 1$. We also added a detuning $\delta_a = -\delta_b = 1.41$

$$L_{\text{relax}}(\rho_S) = \frac{\gamma_{ab}}{2}(2\sigma_{ba}\rho_S\sigma_{ab} - \sigma_{ab}\sigma_{ba}\rho_S - \rho_S\sigma_{ab}\sigma_{ba}) \qquad (16.11)$$

$$+ \frac{\gamma_{bc}}{2}(2\sigma_{cb}\rho_S\sigma_{bc} - \sigma_{bc}\sigma_{cb}\rho_S - \rho_S\sigma_{bc}\sigma_{cb})$$

$$+ \frac{\gamma_a}{2}(2a\rho_S a^\dagger - a^\dagger a\rho_{sys} - \rho_{sys}a^\dagger a)$$

$$+ \frac{\gamma_b}{2}(2b\rho_{sys}b^\dagger - b^\dagger b\rho_{sys} - \rho_{sys}b^\dagger b)\,.$$

In Fig. 16.2, we show $\langle n \rangle$ and the Mandel parameter $Q_{\text{Mandel}} \equiv (\frac{\langle(\Delta n)^2\rangle - \langle n\rangle}{\langle n\rangle})$ as a function of time. The results were obtained averaging over 50 trajectories. We notice that for a certain set of parameters we get a Q value very close to -1, implying that we are generating an almost pure Fock state.

16.2 The Stochastic Schrödinger Equation

If an open system with a density operator ρ starts initially as a pure state and evolves into a mixed state, as a result of the interaction with the Reservoir, there can be no deterministic equation for $\mid \phi \rangle$, but one could define a stochastic equation, as one would expect, given the probabilistic nature of the interaction with the environment.

Gisin and Percival [6, 7, 8, 9] proposed thefollowing equation

$$\mid d\phi \rangle = \mid v \rangle \delta t + \sum_j \mid u_j \rangle d\xi_j\,, \qquad (16.12)$$

where the first term on the right hand side of (16.12), represents the drift and the second one the diffusion. Also, $d\xi_j$ is a complex stochastic Wiener process, such that

$$M(Re(d\xi_j)Re(d\xi_k)) = M(Im(d\xi_j)Im(d\xi_k)) = \delta_{jk}\delta t\,, \qquad (16.13)$$
$$MRe(d\xi_j) = M(Im(d\xi_j)Re(d\xi_k)) = 0\,,$$

which is equivalent to writing

$$M((d\xi_j)^*(d\xi_k)) = 2\delta_{jk}\delta t\,. \qquad (16.14)$$

For normalization purposes, we set $\langle \phi \mid u_j \rangle = 0$ and take the mean value of both $\mid d\phi \rangle$ and $\mid d\phi \rangle\langle d\phi \mid$

$$M \mid d\phi \rangle = \mid v \rangle \delta t\,, \qquad (16.15)$$
$$M \mid d\phi \rangle\langle d\phi \mid := 2\sum_j \mid u_j \rangle\langle u_j \mid \delta t\,.$$

We notice that (16.15) was obtained to order δt and also using (16.14), that is $d\xi_k$ is of order $\sqrt{\delta t}$, and therefore, we have to keep second-order differentials. This is the characteristic of the Ito algebra.

Now, we can write

$$d\rho = M(| \phi\rangle\langle d\phi | + | d\phi\rangle\langle\phi | +. | d\phi\rangle\langle d\phi |) , \qquad (16.16)$$

$$\frac{d\rho}{dt} = | \phi\rangle\langle v | + | v\rangle\langle\phi | +2\sum_j | u_j\rangle\langle u_j | . \qquad (16.17)$$

Now, we try to obtain the diffusion and drift term and relate them to the Master equation.

Multiplying (16.17) from the left and right sides by the projector $(1- | \phi\rangle\langle\phi |)$, we get

$$(1- | \phi\rangle\langle\phi |)\frac{d\rho}{dt}(1- | \phi\rangle\langle\phi |) = 2\sum_j | u_j\rangle\langle u_j | . \qquad (16.18)$$

Now, for the drift part, we take (16.17) and multiply it by $\frac{1}{2}\langle\phi |$ from the left and $| \phi\rangle$ from the right. We readily get

$$\frac{1}{2}\langle\phi | \frac{d\rho}{dt} | \phi\rangle = Re\langle\phi | v\rangle , \qquad (16.19)$$

and multiplying (16.17) by $| \phi\rangle$, we also get

$$\frac{d\rho}{dt} | \phi\rangle = | \phi\rangle\langle v | \phi\rangle + | v\rangle . \qquad (16.20)$$

By combining (16.19, 16.20), we finally get

$$| v\rangle = \frac{d\rho}{dt} | \phi\rangle - (\frac{1}{2}\langle\phi | \frac{d\rho}{dt} | \phi\rangle + ic) | \phi\rangle , \qquad (16.21)$$

where c is a non-physical imaginary phase determined by the convention that this equation has to agree with the conventional Schrödinger equation in the absence of coupling to the environment, and corresponds to $Im\langle\phi | v\rangle$.

As we can see, we have obtained both the diffusion (16.18) and the drift (16.21) in terms of $\frac{d\rho}{dt}$, so all that is left is to replace the Master equation (16.1, 16.2) in both terms, to arrive to the Stochastic Schrödinger equation

$$d | \phi\rangle = -\frac{i}{\hbar}H | \phi\rangle\delta t \qquad (16.22)$$

$$+ \sum_m \left(\langle C_m^\dagger\rangle_\phi C_m - \frac{1}{2}C_m^\dagger C_m - \frac{1}{2}\langle C_m^\dagger\rangle_\phi\langle C_m\rangle_\phi \right) | \phi\rangle\delta t$$

$$+ \frac{1}{\sqrt{2}}\sum_m (C_m - \langle C_m\rangle_\phi) | \phi\rangle d\xi_m .$$

An equivalent Stochastic Schrödinger equation, interpreted as homodyne measurement, for $T = 0$, can be derived [10], with a noise that is real rather than complex

$$\mathrm{d} \mid \phi\rangle \ = \ -\frac{i}{\hbar}H \mid \phi\rangle \delta t \tag{16.23}$$
$$+ \ \{(-\frac{\gamma}{2}a^\dagger a + 2\gamma\langle X(t)\rangle_\phi)\delta t$$
$$+ \ \sqrt{\gamma}\Delta W(t)a\} \mid \phi\rangle \,,$$

where the ΔW is a Wiener increment, satisfying

$$\langle \Delta W \rangle = 0 \,, \tag{16.24}$$

$$\langle (\Delta W)^2 \rangle = \Delta t \,. \tag{16.25}$$

In the next few sections, we will generalize these arguments for $T \neq 0$. We will also show a physical realization, in the context of cavity QED, of both the Monte Carlo and Stochastic Schrödinger methods.

16.3 Stochastic Schrödinger Equations and Dissipative Systems

As we mention at the begining of this chapter, a wide class of master equations describing the evolution of dissipative quantum systems can be written in the Lindblad form [12]

$$\dot{\rho}_S = \mathcal{L}\rho_S \,, \tag{16.26}$$

where

$$\mathcal{L} = \mathcal{L}_0 + \sum_n \mathcal{L}_n \,, \tag{16.27}$$

$$\mathcal{L}_0\rho_S = \frac{i}{\hbar}[\rho_S, H_s] \,, \tag{16.28}$$

$$\mathcal{L}_n\rho_S = -\frac{1}{2}[C_n^\dagger C_n\rho_S + \rho_S C_n^\dagger C_n] + C_n\rho_S C_n^\dagger \,, \tag{16.29}$$

ρ_S is the reduced density operator for the "small" system S (obtained by tracing out the degrees of freedom of the reservoir R from the density operator for the full system $S + R$), and H_S describes the Hamiltonian evolution of the small system S in the interaction picture. The operators C_n act on the space of states of the small system S and express the interaction of S with the reservoir R. The number of them depends on the nature of the problem. We follow here the ([13]).

An example of such an equation is the master equation for a field in a lossy cavity, at temperature T, given in the interaction picture by

$$\frac{d\rho_f}{dt} = \gamma\langle n\rangle_{\text{th}}(a^\dagger\rho_f a - \frac{1}{2}aa^\dagger\rho_f - \frac{1}{2}\rho_f aa^\dagger)$$
$$+ \gamma(1 + \langle n\rangle_{\text{th}})(a\rho_f a^\dagger - \frac{1}{2}a^\dagger a\rho_f - \frac{1}{2}\rho_f a^\dagger a) , \tag{16.30}$$

where a and $a^\dagger$ are the photon annihilation and creation operators, respectively, $\langle n\rangle_{\text{th}}$ is the average number of thermal photons, given by Planck's distribution, and $\gamma = 1/t_{\text{cav}}$, where t_{cav} is the damping time. In this case, one could set

$$C_1 \equiv \sqrt{\gamma(1 + \langle n\rangle_{\text{th}})}a, \qquad C_2 \equiv \sqrt{\gamma\langle n\rangle_{\text{th}}}a^\dagger . \tag{16.31}$$

A formal solution of (16.26) is

$$\rho(t) = \exp\left(\mathcal{L}t\right)\rho(0) . \tag{16.32}$$

Let us define

$$J_n\rho = C_n\rho C_n^\dagger \tag{16.33}$$

and write

$$\rho(t) = \exp\left\{\mathcal{L}_0 + \sum_n [J_n + (\mathcal{L}_n - J_n)]\, t\right\}\rho(0) . \tag{16.34}$$

Note that

$$(\mathcal{L}_n - J_n)\rho_S = -\frac{1}{2}\left(C_n^\dagger C_n\rho_S + \rho_S C_n^\dagger C_n\right) . \tag{16.35}$$

Applying Dyson's expansion to (16.34), we get

$$\rho(t) = \sum_{m=0}^{\infty} \int_0^t dt_m \int_0^{t_m} dt_{m-1} \cdots \int_0^{t_2} dt_1$$
$$\{S(t - t_m)(\sum_n J_n)S(t_m - t_{m-1})$$
$$\times \ldots (\sum_n J_n)S(t_1)\}\rho(0) , \tag{16.36}$$

where

$$S(t) = \exp\left\{\left[\mathcal{L}_0 + \sum_n (\mathcal{L}_n - J_n)\right]t\right\} . \tag{16.37}$$

Equation (16.36) may be rewritten in the following way:

$$\rho(t) = \sum_{m=0}^{\infty} \sum_{\{n_i\}} \int_0^t dt_m \int_0^{t_m} dt_{m-1} \cdots \int_0^{t_2} dt_1$$
$$\{S(t - t_m)J_{n_m}S(t_m - t_{m-1})$$
$$\times \ldots J_{n_1}S(t_1)\}\rho(0) . \tag{16.38}$$

Each term in the above double sum can be considered as a quantum trajectory, the reduced density operator at time t being given by the sum over all possible quantum trajectories . For each of these trajectories, (16.38) shows that the evolution of the system can be considered as a succession of quantum jumps, associated to the operators J_n, interspersed by smooth time evolutions, associated with the operators $S(t)$. The probability of each trajectory is given by the trace of the corresponding term in (16.38).

From (16.35) and (16.37), we can write

$$S(t)\rho = N(t)\rho N(t)^{\dagger} , \tag{16.39}$$

where

$$N(t) = \exp\left[-\frac{i}{\hbar}H_S t - \frac{t}{2}\sum_n \left(C_n^{\dagger}C_n\right)\right] . \tag{16.40}$$

Therefore, if ρ is a pure state, then $S(t)\rho$ is also a pure state. The same is true for $J_n\rho$, with J_n defined by (16.33). This implies that a pure state remains pure, when a single quantum trajectory is considered. Note also that the evolution between jumps is given by the non-unitary operator $N(t)$.

It is clear from (16.34) that different choices of the jump operators are possible. These different choices correspond to different decompositions in terms of quantum trajectories of the time evolution of the density operator ρ_S and, eventually, to different experimental schemes leading to the continuous monitoring of the evolution of the system. It is precisely due to this continuous monitoring that an initial pure state remains pure, since no information is lost in this situation: for a field in a cavity, this continuous monitoring amounts to accounting for every photon gained or lost by the field, due to its interaction with the reservoir. ([13]).

We will discuss now two different realizations of the reservoir, for a field in a cavity, which will lead to a Monte Carlo quantum jump approach, for the first realization, and to a Schrödinger equation with stochastic terms, for the second one.

16.4 Simulation of a Monte Carlo SSE

We exhibit in this section a physical realization of the Monte Carlo method ([13]). The corresponding experimental scheme is shown in Fig. 16.3.

A monokinetic atomic beam plays the role of a reservoir R and crosses a lossless cavity, interacting with one mode of the electromagnetic field. The cavity mode plays the role of a small system S. The atoms, regularly spaced along the atomic beam, are prepared in one of two Rydberg states: an upper state $|a\rangle$ or a lower state $|b\rangle$. The transition frequency ω between these two

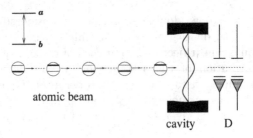

Fig. 16.3. Physical realization of a quantum jump trajectory. A beam of two-level atoms crosses a resonant cavity

states is assumed to be resonant with the cavity mode. A similar model of reservoir was adopted in Sect. 16.1 of [14].

The state of the atoms is measured by a detector just at the exit of the cavity. The ratio between the flux of upper state atoms r_a and the lower state atoms r_b before their entrance into the cavity is chosen so that

$$\frac{r_a}{r_b} = e^{-\hbar\omega/k_B T} \equiv \frac{\langle n \rangle_{\text{th}}}{1 + \langle n \rangle_{\text{th}}} , \qquad (16.41)$$

where $\hbar\omega$ is the difference in energy between $|a\rangle$ and $|b\rangle$, and, as will be shown in the next paragraphs, T is the reservoir temperature.

We analyze now the time evolution of the state vector $|\Psi(t)\rangle$ of S, under the continuous measurement of the atoms after they leave the cavity. We also assume that one knows the state of each atom before it interacts with the cavity. This may be achieved by selectively exciting the atoms to $|a\rangle$ or $|b\rangle$, according to the proportion given by (16.41). We will adopt the following simplifying assumptions: (a) the atom-field interaction time τ is the same for all atoms; (b) the spatial profile of the electric field is constant; (c) the cavity is perfect, i.e., the field state is changed only by the atoms; (d) the atom-field coupling constant g and the interaction time τ are both small, so that the atomic state rotation is very small; (e) the rotating-wave and dipole approximations will be used; and (f), according to the statements (d) and (e), quantum cooperative effects will be neglected. In this case the interaction Hamiltonian in the interaction picture will be

$$H = \hbar g \left(|b\rangle\langle a| a^\dagger + |a\rangle\langle b| a \right) . \qquad (16.42)$$

The operators a and $a^\dagger$ are annihilation and creation operators, acting on the space of states of the field mode. Just before the i-th atom enters the cavity, the state describing the combined system (atom i + field) is given by

$$|\Psi_{a-f}(t_i)\rangle = |\Psi(t_i)\rangle \otimes |\Psi_a(t_i)\rangle . \qquad (16.43)$$

Here $|\Psi_a(t_i)\rangle = |a\rangle$ or $|\Psi_a(t_i)\rangle = |b\rangle$, depending on the state to which the atom was excited before entering the cavity.

At time $t_i + \tau$, the atom-field state vector, up to second order in τ, is given by

$$|\tilde{\Psi}_{a-f}(t_i + \tau)\rangle = \left(1 - ig\tau|b\rangle\langle a| \, a^\dagger - ig\tau|a\rangle\langle b| \, a \right.$$
$$\left. - \frac{g^2\tau^2}{2}|b\rangle\langle b| \, a^\dagger a - \frac{g^2\tau^2}{2}|a\rangle\langle a| \, aa^\dagger \right) |\Psi_{a-f}(t_i)\rangle \,, \quad (16.44)$$

where the tilde indicates that the state vector is not normalized. The expansion (16.44) should be very good in view of condition (d). We assume that $(r_a + r_b)\tau < 1$, so that there is at most one atom inside the cavity at each instant of time. After this atom exits the cavity and is detected, one of the following four cases will be realized:

i. The atom enters the cavity in state $|b\rangle$ and is detected in the same state. In this case, according to (16.44), the state of S at time $t = t_i + \tau$ will be given by

$$|\tilde{\Psi}(t_i + \tau)\rangle = \left(1 - \frac{g^2\tau^2}{2}a^\dagger a\right)|\Psi(t_i)\rangle \,. \quad (16.45)$$

ii. The atom enters the cavity in state $|a\rangle$, and it is detected in the same state $|a\rangle$. In this case,

$$|\tilde{\Psi}(t_i + \tau)\rangle = \left(1 - \frac{g^2\tau^2}{2}aa^\dagger\right)|\Psi(t_i)\rangle \,. \quad (16.46)$$

iii. The atom enters the cavity in the state $|b\rangle$, and it is detected in the state $|a\rangle$. In this case,

$$|\tilde{\Psi}(t_i + \tau)\rangle = -ig\tau a\,|\Psi(t_i)\rangle \,. \quad (16.47)$$

iv. The atom enters the cavity in the state $|a\rangle$ and it is detected in the state $|b\rangle$. Then,

$$|\tilde{\Psi}(t_i + \tau)\rangle = -ig\tau a^\dagger\,|\Psi(t_i)\rangle \,. \quad (16.48)$$

Note that in the cases i and ii a small change in the state of "S" takes place, whereas in the cases iii and iv a big change may happen (quantum jump). However, these last two cases are very rare, due to the small change of the atomic state during the interaction time.

We consider now the change of $|\Psi\rangle$ from t to $t + \delta t$, where the time interval δt is large enough so that many atoms go through the cavity during this time interval ($n_a = r_a\delta t \gg 1$, $n_b = r_b\delta t \gg 1$), and also much smaller than $t_{\text{cav}}/\langle n\rangle_{\text{th}}\langle n\rangle$, where $\langle n\rangle$ is the average number of photons in the state. This last condition, as it will be seen later, implies that the probability of a quantum jump during δt is very small. In most of the time intervals δt, the atoms will be detected at the same state they come in, since the transition probability is very small. The evolution of $|\Psi\rangle$ during these intervals will be given by

$$|\tilde{\Psi}(t+\delta t)\rangle = \left(1 - \frac{g^2\tau^2}{2}aa^\dagger\right)^{n_a}$$

$$\times \left(1 - \frac{g^2\tau^2}{2}a^\dagger a\right)^{n_b}|\Psi(t)\rangle$$

$$= \left(1 - \frac{n_a g^2\tau^2}{2}aa^\dagger - \frac{n_b g^2\tau^2}{2}a^\dagger a\right)|\Psi(t)\rangle . \qquad (16.49)$$

This result does not depend on the ordering of the upper-state and lower-state atoms. We also note that in the interaction picture, the state vector does not evolve when there is no atom inside the cavity, since the only source of field dissipation is the interaction with the atomic beam.

Equation (16.49) displays the interesting property that the wave-function of the system (and, consequently, the mean energy) may change even when there is no exchange of energy between the system and the measurement apparatus (represented by the atoms in the present case). An easy way to understand this effect physically is to imagine that all atoms are sent into the cavity in the lower state, and are detected in the same state after exiting the cavity, for a given realization of the system, which starts with a coherent state in the cavity. Then, even though there is no exchange of energy between the atoms and the field in the cavity, as time evolves the ground state component of the initial state should also increase, since the results of the measurements lead to an increasing probability that there is a vacuum state in the cavity. In other words, the fact that there is no quantum jump, for that specific trajectory, provides us with information about the quantum state of the system, and this information leads to an evolution of the state. ([13]). This is closely related to the quantum theory of continuous measurement [15, 16] and also to quantum non-demolition measurement schemes proposed recently [17]. This problem is also very similar to that of a Heisenberg microscope in which even the unsuccessful events of light scattering produce a change in the quantum-mechanical state of the particle [18].

We introduce now the following definitions:

$$\gamma \equiv (r_b - r_a)g^2\tau^2 = \frac{r_b}{1+\langle n\rangle_{th}}g^2\tau^2 = \frac{r_a}{\langle n\rangle_{th}}g^2\tau^2 , \qquad (16.50)$$

$$C_1 \equiv \sqrt{\gamma(1+\langle n\rangle_{th})}\,a, \qquad C_2 \equiv \sqrt{\gamma\langle n\rangle_{th}}\,a^\dagger . \qquad (16.51)$$

Using these definitions and (16.41), (16.49) may be rewritten in the following way:

$$|\tilde{\Psi}(t+\delta t)\rangle = \left[1 - \frac{\delta t}{2}\sum_m C_m^\dagger C_m\right]|\Psi(t)\rangle . \qquad (16.52)$$

If an atom enters the cavity in state $|a\rangle$ and is detected in the state $|b\rangle$, the state vector of S suffers a "quantum jump," and one photon is added to that system. On the other hand, a de-excitation in S occurs if an atom which entered in $|b\rangle$ is detected in the state $|a\rangle$. The probability of these events to occur may be calculated by using (16.51) and (16.47) or (16.48); thus, the probability of an excitation (action of $a^\dagger$) to occur between t and $t + \delta t$ is given by

$$\delta p_1 = \delta t \langle \Psi(t)| C_1^\dagger C_1 |\Psi(t)\rangle \; . \tag{16.53}$$

The probability of a de-excitation (action of a) during this time interval is

$$\delta p_2 = \delta t \langle \Psi(t)| C_2^\dagger C_2 |\Psi(t)\rangle \; . \tag{16.54}$$

The probabilities δp_1 and δp_2 are very low, so that the joint probability of having one excitation and one de-excitation during the same time interval δt is negligible. One may therefore write

$$|\tilde{\Psi}(t + \delta t)\rangle = C_1^{\delta N_1} C_2^{\delta N_2} \left[1 - \frac{\delta t}{2} \sum_m C_m^\dagger C_m \right] |\Psi(t)\rangle \; . \tag{16.55}$$

where δN_1 and δN_2 are equal to one or zero, with probabilities δp_1 and δp_2 for δN_1 and δN_2 to be equal to one, respectively. This may be represented by writing the statistical mean $M(\delta N_m) = \langle C_m^\dagger C_m \rangle \delta t$. Also, $\delta N_m \delta N_n = \delta N_m \delta_{nm}$. One should note that the instants of time in which the quantum jumps occur during the time interval δt are irrelevant, since the jump operators can be commuted through the no-jump evolution, the commutation producing an overall phase that goes away upon renormalization of the state. This can be easily seen by rewriting the no-jump evolution, during a time interval $\delta t_j < \delta t$, as an exponential

$$1 - \frac{\delta t_j}{2} \sum_m C_m^\dagger C_m = \exp\left(-\frac{\delta t_j}{2} \sum_m C_m^\dagger C_m \right)$$
$$+ O[(\delta t_j)^2] \; , \tag{16.56}$$

and using that

$$C_i e^{\left(-\frac{\delta t_j}{2} \sum_m C_m^\dagger C_m \right)} = e^{\left(-\frac{\delta t_j}{2} \sum_m C_m^\dagger C_m \right)} C_i e^{\lambda_i} \; , \tag{16.57}$$

where $\lambda_1 = -(\delta t_j/2)\gamma(1 + \langle n \rangle_{\text{th}})$ and $\lambda_2 = (\delta t_j/2)\gamma \langle n \rangle_{\text{th}}$.

The results of the measurement may be simulated by picking up random numbers. The state vector in (16.55) may be normalized as follows:

$$\begin{aligned}
|\psi(t + \delta t)\rangle &= \{ \frac{C_1}{\sqrt{C_1^\dagger C_1}} \delta N_1 + \frac{C_2}{\sqrt{C_2^\dagger C_2}} \delta N_2 \\
&\quad + (1 - \delta N_1)(1 - \delta N_2)\left(1 - \frac{\delta t}{2} \sum_m C_m^\dagger C_m \right) \\
&\quad \times \left(1 - \delta t \sum_m \langle C_m^\dagger C_m \rangle \right)^{-\frac{1}{2}} \} \, |\psi(t)\rangle \; .
\end{aligned} \tag{16.58}$$

In the above equation, the first two terms represent the possible jumps, each normalized, as in the Monte Carlo method, and the last term is the no-jump evolution contribution, normalized with the corresponding prefactor that rules out the jumps. From (16.58) one gets for $| \, d\psi(t)\rangle \equiv | \, \psi(t + \delta t)\rangle - | \, \psi(t)\rangle$

$$
| \, d\psi(t)\rangle = \left\{ \sum_m \left[\frac{C_m}{\sqrt{C_m^\dagger C_m}} - 1 \right] \delta N_m \right.
$$
$$
\left. - \frac{\delta t}{2} \sum_m (C_m^\dagger C_m - \langle C_m^\dagger C_m\rangle) \right\} | \, \psi(t)\rangle \, . \tag{16.59}
$$

16.5 Simulation of the Homodyne SSDE

We show now that, by a suitable modification of the atomic configuration, it is also possible to interpret physically diffusion-like Schrödinger equations in terms of continuous measurements made on atoms ([13]), which cross the cavity containing the field. The corresponding scheme is shown in the Fig. 16.4: a beam of three-level atoms with a degenerate lower state (states b and c) crosses the cavity, the field in the cavity being resonant with a transition between one of the two lower levels (say, level b) and the upper atomic state a, whereas a strong classical field connects the other lower state with the upper level (one may assume that both fields are circularly polarized, so that the cavity field cannot connect a and c, whereas the strong field does not induce transitions between a and b).

We also assume that the atom is prepared in either a coherent superposition of the two lower levels:

$$
| \, \psi_{\text{atom}}\rangle = \frac{1}{\sqrt{2}}(| \, b\rangle + | \, c\rangle) \, , \tag{16.60}
$$

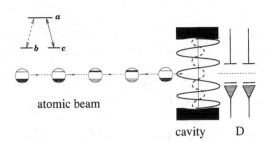

atomic beam

cavity   D

Fig. 16.4. Physical realization of the homodyne stochastic Schrödinger trajectory. A beam of three-level atoms crosses a resonant cavity, being subjected to an external classical field

or in the upper one, following a Boltzmann distribution corresponding to a temperature T for the atoms, which act as a reservoir for the quantum field in the cavity.

In the interaction picture, one can write

$$
\begin{aligned}
H &= \hbar g_{ac}(\varepsilon \mid a\rangle\langle c \mid +\varepsilon \mid c\rangle\langle a \mid) \\
&+ \hbar g_{ab}(a^\dagger \mid b\rangle\langle a \mid +a \mid a\rangle\langle b \mid) .
\end{aligned}
\tag{16.61}
$$

We assume for simplicity that $g_{ac} = g_{ab} = g$, and that ε is real. The time evolution of the wave function to second order in the coupling constant is

$$
\mid \psi(t+\tau)\rangle = \left[1 - \frac{iH\tau}{\hbar} - \frac{H^2\tau^2}{2\hbar^2}\right] \mid \psi(t)\rangle .
\tag{16.62}
$$

As in the previous model, there are two possible quantum jump processes. The first one corresponds to the atom entering the cavity in the coherent superposition of lower states and being detected in the upper state. After the measurement, the state of the field is given by

$$
\mid \psi(t+\tau)\rangle_f^{(b,c\to a)} = \frac{-ig\tau}{\sqrt{2}}(\varepsilon + a) \mid \psi(t)\rangle_f .
\tag{16.63}
$$

The corresponding probability of detecting an atom in $\mid a\rangle$, after a time interval δt, staring from the initial superposition state, is given by

$$
\delta p_1 = n_b \frac{g^2\tau^2}{2} \langle \psi_f(t) \mid (\varepsilon + a^\dagger)(\varepsilon + a) \mid \psi_f(t)\rangle ,
\tag{16.64}
$$

where $n_b \equiv r_b\delta t$, r_b being the rate of atoms injected in the superposition of the lower states.

The second jump process corresponds to the atom entering the cavity in the upper state $\mid a\rangle$, and being detected in the superposition of lower states. Then, the state of the field after the measurement is

$$
\mid \psi(t+\tau)\rangle_f^{(a\to b,c)} = \frac{-ig\tau}{\sqrt{2}}(\varepsilon + a^\dagger) \mid \psi(t)\rangle_f .
\tag{16.65}
$$

The corresponding probability is given by

$$
\delta p_2 = n_a \frac{g^2\tau^2}{2} \langle \psi_f(t) \mid (\varepsilon + a)(\varepsilon + a^\dagger) \mid \psi_f(t)\rangle ,
\tag{16.66}
$$

where $n_a = r_a\delta t$ is the number of atoms that enter the cavity in state $\mid a\rangle$, during the time interval δt.

This analysis suggests that the quantum jump operators corresponding to these two processes should be, respectively, are

$$C_1 = \sqrt{\gamma(1+\langle n\rangle_{\text{th}})}(\varepsilon + a) \, ,$$
$$C_2 = \sqrt{\gamma\langle n\rangle_{\text{th}}}(\varepsilon + a^\dagger) \, , \tag{16.67}$$

where

$$\gamma \equiv (r_b - r_a)\frac{g^2\tau^2}{2} = \frac{r_b}{1+\langle n\rangle_{\text{th}}}\frac{g^2\tau^2}{2} = \frac{r_a}{\langle n\rangle_{\text{th}}}\frac{g^2\tau^2}{2} \, . \tag{16.68}$$

Formally, these jump operators are retrieved by rewriting the master equation (16.30) in the following equivalent form

$$\frac{d\rho_f}{dt} = (J_1 + J_2)\rho_f - \frac{\gamma(1+\langle n\rangle_{\text{th}})}{2}\left[(a^\dagger a + 2\varepsilon a + \varepsilon^2)\rho_f \right.$$
$$\left. + \rho_f(a^\dagger a + 2\varepsilon a^\dagger + \varepsilon^2)\right] - \frac{\gamma\langle n\rangle_{\text{th}}}{2}\left[(aa^\dagger + 2\varepsilon a^\dagger + \varepsilon^2)\rho_f \right.$$
$$\left. + \rho_f(aa^\dagger + 2\varepsilon a + \varepsilon^2)\right] \tag{16.69}$$

with

$$J_i = C_i\rho C_i^\dagger, \qquad i = 1,2 \tag{16.70}$$

being associated with the jumps, the operators C_i being now given by (16.67).

We derive now the Stochastic Schrödinger Equation that describes the present measurement scheme.

With the above jump operators, and using the expansion given by (16.38), we show in the Appendix E that the joint probability of getting m_1 and m_2 jumps corresponding respectively to the first and second processes described above is given by the following expression

$$P_{m_1,m_2}(\Delta t) = \left(\exp\mu_1\frac{(\mu_1)^{m_1}}{m_1!}\right)\left(\exp\mu_2\frac{(\mu_2)^{m_2}}{m_2!}\right)$$
$$\times Tr\left\{\exp\beta'\left[1 + \frac{1}{\varepsilon}(m_1a + m_2a^\dagger)\right]\rho \right.$$
$$\left. \times \left[1 + \frac{1}{\varepsilon}(m_1a^\dagger + m_2a)\right]\exp\beta^{\dagger\prime}\right\} \, , \tag{16.71}$$

where

$$\mu_1 = \gamma\Delta t\varepsilon^2(1+\langle n\rangle_{\text{th}}) \, , \tag{16.72}$$
$$\mu_2 = \gamma\Delta t\varepsilon^2(\langle n\rangle_{\text{th}}) \, ,$$
$$\beta' = -\frac{\gamma\Delta t}{2}\left[a^\dagger a(2\langle n\rangle_{\text{th}} + 1) + 2\varepsilon a(\langle n\rangle_{\text{th}} + 1) + 2\varepsilon a^\dagger\langle n\rangle_{\text{th}} + \langle n\rangle_{\text{th}})\right] \, .$$

From (16.71) and (16.72), one can readily find $\langle m_i\rangle$ and $\langle m_i^2\rangle$ for $i = 1, 2$. Up to order $\varepsilon^{-3/2}$, one finds

$$\langle m_i \rangle = \mu_i \left(1 + \frac{2}{3}\langle X_1 \rangle\right),$$

$$\langle m_i^2 \rangle = \mu_i , \tag{16.73}$$

with

$$X_1 \equiv \frac{a + a^\dagger}{2} . \tag{16.74}$$

Going back to the definition of $S(t)$, one may write

$$S(\Delta t) = N(\Delta t)\rho N^\dagger(\Delta t) , \tag{16.75}$$

in terms of a smooth evolution operator N that preserves pure states. This operator N is given by (16.40). with the jump operators C_m now given by (16.67). Now, if we consider a sequence of jumps (of the two kinds, in the present analysis) and evolutions, the state vector of the field will evolve according to

$$
\begin{aligned}
\mid \widetilde{\psi} \rangle_f(\Delta t) &= N(\Delta t - t_m)C_2 N(t_m - t_{m-1})C_1 ... \mid \psi \rangle_f(0) \\
&= N(\Delta t)C_2^{m_2}C_1^{m_1} \mid \psi \rangle_f(0) .
\end{aligned} \tag{16.76}
$$

In the last step, in deriving (16.76), we used that the commutators between the jump operators and the no-jump evolution produce overall phases, like in the Monte Carlo evolution given by (16.55).

Now, we consider m_i, $i = 1, 2$ as a couple of random variables with non-zero average, and write them as

$$m_i = \langle m_i \rangle + \Delta W_i \frac{\sigma_i}{\sqrt{\Delta t}} , \tag{16.77}$$

where the ΔW_i are two real and independent Wiener increments, with

$$\langle \Delta W_i^2 \rangle = \Delta t, \quad i = 1, 2 . \tag{16.78}$$

From (16.76, 16.77) and up to order $\varepsilon^{-3/2}$, we get the following Homodyne Stochastic Schrödinger Differential Equation (HSSDE)

$$
\begin{aligned}
\Delta^{m_1, m_2} \mid \widetilde{\psi} \rangle_f(\Delta t) &= \mid \widetilde{\psi} \rangle_f(\Delta t) - \mid \psi \rangle_f(0) \\
&= \Big\{ \Big[-\frac{\gamma}{2}(1 + \langle n \rangle_{\text{th}})a^\dagger a - \frac{\gamma}{2}(\langle n \rangle_{\text{th}})aa^\dagger \\
&\quad + 2\gamma\langle X_1 \rangle(a(1 + \langle n \rangle_{\text{th}}) \\
&\quad + a^\dagger \langle n \rangle_{\text{th}}) \Big] \Delta t + a^\dagger \sqrt{\gamma\langle n \rangle_{\text{th}}}\Delta W_2 \\
&\quad + a\sqrt{\gamma(1 + \langle n \rangle_{\text{th}})}\Delta W_1 \Big\} \mid \psi \rangle_f(0).
\end{aligned} \tag{16.79}
$$

At zero temperature, a typical quantum trajectory in this homodyne scheme is as follows:

a) If one starts from a coherent state, the quantum jumps will only produce a multiplicative factor in the wave function of the field, factor that can be absorbed in the normalization.

On the other hand, during the "no-click" periods, the nature of the coherent state is preserved, changing only the coherent amplitude, all the way to the vacuum.

This situation will be studied in Chap. 18, [19], in the context of the continuous measurement theory, applied to three-level atoms and two resonant fields, with the difference that there the number of detections is a predetermined quantity. However, the net result of the preservation of the coherent nature of the state of the field, along the trajectory, is the same([13]).

b) If we start with a Fock state, the quantum jumps will invariably produce a mixture of various Fock states, whereas the waiting or "no-click" periods will only generate numerical factors in front of those Fock states.

In the finite temperature case, the situation is more complex, since there will be also creation of photons that will disturb an initial coherent state and produce further mixtures in the Fock state case.

A more detailed analysis of these various cases is described in the next section, devoted to the numerical simulation.

16.6 Numerical Results and Localization

We present now the numerical calculations corresponding to the two equations associated with the two measurement schemes discussed above ([13]). We consider in these calculations the general case in which the temperature of the reservoir is taken as different from zero.

16.6.1 Quantum Jumps Evolution

We consider first an example in which the initial state of the system is a Fock state with three photons. We assume that the temperature of the reservoir corresponds to an average number of photons also equal to three. The corresponding evolutions is exhibited in the Fig. 16.5.

The state of the system remains a Fock state, with a number of photons that keep jumping between several values, in such a way that the average number of photons is equal to three. We have verified that the probability distribution for the number of photons is a Bose–Einstein distribution, as long as the observation is done over a sufficiently large time.

Figure 16.6 displays two different views of the evolution of the photon number population $|a_n|^2$ of an initial coherent state.

These figures clearly exhibit the dual nature of the system dynamics, with quantum jumps interspersed by non-unitary evolutions. In the displayed realization, the vacuum component of the state increases until the first quantum

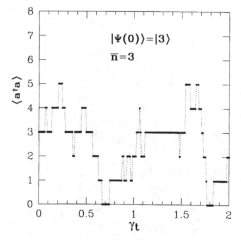

Fig. 16.5. Quantum jump for an initial Fock state with $n = 3$, the number of thermal photons being also equal to three. (After [13])

jump occurs. This jump corresponds to the addition of a thermal photon to the system, leading to the disappearance of the vacuum component. The second jump corresponds to the absorption of a photon from the cavity field, leading to the reappearance of the vacuum state. The combination of the non-unitary evolution with the quantum jumps finally leads to a Fock state, which under the action of the reservoir keeps jumping, in such a way that the photon number distribution over a long time span reproduces the Bose–Einstein distribution. This process is illustrated in Fig. 16.7, which displays the time evolution of the Q distribution for the field, defined for each realization as $Q = |\langle \alpha|\psi\rangle|^2/\pi$, where $|\alpha\rangle$ is a coherent state with amplitude α.

The initial Q distribution is a Gaussian, corresponding to the initial coherent state $|\alpha_0\rangle$, with $\alpha_0 = \sqrt{15/2}(1 + i)$. This distribution evolves into the one corresponding to a Fock state, with a number of photons that keep jumping around the thermal value $\langle n\rangle_{\text{th}} = 2$, in the same way as shown in Fig. 16.5.

16.6.2 Diffusion-like Evolution

We consider now the evolution corresponding to the situation displayed in Fig. 16.4. We consider as initial state the same coherent state as in Fig. 16.8, the reservoir temperature being also the same as before ($\langle n\rangle_{\text{th}} = 2$). In this case, the system evolves according to the homodyne stochastic Schrödinger equation given by (16.79). After some time, the Q function approaches a distorted Gaussian, with a mild amount of squeezing along the direction of the axis corresponding to the real part of α. The centre of this Gaussian keeps diffusing in phase space, so that after a long time span the time-averaged distribution coincides with the Bose–Einstein distribution.

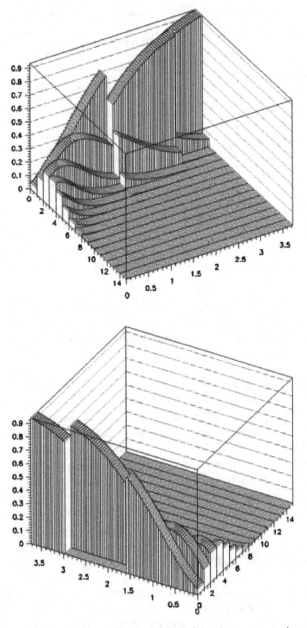

Fig. 16.6. Two views of the evolution of an initial coherent state (average photon number equal to three), in the quantum jump approach. The temperature of the reservoir corresponds to a number of thermal photons equal to 0.2. At around $\gamma t = 1.52$, a photon is absorbed by the cavity mode, whereas around $\gamma t = 3$, a photon is lost by the cavity. Before the first jump, the amplitude of the coherent state decreases exponentially. After some jumps, the state becomes a jumping Fock state. (After [13])

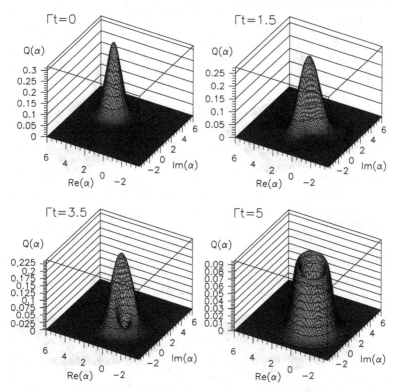

Fig. 16.7. Evolution of the Q function, for the quantum jump approach, and an initial coherent state, with $\alpha_0 = \sqrt{\frac{15}{2}}(1+i)$. The reservoir temperature corresponds to two thermal photons (average). The initial Gaussian, corresponding to a coherent state, evolves into a distribution corresponding to a jumping Fock state. (After [13])

16.6.3 Analytical Proof of Localization

For the quantum jump situation, it is actually possible to demonstrate that the system evolves towards a Fock state, for non-zero temperatures.

We first define two kind of variances, for an arbitrary operator O.

For the Hermitian case

$$\langle \Delta O^2 \rangle = \langle O^2 \rangle - \langle O \rangle^2, \tag{16.80}$$

and for the non-Hermitian case

$$\begin{aligned}
\mid \Delta O \mid^2 &= (O^\dagger - \langle O^\dagger \rangle)(O - \langle O \rangle) \\
&= O^\dagger O - \langle O^\dagger \rangle O - O^\dagger \langle O \rangle - \langle O^\dagger \rangle \langle O \rangle, \tag{16.81}
\end{aligned}$$

so that

$$\langle \mid \Delta O \mid^2 \rangle = \langle O^\dagger O \rangle - \langle O^\dagger \rangle \langle O \rangle. \tag{16.82}$$

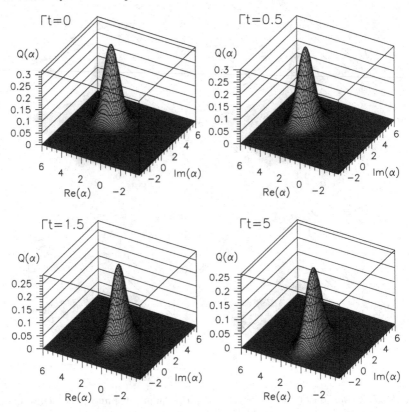

Fig. 16.8. Evolution of the Q function, for the quantum jump approach, and an initial coherent state, with $\alpha_0 = \sqrt{\frac{15}{2}}(1+i)$. The reservoir temperature corresponds to two thermal photons (average). The initial Gaussian, corresponding to a coherent state, evolves into a distorted Gaussian, whose centre diffuses in phase space. (After [13])

In particular, we are interested in two quantities

$$Q_1 = \langle |\, \Delta a\, |^2 \rangle, \tag{16.83}$$
$$Q_2 = \langle |\, \Delta n\, |^2 \rangle, \tag{16.84}$$

that measure the distance of the state from being a coherent or a Fock state, respectively.

We start with the quantum jump equation

$$
\begin{aligned}
|\, d\psi \rangle = & -\frac{i}{\hbar} H \,|\, \psi \rangle \delta t \\
& -\frac{1}{2} \sum_m (C_m^\dagger C_m - \langle C_m^\dagger \rangle \langle C_m \rangle) \,|\, \psi \rangle \delta t \\
& + \sum_m (\frac{C_m}{\sqrt{C_m^\dagger C_m}} - 1) \,|\, \psi \rangle \delta N_m,
\end{aligned}
\tag{16.85}
$$

with

$$M(\delta N_m) = \langle C_m^\dagger C_m \rangle \delta t, \tag{16.86}$$

$$\delta N_m \delta N_m = \delta N_n \delta_{n,m}. \tag{16.87}$$

We will calculate, using Ito's rule of calculus, Q_1 and Q_2 for $T = 0$ ($C = \sqrt{\Gamma}a$) and $T > 0$ ($C_1 = \sqrt{(\langle n \rangle_{th} + 1)\gamma}a$, $C_2 = \sqrt{\gamma \langle n \rangle_{th}}a^\dagger$).

We first develop some general expressions, which will be applied to calculate the above variances.

$$d\langle O \rangle = \langle d\psi \mid O \mid \psi \rangle + \langle \psi \mid O \mid d\psi \rangle + \langle d\psi \mid O \mid d\psi \rangle$$

$$= -\frac{i}{\hbar}\langle [O, H] \rangle \delta t - \frac{1}{2}\langle \{O, C^\dagger C\} \rangle \delta t + \langle O \rangle \langle C^\dagger C \rangle \delta t$$

$$+ \frac{(\langle C^\dagger O C \rangle - \langle C^\dagger C \rangle \langle O \rangle)}{\langle C^\dagger C \rangle} \delta N, \tag{16.88}$$

and similarly for the case in which several jump operators are present.

For the variance of a non-Hermitian operator, we have

$$d(\langle \mid \Delta O \mid^2 \rangle) = d\langle O^\dagger O \rangle - \langle O \rangle d\langle O^\dagger \rangle - \langle O^\dagger \rangle d\langle O \rangle$$

$$- d\langle O^\dagger \rangle d\langle O \rangle. \tag{16.89}$$

After a simple calculation, one gets

$$d(\langle \mid \Delta O \mid^2 \rangle) = -\frac{i}{\hbar}\langle [\mid \Delta O \mid^2, H] \rangle \delta t$$

$$- \frac{1}{2}\langle \{\mid \Delta O \mid^2, C^\dagger C\} \rangle \delta t$$

$$+ \langle \mid \Delta O \mid^2 \rangle \langle C^\dagger C \rangle \delta t - \langle \mid \Delta O \mid^2 \rangle \delta N$$

$$+ \frac{\langle C^\dagger O^\dagger O C \rangle \langle C^\dagger C \rangle - \langle C^\dagger O^\dagger C \rangle \langle C^\dagger O C \rangle}{\langle C^\dagger C \rangle \langle C^\dagger C \rangle} \delta N. \tag{16.90}$$

In the Hermitian case, on the other hand, we get

$$d(\langle \Delta O^2 \rangle) = -\frac{i}{\hbar}\langle [\Delta O^2, H] \rangle \delta t - \frac{1}{2}\langle \{\Delta O^2, C^\dagger C\} \rangle \delta t$$

$$+ \langle \Delta O^2 \rangle \langle C^\dagger C \rangle \delta t - \langle \Delta O^2 \rangle \delta N$$

$$+ \frac{\langle C^\dagger O^2 C \rangle \langle C^\dagger C \rangle - \langle C^\dagger O C \rangle \langle C^\dagger O C \rangle}{\langle C^\dagger C \rangle \langle C^\dagger C \rangle} \delta N. \tag{16.91}$$

Now we specialize to several cases

a) $T = 0, O = a; C = \sqrt{\gamma}a; H = \hbar\omega a^\dagger a$.

Using the above general expressions, we write

$$
\begin{aligned}
d(\langle |\, \Delta a\, |^2 \rangle) = [&-\gamma\langle a^\dagger aa^\dagger a\rangle - 2\gamma\langle a^\dagger a\rangle\langle a^\dagger\rangle\langle a\rangle \\
&+\gamma\langle a^\dagger a\rangle\langle a^\dagger a\rangle + \frac{\gamma}{2}\langle a^\dagger a^\dagger a\rangle\langle a\rangle + \frac{\gamma}{2}\langle a^\dagger aa^\dagger\rangle\langle a\rangle \\
&+\frac{\gamma}{2}\langle aa^\dagger a\rangle\langle a^\dagger\rangle + \frac{\gamma}{2}\langle a^\dagger aa\rangle\langle a^\dagger\rangle]\delta t \\
&-\langle a^\dagger a\rangle\delta N + \langle a^\dagger\rangle\langle a\rangle\delta N \\
&+\frac{\langle a^\dagger a^\dagger aa\rangle\langle a^\dagger a\rangle - \langle a^\dagger a^\dagger a\rangle\langle a^\dagger aa\rangle}{\langle a^\dagger a\rangle\langle a^\dagger a\rangle}\delta N.
\end{aligned}
\tag{16.92}
$$

The above results are strictly neither positive nor negative, so we cannot draw any conclusion, however, for the statistical mean

$$
\begin{aligned}
M\frac{d(\langle |\, \Delta a\, |^2\rangle)}{dt} = &-\gamma\langle |\, \Delta a\, |^{\langle 2\rangle}\rangle \\
&-\frac{\gamma\langle(\Delta a^\dagger)a^\dagger a\rangle\langle a^\dagger a\Delta a\rangle}{\langle a^\dagger a\rangle} \le 0,
\end{aligned}
\tag{16.93}
$$

so, in the mean, the system goes to a coherent state, which, in this case, is the vacuum.

b) $T > 0; O = a; C_1 = \sqrt{(\langle n\rangle_{\text{th}} + 1)\gamma}a, C_2 = \sqrt{\gamma\langle n\rangle_{\text{th}}}a^\dagger, H = \hbar\omega a^\dagger a$

The reader can easily verify, with a little algebra, that, in this case, neither $d(\langle |\, \Delta a\, |^2\rangle)$ or $Md(\langle |\, \Delta a\, |^2\rangle)$ are strictly negative.

c) $T > 0, O = a^\dagger a; C_1 = \sqrt{(\langle n\rangle_{\text{th}} + 1)\gamma}a, C_2 = \sqrt{\gamma\langle n\rangle_{\text{th}}}a^\dagger, H = \hbar\omega a^\dagger a$

In this case, as shown in Appendix F, $d\langle(\Delta a^\dagger a)^2\rangle$ is not negative, but $Md\langle(\Delta a^\dagger a)^2\rangle$ is

$$
\begin{aligned}
M\frac{d\langle(\Delta a^\dagger a)^2\rangle}{dt} = &-\gamma(\langle n\rangle_{\text{th}} + 1)\frac{\langle(\Delta a^\dagger a)a^\dagger a\rangle\langle a^\dagger a(\Delta a^\dagger a)\rangle}{\langle a^\dagger a\rangle} \\
&-\gamma(\langle n\rangle_{\text{th}})\frac{\langle(\Delta aa^\dagger)aa^\dagger\rangle\langle aa^\dagger(\Delta aa^\dagger)\rangle}{\langle aa^\dagger\rangle} \le 0.
\end{aligned}
\tag{16.94}
$$

So Q_2 is strictly diminishing in the mean, even at $T > 0$. Since Q_1 is not, the final state will not be the vacuum. It is easy to show from (16.94) that $M[d\langle(\Delta a^\dagger a)^2\rangle/dt] = 0$ if and only if the state of the system is a Fock state. This result shows therefore that any initial state approaches eventually a Fock state $|\, n\rangle$, with n fluctuating with a mean $\langle n\rangle_{\text{th}}$.

16.7 Conclusions

The dynamics of dissipative quantum systems is often described through Master Equations for the reduced density matrix, obtained by tracing the degrees of freedom of the reservoir and making the usual Markov–Born approximation.

However, in recent years, monitoring single quantum systems has become a reality in Paul traps, micromasers, etc., so new methods have been searched, through the evolution of state vectors.

The two methods discussed in this chapter are the Monte Carlo wavefunction or quantum jump method, involving random finite discontinuities, and the Stochastic Schrödinger equation characterized by a diffusive term added to the equation for the state vector, generally associated to a homodyne measurement .

We propose here a physical interpretation of the Quantum Jump approach and the Homodyne Stochastic Schrödinger Differential Equation, using as an example the damping of one field mode in a cavity at temperature T.

This field-damping mechanism can be modeled as an atomic beam, whose upper and lower population ratio is given by the Boltzmann factor, crossing a lossless cavity.

The quantum jump trajectory can be interpreted as a continuous monitoring of the outgoing two-level atoms, which are resonant with the cavity mode. We show both numerically and analytically that this continuous measurement on the reservoir leads, for each trajectory, to a pure Fock state. At a later time and due to the non-zero temperature, a thermal photon may produce a jump to a different Fock state, thus leading, as time goes on, to a series of Fock states, whose statistics will reproduce the thermal distribution.

In the case of the Homodyne Stochastic Schrödinger Differential Equation, the proposed damping mechanism consists of a three-level atomic beam, with a split ground state, whose population ratio of the upper and lower levels is given by the Boltzmann factor. The atoms cross again a lossless cavity, being resonant with the mode of the field under consideration. A second field is externally applied, with the same frequency but different polarization, so that each of the two fields connects the upper atomic state with a different lower sub-level. If this external field is a strong classical field, we show analytically that the Stochastic Schrödinger Equation describing the behaviour of the quantum field in the cavity corresponds precisely to the Homodyne Stochastic Schrödinger equation.

The beam is then continuously monitored as it exits the cavity. Numerically, one observes, for low temperatures, that the state of the field goes to a mildly squeezed state, centred around a value of α which diffuses in phase space, in such a way that the time-averaged distribution again reproduces the thermal state.

Recently, the monte carlo simulation has been used to describe spontaneous emission [20], two-photon processes [21]. Also, there has been serval publications related to quantum diffusion and localization [22, 23, 24, 25, 26, 27]

Problems

16.1. The Lindblad form of the master equation is not unique. Show that if we transform

$$D_m = T^\dagger C_m T \ ,$$

where T is a unitary transformation, the Master equation is unchanged. However, the nature of the jumps have changed, since now the system may jump to one of the states

$$D_m \mid \phi(t)\rangle \ ,$$

with a probability

$$\delta p_m^D = \langle \phi \mid D_m^\dagger D_m \mid \phi \rangle \delta t \ .$$

16.2. Derive a Linear Stochastic equation equivalent to the Master equation. Hint: See [11].

References

1. Carmichael, H.J.: An Open System Approach to Quantum Optics. Lecture Notes in Physics, Springer, Berlin (1993). See also Tian, L., Carmichael, H.J.: Phys. Rev. A, **46**, 6801 (1992)
2. Dalibard, J., Castin, Y., Mölmer, K.: Phys. Rev. Lett., **68**, 580 (1992)
3. Mölmer, K., Castin, Y., Dalibard, J.: J.O.S.A.A., **10**, 524 (1993)
4. Haake, F., Kolobov, M.J., Fabre, C., Giacobino, E., Reynaud, S.: Phys. Rev. Lett., **71**, 995 (1993)
5. Pellizzari, T., Ritsch, H.: Phys. Rev. Lett., **72**, 3973 (1994)
6. Gisin, N., Percival, I.C.: J. Phys. A, **25**, 5677 (1992)
7. Gisin, N., Percival, I.C.: J. Phys. A, **26**, 2233 (1993)
8. Gisin, N., Percival, I.C.: J. Phys. A, **26**, 2245 (1993)
9. Also, for earlier work, see Bohm, D., Bub, J.: Rev. Mod. Phys., **38**, 453 (1966); Pearl, D.: Phys. Rev. D, **13**, 857 (1976); Diosi, L.J.: J. Phys. A, **21**, 2885 (1988)
10. Wiseman, H.M., Milburn, G.J.: Phys. Rev. A, **47**, 642 (1993). See also: Wiseman, H.M., Milburn, G.J.: Phys. Rev. A, **47**, 1652 (1993)
11. Goetsch, P., Graham, R., Haake, F.: Quant. Semiclass. Opt., **8**, 1571 (1996)
12. Lindblad, G.: Commun. Math. Phys., **48**, 119 (1976)
13. Kist, T.B., Orszag, M., Brun, T., Davidovich, L.: Quant. Semiclass. Opt., **1**, 251 (1999)
14. Sargent III, M., Scully, M. O., Lamb, W. E., Jr.: Laser Physics. Addison-Wesley Publishing Company, New York (1974)
15. Davies, E.B., Srinivas, M.D.: Opt. Acta, **28**, 981 (1981)
16. Ogawa, T., Ueda, M., Imoto, N.: Phys. Rev. Lett., **66**, 1046 (1991); Phys. Rev. A, **43**, 6458 (1991); Ueda M., et al., Phys. Rev. A, **46**, 2859 (1992)

17. Brune M., Haroche, S., Raimond, J.M., Davidovich, L., and Zagury, N.: Phys. (Paris), **2**, 659 (1992)
18. Dicke, R.H.: Am. J. Phys., **49**, 925 (1981)
19. Agarwal, G.S., Graf, M., Orszag, M., Scully, M.O., Walther, H.: Phys. Rev. A, **49**, 4077 (1994)
20. Dum, R., Zoller, P., Ritsch, H.: Phys. Rev. A, **45**, 4879 (1992)
21. Garraway, B.M., Knight, P.L.: Phys. Rev. A, **49**, 1266(1994)
22. Percival, I.: Quantum State Diffusion Cambridge University Press, Cambridge (1998)
23. Gisin, N.: J. Mod. Opt., **40**, 2313 (1993)
24. Gisin, N., Percival, I.C.: In: *Experimental Metaphysics* Kluwe, Dordrecht (1997)
25. Percival, I.C., Struntz, W.T.: J. Phys. A, **31**, 1815 (1998)
26. Rigo, M., Mota-Hurtado, F., Alber, G., O´Mahony, P.F.: Phys. Rev. A, **55**, 1165 (1997)
27. Molmer, K., Castin, Y.: Quantum. Semiclass. Opt., **8**, 49 (1996)

Further Reading

Quantum Jumps

- Cohen-Tannoudji, C., Dalibard, J.: Europhys. Lett., **1**, 441 (1986)
- Zoller, P., Marte, M., Walls, D.F.: Phys. Rev. A, **35**, 198 (1987)

Quantum Jumps in Lasing Without Inversion

- Cohen-Tannoudji, C., Zambon, B., Arimondo, E.: Compte Rendu de l´Academie des Sciences, Serie II, **314**, 1139 (1992)
- Cohen-Tannoudji, C., Zambon, B., Arimondo, E.: Compte Rendu de l´Academie des Sciences, SerieII, **314**, 1293 (1992)
- Cohen-Tannoudji, C., Zambon, B., Arimondo, E.: JOSA.B, **10**, 2107 (1993)

Quantum Jumps in Laser Cooling

- Castin, Y., MØlmer, K.: Phys. Rev. Lett., **74**, 3772 (1995)
- Marte, M., Dum, R., Taieb, R., Lett, P.D., Zoller, P.: Phys. Rev. Lett., **71**, 1335 (1993)
- Marte, M., Dum, R., Taieb, R., Zoller, P.: Phys. Rev. A, **47**, 1378 (1993)

Quantum Jumps and Quantum Zeno Effect

- Mistra, B., Sudarshan, E.C.G., J. Math. Phys, **18**, 756 (1977)

Quantum Jumps and Resonance Fluorescence

- Power, W.L., J. Mod. Optics, **42**, 913 (1995)
- Power, W.L., Knight, P.L.: Phys. Rev. A, **53**, 1052 (1996)
- Beige, A., Hegerfeld, G.C.: Phys. Rev. A, **53**, 53 (1996)
- Beige, A., Hegerfeld, G.C.: Quant. Optics, **8**, 999 (1996), see also:
- Frerichs, V., Schenzle, A., Phys. Rev. A, **44**, 1962 (1991)
- Gardiner, C.W., Parkins, A.S., Zoller, P.: Phys. Rev. A, **46**, 4363 (1992)

Review Papers

- Zoller, P., Gardiner, C.W.: Lecture Notes of Les Houches Summer School on Quantum Fluctuations Elsevier, (1995)
- Knight, P.L., Garraway, B.M.: In: Opp., G.L., Barnett, S.M., Wilkinson, M. (eds) Quantum dynamics of simple systems, Proceedings of the 44-th scotish universities summer school in physics., Institute of Physics.Pub (1996)
- Molmer, K., Castin, Y.: Quant. Opt., **8**, 49 (1996)
- Plenio, M.B., Knight, P.L.: Los Alamos Preprint quant-ph\9702007

17. Atom Optics

This chapter is an introduction to Atom Optics.

Atom optics [1], in analogy with electron or neutron optics, deals with manipulation of matter waves. As such, they are characterized by a wavelength, which is the de Broglie wavelength $\lambda_{dB} \equiv \frac{h}{p}$ and the momentum $p = mv$.

The momentum of a typical atom is larger than that of a typical photon, absorbed or emitted by that atom.

There are several advantages of using atoms instead of photons for optical experiments.

1) Atoms have a non-zero rest mass, which is interesting when, for example we want to detect gravitational waves.
2) Atoms, as opposed to neutrons or electrons, are less susceptible to stray fields, but cannot be manipulated as easily as the charged particles.
3) Atoms have variable velocities, and as a result, one can in principle, control its de Broglie wavelength.
4) Atoms are easy and cheap to produce, as compared, for instance, with neutrons.
5) A very important aspect of the atom optics is that **atoms have internal structure, which can be probed and modified using light.**

17.1 Optical Elements

In general, a typical atom optics experiment consists of a source, optical elements and a detector.

Sources, in general, provide a well-collimated, monochromatic atomic beam.

Sources can be fast or slow. Among the sources of slow atoms are thermal expansions, with a Maxwellian velocity distribution. This type of sources are in general easy to operate and have a large flux. Within the fast type, when the reservoir pressure is increased, supersonic sources can be created, with a narrow longitudinal velocity distribution, typically a Gaussian distribution.

For some experiments, a better controlled atomic source is required. These slow beams are produced loading atoms from thermal sources into an atomic trap and then release them in a controlled fashion. As these atoms are extremely cold (30 μK), we may have a small velocity spread and large de Broglie wavelength.

A large number of different schemes have been used for the detection of atomic beams. Neutral atom can be collected using hot wire detectors, which absorb the atom briefly and ionize it. The ions are then detected as a current proportional to the incident atomic flux.

On the other hand, atoms in a metastable state can be detected by ionization followed by Auger neutralization.

Another versatile detection method in atom optics is laser-induced fluorescence.

In recent years, a large amount of effort has been put into developing optical elements, such as mirrors (for example, reflection of sodium atoms from evanescent waves [2, 3]), lenses, beam splitters etc.

An interesting note is that although the atom optics experiments belong to the decades of the 1980s and 1990s, the diffraction of atoms was actually performed as early as 1929, by Stern et al. [4].

17.2 Atomic Diffraction from an Optical Standing Wave

The diffraction of an atomic beam by an optical standing wave can be easily visualized as a follows: the standing wave acts as a phase grating for the atoms, splitting the incoming plane wave in a series of plane waves, separated by an integer number of photon momentum units $\hbar k$. (See Fig. 17.1)

In the present theoretical treatment, we will be using the Raman–Nath approximation that consists in neglecting the transverse Kinetic Energy of

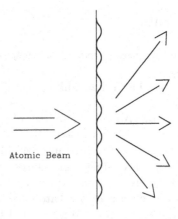

Fig. 17.1. Atomic beam crossing and being deflected by a standing wave

the atoms. We will also assume near resonance between the field and a couple of atomic levels. One of the first experimental observation of this effects was achieved by Moskowitz et al. [5], in 1983. An improved version was done by Martin et al. in 1987 [6]. In this experiment, low velocity sodium atoms $(2\frac{m}{s})$ are diffracted by a near resonant standing wave of light. There have been several other experiments [7, 8, 9, 10].

17.2.1 Theory

The deflection of atoms by standing waves has been, for some years, a subject of considerable interest, in particular, in connection with atomic interferometers.

When the standing wave is intense, classical fields are adequate. However, with the modern experimental tools, we may observe in the near future, diffraction from a few photons, where quantum effects are important. Also, spontaneous emission plays a role. In this particular treatment, we will assume the detuning to be sufficiently large, as to neglect this effect altogether.

Now, consider a collimated atomic beam traveling in the y direction (see Fig. 17.2).

The individual atoms are deflected by the photons. The field induces absorption and emission. As a result of this interaction, the atomic transverse momentum spreads. This behavior can be understood in terms of travelling waves. The atom absorbs a photon, thus gaining $\hbar k$ transverse momentum from one of the travelling waves and can emit a photon into the other travelling wave, thus **changing its own momentum in $2\hbar k$**. This is shown in the Fig. 17.3.

Actually, the above argument is only approximate, because there is a difference between a standing wave and two travelling waves. In principle, in two travelling waves, the momentum exchange between the field and the atom can only be finite, limited by the number of photons available. On the other

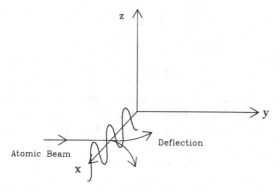

Fig. 17.2. Atomic diffraction by a standing wave light field

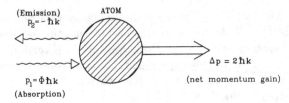

Fig. 17.3. Change of the atom's momentum after an absorption and emission event. As a result, the atom gains two units of the photon's momentum

hand, in the standing wave, there is an inseparable quantum unit, with zero average momentum. Here an important role is played by the fixed mirrors, which act as an infinite sink of momentum, and the amount of momentum exchanged between the standing wave and the atom is not limited anymore. An interesting discussion on this point is found in Shore et al. [11].

The Hamiltonian of the system is

$$H = \frac{p^2}{2m} + \hbar\omega a^\dagger a + \hbar\omega_o \sigma_z + \hbar g(a^\dagger \sigma_- + a\sigma_+)\cos kx , \qquad (17.1)$$

where the first term in (17.1) represents the atomic center of mass motion; the second and third terms are the free field and the internal energy of the two-level atom respectively. The last term represents the atom-standing wave interaction, with the $\sigma_z, \sigma_+, \sigma_-$ are the usual Pauli spin matrices.

The total kinetic energy can be split in a transverse and a longitudinal part

$$\frac{p^2}{2m} = \frac{p_x^2}{2m} + \frac{p_y^2}{2m} , \qquad (17.2)$$
$$p_y \gg p_x ,$$

and the transverse kinetic energy can be written as

$$\frac{p_x^2}{2m} = \hbar\left(\frac{\hbar k^2}{2m}\right) S_z^2 , \qquad (17.3)$$

where we defined

$$S_z \equiv \frac{p_x}{\hbar k} , \qquad (17.4)$$

where S_z is just the transverse momentum change, in units of a photon momentum. Normally (although this is not really necessary) one assumes that initially the transverse momentum is zero, and we define an $|m\rangle$ basis, with m integer, such that

$$S_z \,|\, m\rangle = m \,|\, m\rangle . \qquad (17.5)$$

We define also

$$S^{\pm} \equiv \exp(\pm ikx) \, . \tag{17.6}$$

It is simple to see that

$$[S^+, S^-] = 0 \, , \tag{17.7}$$

$$[S_z, S^{\pm}] = \pm S^{\pm} \, .$$

As we can see, the S^+, S^- operators are the step operators for the center of mass momentum of the atom, such that:

$$S^+ \mid m\rangle = \mid m + 1\rangle \, , \tag{17.8}$$

$$S^- \mid m\rangle = \mid m - 1\rangle \, .$$

The Hamiltonian can be written now as:

$$H = \hbar \epsilon_R S_z^2 + \frac{p_y^2}{2m} + \hbar \omega a^\dagger a + \frac{\hbar \omega_o \sigma_z}{2} + \frac{\hbar g}{2}(a^\dagger \sigma_- + a\sigma_+)(S^+ + S^-) \, . \tag{17.9}$$

Now, the quantity $\hbar \omega(a^\dagger a + \sigma_z)$ is a constant of motion, because it commutes with the total Hamiltonian. Also, the longitudinal kinetic energy is very large and can be considered approximately as a constant, so these three terms can be eliminated from the energy, getting

$$H = \hbar \epsilon_R S_z^2 + \frac{\hbar \Delta \sigma_z}{2} + \frac{\hbar g}{2}(a^\dagger \sigma_- + a\sigma_+)(S^+ + S^-) \, . \tag{17.10}$$

Now, we use the $\mid n\rangle \mid m\rangle$ basis, and because $\sigma_z + a^\dagger a$ is a constant of motion, we write the wave function as:

$$\mid \Psi\rangle = \sum_{m=-\infty}^{m=+\infty} \left[C_{nm}^+ \mid n\rangle \mid m\rangle \begin{pmatrix} 1 \\ 0 \end{pmatrix} + C_{n+1,m}^- \mid n+1\rangle \mid m\rangle \begin{pmatrix} 0 \\ 1 \end{pmatrix} \right] \, . \tag{17.11}$$

We now write the Schrödinger equation

$$i\hbar \frac{\partial \mid \Psi\rangle}{\partial t} = H \mid \Psi\rangle = \tag{17.12}$$

$$i\hbar \sum_{m=-\infty}^{m=+\infty} \left[\frac{dC_{nm}^+}{dt} \mid n\rangle \mid m\rangle \begin{pmatrix} 1 \\ 0 \end{pmatrix} + \frac{dC_{n+1,m}^-}{dt} \mid n+1\rangle \mid m\rangle \begin{pmatrix} 0 \\ 1 \end{pmatrix} \right]$$

$$= \frac{\hbar \Delta}{2} \sum_{m=-\infty}^{m=+\infty} \left[C_{nm}^+ \mid n\rangle \mid m\rangle \begin{pmatrix} 1 \\ 0 \end{pmatrix} - C_{n+1,m}^- \mid n+1\rangle \mid m\rangle \begin{pmatrix} 0 \\ 1 \end{pmatrix} \right]$$

$$+ \hbar \epsilon_R \sum_{m=-\infty}^{m=+\infty} m^2 \left[C_{nm}^+ \mid n\rangle \mid m\rangle \begin{pmatrix} 1 \\ 0 \end{pmatrix} + C_{n+1,m}^- \mid n+1\rangle \mid m\rangle \begin{pmatrix} 0 \\ 1 \end{pmatrix} \right]$$

$$+ \frac{\hbar g}{2} \sum_{m=-\infty}^{m=+\infty} \left[\begin{array}{l} C_{nm}^+ \sqrt{n+1} \mid n+1\rangle(\mid m+1\rangle + \mid m-1\rangle) \begin{pmatrix} 0 \\ 1 \end{pmatrix} \\ + C_{n+1,m}^- \sqrt{n+1} \mid n\rangle(\mid m+1\rangle + \mid m-1\rangle) \begin{pmatrix} 1 \\ 0 \end{pmatrix} \end{array} \right] \, .$$

Finally, by comparing $\begin{pmatrix} 1 \\ 0 \end{pmatrix}$ and $\begin{pmatrix} 0 \\ 1 \end{pmatrix}$ terms, we get

$$i\frac{dC_{nm}^+}{dt} = (\frac{\Delta}{2} + \epsilon_R m^2)C_{nm}^+ + \frac{g}{2}\sqrt{n+1}\left[C_{n+1,m-1}^- + C_{n+1,m+1}^-\right] , \quad (17.13)$$

$$i\frac{dC_{n+1,m}^-}{dt} = (-\frac{\Delta}{2} + \epsilon_R m^2)C_{n+1,m}^- + \frac{g}{2}\sqrt{n+1}\left[C_{n,m-1}^+ + C_{n,m+1}^+\right] .$$

Equations (17.13) are quite general and exact.

17.2.2 Particular Cases

a) $\Delta = 0, p_x^2 \approx 0$.

This is the Raman–Nath regime with no detuning. If we also assume that $n \gg 1$, so that $g\sqrt{n+1} = $ constant, then (17.13) reduce to

$$i\frac{dC_m}{dt} = \frac{g}{2}\sqrt{n+1}\left[C_{m-1} + C_{m+1}\right] . \quad (17.14)$$

The difference-differential equation for the Bessel functions is

$$2\frac{d}{dz}J_n(z) = J_{n-1}(z) - J_{n+1}(z) . \quad (17.15)$$

Thus, by direct comparison between (17.14) and (17.15) , we get

$$C_m = (-i)^m J_m(\Omega t) , \quad (17.16)$$

with $\Omega \equiv g\sqrt{n+1}$, or

$$P_{n,m}(t) = J_m^2\left(g\sqrt{n+1}t\right) . \quad (17.17)$$

If instead of having initially a Fock state, we have a general superposition of Fock states, distributed with probability W_n, we get

$$P_m(t) = \sum_{n=0}^{n=\infty} W_n J_m^2\left(g\sqrt{n+1}t\right) . \quad (17.18)$$

Equation (17.18) gives the probability distribution for the transverse momentum of an atom after an interaction time t with a standing wave light field for any given initial field distribution. The momentum distribution of the atom is a signature of the field this atom interacted with.

Figures 17.4 and 17.5 show $P_m(t)$ for $gt = 10$ and $gt = 100$, for a Fock state with n=9. We notice that the maximum of the Bessel function $J_m\left(g\sqrt{n+1}t\right)$ happens when $m \approx g\sqrt{n+1}t$ and then it sharply drops to zero, as we see from these figures [12].

We also show the momentum distribution for a squeezed state in Figs. 17.6 and 17.7.

We notice that the case $\Delta = 0$ is not very realistic, because spontaneous emission has not been considered [13, 14].

However, for large Δ, the model, again, is reasonable.

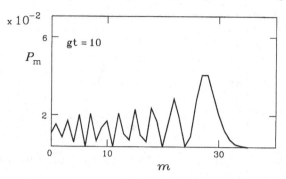

Fig. 17.4. Momentum distribution of atoms scattered off a Fock state $n = 9$, and $gt = 10$ (after [12])

b) $\Delta \gg g\sqrt{n}, \epsilon_R m^2$.

If we consider the Raman–Nath approximation $\left(\frac{p_x^2}{2m} \sim 0\right)$, it is not difficult to show that for a large detuning, one can write an approximate effective Hamiltonian:

$$V_{\text{eff}} = \frac{\hbar g^2}{\Delta}\sigma_z a^\dagger a \left(\frac{S^+ + S^-}{2}\right), \qquad (17.19)$$

where, this time $S^\pm \equiv \exp \pm 2ikx$, $S_z \equiv \frac{p_x}{2\hbar kx}$. Notice, that this definition is similar to that of the previous case, **except that the transitions are in steps of two photon momentum units.**

We also notice in this case, that both σ_z and $a^\dagger a$ are constants of motion, and therefore, there is only one index left $m = \frac{p_x}{2\hbar k}$, and

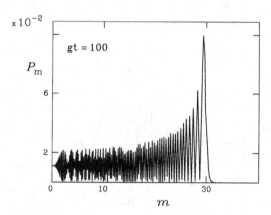

Fig. 17.5. Momentum distribution of atoms scattered of a Fock state with $n = 9$ for $gt = 100$ (after [12])

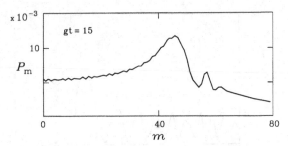

Fig. 17.6. Momentum distribution for the atoms interacting with a squeezed state with $\alpha^2 = 9$ and $r = 50$ and $gt = 15$ (after [12])

$$| \Psi \rangle = \sum_m C_m \mid m \rangle \begin{pmatrix} 1 \\ 0 \end{pmatrix} , \tag{17.20}$$

and Schrödinger's equation can be written as

$$2i\frac{dC_m}{d\tau} = \frac{\mid g \mid^2 n}{2\Delta}(C_{m-1} + C_{m+1}) , \tag{17.21}$$

where $\tau \equiv \frac{|g|^2 t n}{2\Delta}$ is a one-dimensional scaled time and the procedure to arrive to the (17.21) is the same to the one used in the previous section. The solution of (17.21) is

$$C_{n,m} = (-i)^m J_m \left(\frac{\mid g \mid^2 n}{2\Delta}\tau \right) , \tag{17.22}$$

where the formula (17.22) is only valid for **even** m.

In the Fig. 17.8, we show a comparison of the theoretical predictions presented here with the experimental observation.

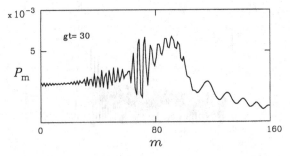

Fig. 17.7. Momentum distribution for the atoms interacting with a squeezed state with $\alpha^2 = 9$ and $r = 50$ and $gt = 30$ (after [12])

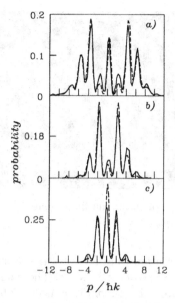

Fig. 17.8. Diffraction patterns for different velocities v_x, which can be done experimentally by tilting the standing wave with respect to the atomic beam. a) $v_x = 0.06\frac{m}{s}$, b) $v_x = 1.22\frac{m}{s}$, c) $v_x = 1.68\frac{m}{s}$. (Solid line experimental, dashed line theory, after [6])

17.3 Atomic Focusing

Lenses are important elements in many optical systems. In atomic optics, a possible application to lenses for atoms are found in the fields of microscopy and lithography.

The typical resolution of a diffraction limited microscope is determined by the wavelength. The instant success, for instance, of electron microscopy is that wavelengths much smaller than optical can be achieved. A particle with a kinetic energy E, has a de Broglie wavelength $\lambda_{dB} = \frac{h}{\sqrt{2mE}}$. Atomic resolution is possible, but in the Kev range, which may damage the sample.

On the other hand, for atoms, with much larger mass, the same resolution is possible but with **much lower energies.**

Another important application of lenses is in **atomic lithography**, where atoms are deposited into a surface with a very high resolution. An example of such an experiment is the one by McClelland et al. [15] (see Fig. 17.9).

17.3.1 The Model

We consider a collimated beam of atoms of mass m moving in the x-z plane along $x = \kappa > 0$. We will assume that the atoms are prepared in such a way that they can be modelled by two-level atoms [16, 17]. The interaction

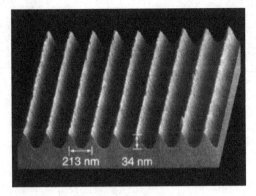

Fig. 17.9. A typical image of chromium lines created by atomic deposition. The image shows a 2 μm by 2 μm region (after [15])

region $-L<z<0$ (or interaction time $T = \frac{L}{v_z}$), the atoms cross an orthogonal one-mode standing light wave detuned by Δ. The longitudinal velocity v_z of the atoms along the beam axis (z-axis) is considered to be sufficiently large, so that the spatial dependence of the field in z can be replaced by a time dependence $t = \frac{z}{v_z}$.

The Hamiltonian is [18]

$$H = \frac{p^2}{2m} + \hbar\omega a^\dagger a + \frac{\hbar\omega_o}{2}\sigma_z + \hbar[a^\dagger\sigma_- g(x) + a\sigma_+ g(x)] \, . \qquad (17.23)$$

In (17.23), $g(x)$ is the space-dependent dipole of the atomic transition. In the limit of high detuning, and keeping the kinetic energy, we have

$$H_{\text{eff}} = \frac{p_x^2}{2m} + \frac{\hbar \mid g(x) \mid^2}{\Delta}\sigma_z a^\dagger a \, . \qquad (17.24)$$

In the above limit, again, $\sigma_z, a^\dagger a$ are constants, and we neglect the effects of the spontaneous emission.

We consider a relatively narrow atomic beam, so that the usual sinusoidal coupling constant can be expanded

$$\mid g(x) \mid^2 = \mid G\sin(\frac{2\pi x}{\lambda}) \mid^2 \cong \left(\frac{2\pi G}{\lambda}\right)^2 x^2 \, . \qquad (17.25)$$

We now expand the quantum mechanical states in the following basis

$$\sum_{j=-,+}\sum_{n=0}^{\infty}\int_{-\infty}^{+\infty} dx' \mid j, n, x', t\rangle\langle j, n, x', t \mid = 1 \, , \qquad (17.26)$$

$$\langle j, n, x', t \mid k, m, x'', t\rangle = \delta_{jk}\delta_{nm}\delta(x' - x'') \, .$$

We notice that if the atoms are in the lower state at $t = 0$, they remain there all the time. Thus, for $\Delta < 0$, we get a harmonic oscillator

$$ i\hbar \frac{d}{dt} \langle -, n, x', t \mid \Psi \rangle = \langle -, n, x', t \mid \frac{p_x^2}{2m} + \frac{m\omega_n^2 x^2}{2} \mid \Psi \rangle \,, \tag{17.27} $$

for each n, with a frequency

$$ \omega_n^2 = -\frac{2\hbar}{\Delta m} \left(\frac{2\pi G}{\lambda} \right)^2 n \,. \tag{17.28} $$

17.3.2 Initial Conditions and Solution

In a real experiment, the lateral velocity of the atoms, before entering the interaction region is, in general, not exactly zero. Also, the orthogonal alignment between the beam and the standing wave has certain deviation from orthogonality, etc. All these effects can be included in the fluctuation of the lateral position κ as well as the initial lateral momentum p. Therefore, we will model all these effects by a Gaussian

$$ \langle -, n, x', t = -T \mid \Psi \rangle = w_n \frac{1}{\sqrt{\sqrt{\pi}d}} \exp\left[-\frac{1}{2} \left(\frac{x' - \kappa}{d} \right)^2 - \frac{i}{\hbar} p \left(x' - \kappa \right) \right] \,, \tag{17.29} $$

where w_n is the initial field amplitude, assumed to be a pure state $\sum w_n \mid n \rangle$.

During the time the atom is going through the interaction region (from $t = -T$ to $t = 0$), the dynamics is that of a Gaussian wave packet in a harmonic potential [19]. After traversing the interaction region, the atoms become free again (from $t = 0$ to $t > 0$). By applying the free propagator to the previous result, one gets a time-dependant Gaussian

$$ \mid \langle -, n, x', t \mid \Psi \rangle \mid^2 = \frac{\mid w_n \mid^2}{\sqrt{\pi} D_n(t)} \exp\left\{ -\left[\frac{x' - x_n'(T) + v_n'(T)t}{D_n(t)} \right]^2 \right\} \,, \tag{17.30} $$

whose width $D_n(t)$ is

$$ D_n(t) = d \left\{ \begin{matrix} \left(\frac{\hbar}{d^2 m\omega_n} \right)^2 [\omega_n t \cos(\omega_n T) + \sin(\omega_n T)]^2 \\ + [\cos(\omega_n T) - \omega_n t \sin(\omega_n T)]^2 \end{matrix} \right\}^{\frac{1}{2}} \,. \tag{17.31} $$

The definitions of $x_n'(T), v_n'(T)$ are

$$ x'(T) = \kappa \frac{\cos(\omega T + \phi)}{\cos(\phi)} \,, \tag{17.32} $$

$$ p'(T) = p \frac{\sin(\omega T + \phi)}{\sin(\phi)} \,, $$

$$ \tan\phi = \frac{p}{m\omega\kappa} \,. $$

The physical picture emerging from these results is quite simple. A classical mass subjected to the harmonic potential, after the interaction time T has as solutions $x'(T), p'(T)$ with initial conditions $x'(0) = \kappa$, $p'(0) = p$.

17.3.3 Quantum and Classical Foci

From (17.30), we see that the classical trajectory of the n-th wave packet is

$$x' = x'_n(T) - v'_n(T)t , \qquad (17.33)$$

and in the paraxial approximation ($\phi \ll 1$), the incoming atoms all intersect at

$$\begin{pmatrix} x_n^{cf} \\ z_n^{cf} \end{pmatrix} = \begin{pmatrix} -\frac{p}{m\omega_n}\csc(\omega_n T) \\ \frac{v_z}{\omega_n}\cot(\omega_n T) \end{pmatrix} , \qquad (17.34)$$

which we call the **classical focus**.

On the other hand, from (17.31), one can write D_n as

$$D_n = d \left[\left(\frac{\cos^2 \varphi_n}{\chi^2 \varphi_n^2} + \sin^2 \varphi_n \right)(\omega_n t - \omega_n t_n^{qf})^2 + \frac{1}{\cos^2 \varphi_n + \chi^2 \varphi_n^2 \sin^2 \varphi_n} \right]^{\frac{1}{2}} , \qquad (17.35)$$

with

$$\omega_n t_n^{qf} \equiv \sin \varphi_n \cos \varphi_n \frac{\chi^2 \varphi_n^2 - 1}{\cos \varphi_n^2 + \chi^2 \varphi_n^2 \sin^2 \varphi_n} , \qquad (17.36)$$

$$\chi \equiv \frac{md^2}{\hbar T}, \varphi_n = \omega_n T .$$

From (17.35) and (17.36), it is clear that the beam converges at the position

$$z_n^{qf} = v_z t_n^{qf} , \qquad (17.37)$$

and we define the quantum focus at this position. The value of D_n at the quantum focus becomes

$$D(t_n^{qf}) = \frac{d}{\sqrt{\cos \varphi_n^2 + \chi^2 \varphi_n^2 \sin^2 \varphi_n}} . \qquad (17.38)$$

If one is restricted to photon numbers $n > \frac{-\Delta\hbar}{2d^4 m}\left(\frac{\lambda}{2\pi G}\right)^2$, then $\chi^2 \varphi_n^2 > 1$, and we have focusing, in the sense that $D(t_n^{qf}) < d$.

17.3.4 Thin Versus Thick Lenses

In many experimental situations, the particle trajectories are only slightly deflected, which in the present notation, means that $\varphi_n \ll 1$, for all relevant n values. This is the **thin lens condition**. According to (17.34), different rays coming with the same p but different κ all intersect at

$$\begin{pmatrix} x_n^{cf} \\ z_n^{cf} \end{pmatrix} \approx \frac{1}{\varphi_n^2} \begin{pmatrix} -\frac{p}{m} \\ v_z \end{pmatrix} T , \qquad (17.39)$$

which implies that the focal length z_n^{cf} goes as $\frac{1}{n}$.

If we assume a 'classical' light, that is a coherent state

$$| w_n |^2 = \frac{\langle a^\dagger a \rangle^n}{n!} \exp -\langle a^\dagger a \rangle ,$$ (17.40)

with large $\langle a^\dagger a \rangle$, the single classical focus corresponding to each n will be distributed over a distance characterized by $\sqrt{\langle a^\dagger a \rangle}$, there is the focal spot along z will have a size of the order of:

$$\frac{-\Delta m v_z \lambda^2}{8\hbar T \pi^2 G^2 \langle a^\dagger a \rangle^{\frac{3}{2}}}$$

centered at

$$z^{cf}_{\langle n \rangle} = \frac{-\Delta m v_z \lambda^2}{8\hbar T \pi^2 G \langle a^\dagger a \rangle}.$$

The spot width in x will be

$$D_{\langle n \rangle}(t^{cf}_{\langle n \rangle}) = \frac{-\Delta \lambda^2}{8\pi^2 G^2 dT \langle a^\dagger a \rangle} ,$$ (17.41)

which does not contain $\hbar$ nor the atomic mass m.

The thin lens is convergent. However, if $\varphi_n > \frac{\pi}{2}$, the classical focus becomes negative, and we speak of a divergent lens.

17.3.5 The Quantum Focal Curve

If we introduce (17.36) in (17.33), we have the quantum focus

$$\begin{pmatrix} x^{qf}_n \\ z^{qf}_n \end{pmatrix} = \begin{pmatrix} \kappa \xi_n \\ \chi L \zeta_n \end{pmatrix} ,$$ (17.42)

where x and z have been parametrized

$$\xi_n = \kappa \frac{c_n}{c_n^2 + l_n^2 s_n^2} ,$$ (17.43)

$$\zeta_n = v_z \frac{c_n s_n}{\omega_n} \frac{l_n^2 - 1}{c_n^2 + l_n^2 s_n^2} ,$$

$$c_n = \cos \omega_n T, s_n = \sin \omega_n T, l_n = \frac{d^2 m \omega_n}{\hbar} .$$

Equation (17.42) is the parametrized focal curve. For large χ, it approaches, with growing n to

$$\left(| \xi_n | - \frac{1}{2} \right)^2 + \zeta_n^2 = 1 .$$ (17.44)

Equation (17.44) describes a double circular lobe (see Fig. 17.10).

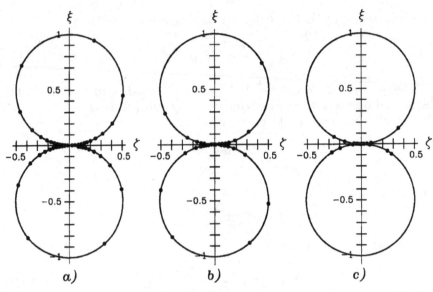

Fig. 17.10. Quantum focal distribution for $\chi = 0.4(a), \chi = 1(b), \chi = 4(c)$. For large χ, all foci concentrate close to the origin (after[18])

Let us assume $\chi \varphi_n \gtrsim 1$, then one can easily check that:

$$| t_n^{cf} | \geq | t_n^{qf} |,$$

and the classical and quantum foci become real, for the same value of n. As they both lie in the trajectory line 17.33, it is evident from the geometry that they should essentially coincide in position, when close to the z-axis.

In this case, and if: $\chi^2 \varphi_n^2 \sin^2 \varphi_n \succeq | \cos \varphi_n |$

$$| \chi_n^{qf} | \ll \kappa . \tag{17.45}$$

Classical and quantum foci will then be equally distributed, at

$$z_n^{qf} \approx v_z T \frac{\cos \varphi_n}{\varphi_n \sin \varphi_n} \approx z_n^{cf},$$

and we will introduce the subscript f to refer to both.

In the case of thin lenses, and more generally, whenever $| \varphi_n - m\pi | << 1$ hods, for a given $m = 0, 1, 2...$, and if (17.45) is satisfied, the focal position becomes

$$z_n^f \approx \frac{L}{\varphi_n(\varphi_n - m\pi)} , \tag{17.46}$$

and

$$z_n^f - z_{n+1}^f \approx \frac{L}{2n} . \tag{17.47}$$

17.3.6 Aberrations

a) Chromatic

The chromatic aberration arises from the velocity spread in the incident beam. In other words, instead of having a plane wave with velocity v_z, we may assume an incoherent superposition of plane waves with different velocities. This implies different interaction times of the atoms. If we consider a velocity shift $v_z \rightarrow v_z + \delta v_z$, it will produce a shift $z_n \rightarrow z_n + \delta z_n$

$$\delta z_n \approx \delta v_z \frac{dz_n}{dv_z} = \delta v_z T \left[\frac{2z_n}{L} + \frac{1}{\sin^2 \varphi_n} \right] , \qquad (17.48)$$

for the n-th focus.

Unfortunately, the quantity in the parenthesis is positive and the lens cannot be made achromatic.

b) Isotopic

When the atomic species used in the beam consists of various isotopes, the exact focal length will vary with the mass. For a mass difference between isotopes δm, the focal shift is:

$$\delta z_n^f \approx \delta m \frac{dz_n^f}{dm} = \frac{L \delta m}{2m} \left(\frac{z_n^f}{L} + \frac{1}{\sin^2 \varphi_n} \right) . \qquad (17.49)$$

The isotopic aberration could be used to create lines made of different isotopes.

c) Spherical

If we want to take into account the anharmonicity of the potential , we have to consider quartic terms or the full sinusoidal potential. Both problems are rather difficult. The sinusoidal potential leads to Mathieu functions.

Problems

17.1. Show that

$$S^+ \mid m \rangle = \mid m + 1 \rangle,$$
$$S^- \mid m \rangle = \mid m - 1 \rangle,$$

where

$$S^\pm = \exp(\pm ikx).$$

17.2. Show that for the Hamiltonian given by (17.9), the quantity

$$C = \hbar \omega (a^\dagger a + \sigma_z),$$

is a constant of motion.

17.3. Justify the fact that neglecting spontaneous emission in atomic diffraction is only reasonable if $\Delta \gg g\sqrt{n}$.

17.4. Prove (17.30, 17.31, 17.32).

References

1. Baldwin, K.H.: Aust. J. Phys., **49**, 855 (1996)
2. Hajnal, J.V., Baldwin, K.G.H., Fisk, P.T.H., Bachor, H.A., Opat, G.I.: Opt. Comm., **73**, 331 (1989)
3. Baldwin, K.G.H., Hajnal, J.V., Fisk, P.T.H., Bachor, H.A., Opat, G.I.: J. Mod. Op., **37**, 1839 (1990)
4. Stern, O.: Naturwissensch, **17**, 391 (1929)
5. Moskowitz, P.E., Gould, P.L., Atlas, S.R., Pritchard, D.E.: Phys. Rev. Lett., **51**, 370 (1983)
6. Martin, P.J., Gould, P.L., Oldaker, B.G., Miklich, A.H., Pritchard, D.E.: Phys. Rev. A, **36**, 2495 (1987)
7. Tanguy, C., Reynaud, S., Cohen-Tannoudji, C., J. Phys. B., **17**, 4623 (1984)
8. Gould, P.L., Ruff, G.A., Pritchard, D.E.: Phys. Rev. Lett., **56**, 827 (1986)
9. Martin, P.J., Oldaker, B.G., Miklich, A.H., Pritchard, D.E.: Phys. Rev. A, **60**, 515 (1988)
10. Gould, P.L., Martin, P.G., Ruff, G.A., Stoner, R.E., Pique, L., Pritchard, D.E.: Phys. Rev. A, **43**, 585 (1991)
11. Shore, B., Meystre, P., Stenholm, S.: J.O.S.A.B, **8**, 903 (1991)
12. Akulin, V.M., Fam, Le Kien, Schleich, W.P.: Phys. Rev. A, **44**, R1642 (1991)
13. Tan, S.M., Walls, D.F.: Phys. Rev. A, **44**, R2779 (1991)
14. Wilkens, M., Schumacher, E., Meystre, P.: Opt. Comm., **86**, 34 (1991)
15. McClelland, J.J., Scholten, R.E., Palm, E.C., Celotta, R.J.: Science, **262**, 877 (1993)
16. Averbukh, I.Sh., Akulin, V.M., Schleich, W.P.: Phys. Rev. Lett., **72**, 437 (1994)
17. Mayr, E., Krähmer, D., Herkommer, A.M., Akulin, V.M., Schleich, W.P.: Acta Polonica, Proceedings of Quantum Optics III. Acta phys. Pol, **86**, 81 (1994)
18. Rohwedder, B., Orszag, M.: Phys. Rev. A, **54**, 5076 (1996)
19. Saxon, D.S.: Elementary Quantum Mechanics. Holden-Day, San Francisco (1968). Gisin, N.: J. Mod. Opt., **40**, 2313 (1993)
20. McClelland, J.J., Gupta, R., Jabbour, Z.J., Celotta, R.J., Aust J.: Phys. **49**, 555 (1996)

18. Measurements, Quantum Limits and All That

In this Chapter, we study the various quantum limits. We also deal with quantum non-demolition (QND) as well as continuous measurements.

18.1 Quantum Standard Limit

18.1.1 Quantum Standard Limit for a Free Particle

We, in this section, study the motion of a free particle, or even better, we monitor its position during a time τ, confining ourselves with only two measurements. [1, 2].

We assume, at $t = 0$, that we measure a free particle's position with an error $(\Delta x_{\text{measure}})_1$, which, according to the uncertainty principle, produces a perturbation in the momentum

$$(\Delta p)_{\text{pert}} \geq \frac{\hbar}{2(\Delta x_{\text{measure}})_1} . \tag{18.1}$$

Now, a second measurement is performed at time $t = \tau$, and the momentum perturbation will produce an additional uncertainty in the position

$$(\Delta x)_{\text{add}} = \frac{(\Delta p)_{\text{pert}}\tau}{m} \geq \frac{\hbar\tau}{2m(\Delta x_{\text{measure}})_1} . \tag{18.2}$$

Now, if these contributions superpose incoherently, and we assume an error in the second measurement $(\Delta x_{\text{measure}})_2$, then

$$(\Delta x(\tau))^2 = (\Delta x_{\text{measure}})_1^2 + (\Delta x_{\text{measure}})_2^2 + (\Delta x_{\text{add}})^2 , \tag{18.3}$$

and replacing (18.2) in (18.3), we can minimize the above expression, making

$$\frac{\mathrm{d}(\Delta x(\tau))^2}{\mathrm{d}(\Delta x_{\text{measure}})_1^2} = 0 . \tag{18.4}$$

The optimum happens for: $(\Delta x_{\text{measure}})_1 = \sqrt{\frac{\hbar\tau}{2m}}$, thus giving an optimum for $(\Delta x(\tau))$ (if $(\Delta x_{\text{measure}})_2^2 = 0$)

$$(\Delta x(\tau)) = \Delta x_{\mathrm{SQL}} = \sqrt{\frac{\hbar\tau}{m}} \ . \tag{18.5}$$

Also, we get for Δp

$$\Delta p_{\mathrm{SQL}} = \sqrt{\frac{m\hbar}{2\tau}} \ . \tag{18.6}$$

On the other hand, if one could prepare the state of the system, then we repeat the argument

$$x(\tau) = x(0) + \frac{p(0)\tau}{m} \ , \tag{18.7}$$

whose variance is

$$(\Delta x(\tau))^2 = (\Delta x(0))^2 + \frac{(\Delta p(0))^2 \tau^2}{m^2} + \langle \Delta x(0)\Delta p(0) + \Delta p(0)\Delta x(0)\rangle \frac{\tau}{m} \ , \tag{18.8}$$

and if we prepared the system in a "contractive state", such that

$$\langle \Delta x(0)\Delta p(0) + \Delta p(0)\Delta x(0)\rangle < 0 \ , \tag{18.9}$$

then one could beat the standard quantum limit. [3]

18.1.2 Standard Quantum Limit for an Oscillator

Consider a harmonic oscillator with a mass m and angular frequency ω, with

$$H_0 = \frac{p^2}{2m} + \frac{1}{2}m\omega^2 x^2 \ , \tag{18.10}$$

and a number of quanta

$$N = \frac{H_0}{\hbar\omega} - \frac{1}{2} \ . \tag{18.11}$$

As the harmonic oscillator is quantum mechanical, one has

$$[x, p] = i\hbar \ , \tag{18.12}$$

$$\Delta x \Delta p \geq \frac{\hbar}{2} \ ,$$

when such an oscillator is in the ground state, the variance of x and p has the minimum allowed by the uncertainty principle

$$\Delta x = \frac{\Delta p}{m\omega} = \sqrt{\frac{\hbar}{2m\omega}} \ , \tag{18.13}$$

which is the half width of the oscillators ground state. Even more generally, in a coherent state, one has the same uncertainties.

18.1.3 Thermal Effects

Now, we may think that these quantum limits are not very relevant if one has large classical thermal fluctuations.

When the measurement time τ is larger than the oscillators' relaxation time τ^*

$$\tau > \tau^* , \tag{18.14}$$

then the criteria to neglect thermal fluctuations is

$$k_B T \leq \frac{\hbar\omega}{2} . \tag{18.15}$$

The above criteria could correspond to extremely low temperatures.

For short measurements ($\tau < \tau^*$), the condition is less stringent on the temperature. As a matter of fact, the above condition is not valid because the energy exchange with the oscillator, on the average, is only a fraction $\frac{\tau}{\tau^*}k_B T$.

One can show that the thermally induced fluctuations for $\tau << \tau^*$ are

$$\Delta x_{\text{thermal}} = \sqrt{\frac{k_B T \tau}{m\omega^2 \tau^*}} . \tag{18.16}$$

The direct comparison of the above limit with the quantum standard limit for the oscillator gives us the criteria to decide whether a system is classical or quantum mechanical. The system behaves quantum mechanically, in the case of short measurements if

$$k_B T \frac{\tau}{\tau^*} << \frac{\hbar\omega}{2} . \tag{18.17}$$

To verify the above discussion, we borrow a result from the chapter on the damped harmonic oscillator

$$\Delta x \Delta p = \frac{\hbar}{2} \{1 + 2\langle n \rangle_{\text{th}} [1 - \exp(-\gamma\tau)]\} , \tag{18.18}$$

and defining $\gamma = \tau^{*-1}$, we have two regimes
a) $\frac{\tau}{\tau^*} >> 1$, then

$$\Delta x \Delta p = \frac{\hbar}{2}(1 + 2\langle n \rangle_{\text{th}}) \tag{18.19}$$

and, for the oscillator to be in a quantum regime, the 1 in the above equation must be much larger than the $\langle n \rangle_{\text{th}}$

$$1 >> 2\langle n \rangle_{\text{th}} = 2\frac{1}{\exp\frac{\hbar\omega}{k_B T} - 1} , \tag{18.20}$$

which implies

$$\frac{\hbar\omega}{2} \gg k_B T .$$

(18.21)

b) $\frac{\tau}{\tau^*} \ll 1$, then

$$\Delta x \Delta p = \frac{\hbar}{2} \left[1 + \left(2\langle n \rangle_{\mathrm{th}} \frac{\tau}{\tau^*} \right) \right] .$$

(18.22)

The quantity in the square parenthesis in (18.22) should be much less than one, for the system behave quantum mechanically. That condition implies

$$k_B T \frac{\tau}{\tau^*} \ll \frac{\hbar\omega}{2} ,$$

(18.23)

proving the above discussion.

18.2 Quantum Non-Demolition (QND) Measurements

18.2.1 The Free System

The problem of measuring classical signals [2] that are very weakly coupled to detectors was originally for interest in the research of gravitational wave detection. In the case of large bar detectors, with masses of the order of 10 tons, the gravitational waves interact so weakly with these detectors that produce a typical displacement of the order of 10^{-19}cm. With such a small signal, the actual position measurement will introduce a momentum uncertainty that will feed back in extra position uncertainty, as discussed in the last section, leading to a quantum standard limit

$$\Delta x_{\mathrm{SQL}} = \sqrt{\frac{\hbar\tau}{m}} \approx 5 \times 10^{-19} cm ,$$

(18.24)

for a typical gravitational wave period of 10^{-3} sec.

As we can see, the minimum uncertainty introduced in the first measurement has made it impossible to detect with certainty whether the gravitational wave has acted on the detector.

On the other hand, if one tries to detect p rather than x, something non-trivial from the experimental point of view, the error in p produces an added uncertainty in x, according to the uncertainty principle. However, this added noise will not feed back to p, as for a free particle, p is a constant of motion. Therefore, a second measurement of p can be made, with the same accuracy as the first one.

This is an example of a "Quantum Non-demolition measurement" (QND), to avoid the "back action" of the measurement on the observed variable.

In the optical domain, QND experiments have been realized [4, 5, 6, 7] based on the Kerr coupling of a signal field to be measured with a probe field, whose phase is changed linearly with the number of photons of the signal.

Here, we will discuss another proposal [8], where photons are stored in a microwave resonant cavity, and they are detected by measuring the phase shift of the electric dipoles of non-resonant Rydberg atoms crossing such cavity.

In this way, weak fields (with a small photon number) can be monitored continuously, with no back action on the number of photons. In this example, the detector is the atomic beam crossing the microwave cavity.

As we already mentioned, the original research on QND measurements was triggered by the desire of monitoring a very weak force, acting on a harmonic oscillator, with an accuracy better than the quantum standard limit.

Braginsky et al. [9] proposed what they baptized as the 'Quantum Non-demolition measurement', where one monitors an observable of the oscillator, which has to be measured many times, with each measurement being completely determined by the result of an initial precise measurement. Such an observable we call 'Quantum Non-demolition observable'.

To fix some ideas, let us assume that we have a system described by a Hamiltonian H_S and we want to measure an observable (Hermitian operator) $\mathbf{A}_S$ that could be, for example, the number of quanta of the harmonic oscillator, to monitor a classical force produced, in this particular example, by the gravitational wave.

In the optical case, $\mathbf{A}_S$ could be the photon number of a field. The measurement of $\mathbf{A}_S$, however, is not made directly, but through the detection of a probe observable $\mathbf{A}_P$, conveniently coupled to the system, during the measurement.

The above definition of a QND observable can be used to derive its condition.

For the moment, we neglect the interaction to the probe, or measuring apparatus.

Now we assume a sequence of measurements of $\mathbf{A}_S$, assuming that we have no control over the state of the system.

Also, we denote by $\mid A_S, \alpha_S\rangle$ the normalized eigenstate of $\mathbf{A}_S(\mathbf{t_0})$, with

$$\mathbf{A}_S \mid A_S, \alpha_S\rangle = A_S \mid A_S, \alpha_S\rangle , \qquad (18.25)$$

where α_S labels the degeneracy index.

As a result of a first measurement, one gets the eigenvalue A_0 of $\mathbf{A}_S(t_0)$, and the eigenstate, after this measurement is

$$\mid \psi(t_0)\rangle = \sum_\alpha C_\alpha \mid A_0, \alpha\rangle . \qquad (18.26)$$

In the interval between the first and the second measurement, in the Heisenberg picture $\mid \psi(t)\rangle = \mid \psi(t_0)\rangle$, that is the state does not change.

If a second measurement at $t = t_1$ is to produce a predictable result, it means that all the states $\mid A_0, \alpha\rangle$ must be eigenstates of $\mathbf{A}_S(\mathbf{t}_1)$, but in general, with different eigenvalue

$$\mathbf{A}_S(t_1) \mid A_0, \alpha\rangle = f_1(A_0) \mid A_0, \alpha\rangle . \tag{18.27}$$

As the above result is true for all the eigenvalues of $\mathbf{A}_S(t_0)$, we must have

$$\mathbf{A}_S(t_1) = f_1(\mathbf{A}_S(t_0)) . \tag{18.28}$$

For a QND measurement at times $t = t_0, t_1, ..., t_n$, one must have

$$\mathbf{A}_S(t_k) = f_k(\mathbf{A}_S(t_0)), k = 1, 2, ..n , \tag{18.29}$$

where f_k is a real function.

For a continuous measurement, or at arbitrary times, one writes

$$\mathbf{A}_S(t) = f(\mathbf{A}_S(t_0); t, t_0) . \tag{18.30}$$

The above condition is satisfied by a constant of motion, which in the absence of interactions satisfies

$$\frac{\mathrm{d}\mathbf{A}_S(t)}{\mathrm{d}t} = -\frac{i}{\hbar} [\mathbf{A}_S(t), H_S] + \frac{\partial \mathbf{A}_S(t)}{\partial t} . \tag{18.31}$$

In a harmonic oscillator, $\mathbf{x}$ and $\mathbf{p}$ are not QND observables; however, $\mathbf{N}$ is conserved. In the case of a free particle, $\mathbf{x}$ is not a QND variable but p is.

Another way of expressing the QND condition (18.30) is

$$[\mathbf{A}_S(t), \mathbf{A}_S(t')] = 0 . \tag{18.32}$$

18.2.2 Monitoring a Classical Force

Once we have defined a continuous QND observable , and a QND measurement, satisfying the condition given in (18.32), we consider its application to monitoring a classical force F(t). The procedure is the following: we make a sequence of QND measurements and detect the changes the classical force produced in the precisely predictable values of the QND variable, in the absence of the force.

We would like to go even further and actually monitor the time dependence of the force with arbitrary accuracy, satisfying the following conditions: [2]

a) The measuring apparatus and its coupling to the system can produce arbitrarily precise measurements.

b) The (k+1)-th measurement at time t_k must be uniquely determined as a result of an initial measurement at time t_0 and the history of F(t) between t_0 and t_k.

For the b) condition to be satisfied, one must have

$$\mathbf{A}(t) = f(\mathbf{A}(t_0); F(t'); t, t_0) , \tag{18.33}$$
$$t_0 < t' < t .$$

In the above condition, $A(t)$ is a Heisenberg operator evolving with a Hamiltonian that includes a coupling term to the apparatus.

c) From the history of the measured values of $A(t)$, one should in principle, derive F(t). This implies that the above condition (18.33) should be invertible.

Now, we concentrate in the measuring apparatus and its interaction with the system.

18.2.3 Effect of the Measuring Apparatus or Probe

We assume that we want to measure a quantum observable A_S of the system S by detecting it indirectly, that is by measuring the change in an observable of a probe A_P, during a time interval T.

We notice that here we talk about measuring a quantum observable A_S, thus generalizing the argument of monitoring a classical force of the previous section.

The Hamiltonian of the system coupled to the probe is

$$H = H_S + H_P + H_I , \tag{18.34}$$

where H_S and H_P are the free terms for the system and probe respectively, and H_I is their interaction.

To do this measurement, the interaction Hamiltonian must depend on A_S, that is

$$\frac{\partial H_I}{\partial A_S} \neq 0 . \tag{18.35}$$

On the other hand, if we are doing this measurement indirectly through another quantum observable A_P, and if furthermore, we want to monitor it with several measurements, then A_P must respond to Heisenberg´s equation, as a dynamic variable, or in other words, we require that

$$[A_P, H_I] \neq 0 . \tag{18.36}$$

Finally, and most importantly, we have the original QND restriction given by (18.32), which, in this model, implies

$$[A_S, H_I] = 0 , \tag{18.37}$$

in the particular case when there is no explicit time dependence of the variables ($\frac{\partial A_S}{\partial t} = 0$).

Equations (18.35, 18.36, 18.37) describe completely a QND measurement, and they will be instrumental in describing a particular QND measurement scheme in an optical system, presented in the next section.

18.3 QND Measurement of the Number of Photons in a Cavity

18.3.1 The Model

An interesting example of a time-independent QND measurement is the one involving cavity quantum electrodynamics. [8]

We assume a beam of three-level atoms interacting non-linearly and non-resonantly with a signal field.

What we want to measure as accurately as possible, is the photon number of the signal field and the probe is the atomic beam.

In order to do this detection efficiently, the atoms are prepared in Rydberg states, that is with large dipole moments, and the signal is a microwave field, in a high Q cavity, nearly resonant with a couple of adjacent atomic energy levels, as shown in the Fig. 18.1, where the three levels are denoted by a,b,c and the atoms cross a microwave cavity with frequency ω, with a relative detuning δ

$$\delta = \omega - \omega_{ab} , \tag{18.38}$$

with $\omega_{ab} = \frac{E_a - E_b}{\hbar}$.

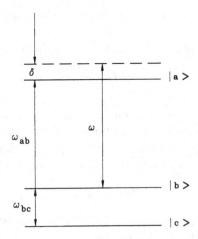

Fig. 18.1. Three-level atom used in the QND measurement of the photon number in the cavity

We also assume that $\frac{g^2 n}{\delta^2} \ll 1$, where g(r) is a position-dependent coupling constant (position of the atom within the cavity).

If one has the combined atom-field state $\mid b, n\rangle$, with an unperturbed energy $E_b + \hbar\omega(n + \frac{1}{2})$, this level suffers an energy Stark shift

$$\hbar\Delta_b = \frac{\hbar g^2(r)n}{\delta} , \tag{18.39}$$

for $E_b < E_a$, and

$$\hbar\Delta_b = -\frac{\hbar g^2(r)(n+1)}{\delta} , \tag{18.40}$$

for $E_b > E_a$.

Here, we consider the level scheme of the Fig. 18.1, with $E_b < E_a$.

18.3.2 The System-Probe Interaction

We will assume that the off-resonant level $\mid c\rangle$ is not affected by this field and will be used for the measurement only.

For the QND measurement purpose, we concentrate, now, in the subspace with the atomic levels $\mid b\rangle$ and $\mid c\rangle$.

The atomic Hamiltonian, in this subspace, is

$$H_{at}^{(b,c)} = \hbar\omega_{bc} \mid b\rangle\langle b \mid , \tag{18.41}$$

and, in the presence of the field, the effective Hamiltonian is (a detailed derivation of such an effective Hamiltonian is found in the chapter of trapped ions)

$$H_I = \hbar\left(\omega_{bc} + \frac{g^2 n}{\delta}\right) \mid b\rangle\langle b \mid . \tag{18.42}$$

As we mentioned before, we consider the atom as probe to measure the photon number of the field.

The Hamiltonian for the S–P coupling, from (18.42) is (noticing that n is the photon number corresponding to the $a_S^\dagger a_S$ field)

$$H_{at}^{(b,c)eff} = \frac{\hbar g^2}{\delta} a_S^\dagger a_S \mid b\rangle\langle b \mid \tag{18.43}$$

$$= \frac{\hbar g^2}{\delta} a_S^\dagger a_S (D_{bc}^\dagger D_{bc}) ,$$

where

$$D_{bc} = \mid c\rangle\langle b \mid .$$

The probe observable is defined as the atomic dipole operator:

$$A_P^{at} = \frac{1}{2i}(D_{bc}^\dagger - D_{bc}) , \tag{18.44}$$

quantity sensitive to the atomic phase, something that one could measure.

If the field contains n photons, the change of atomic phase, after a time interval t, is

$$\Delta\varphi = \left(\omega_{bc} + \frac{g^2 n}{\delta}\right) t \ . \tag{18.45}$$

One can easily check that in this system and probe, all the QND measurement criteria are satisfied.

18.3.3 Measuring the Atomic Phase with Ramsey Fields

Now a device is needed to measure the atomic phase shift $\Delta\varphi$. An interesting possibility is the Ramsey method of two oscillating fields R_1 and R_2 as shown in the Fig. 18.2.

Before entering the cavity, each atom is prepared by a laser in a Rydberg level $\mid b\rangle$. Then, the atom interacts with the first Ramsey field (R_1), which is a microwave field at frequency ω_r, quasiresonant with the b–c transition.

The atom, after this interaction, leaves in a linear superposition of the $\mid b\rangle$ and $\mid c\rangle$ states .Then it crosses the cavity, and outside the cavity, it interacts with a second Ramsey microwave field (R_2), with frequency ω_r.

In the absence of photons, the atomic dipole phase shift introduced by the two Ramsey fields is

$$\varphi_0 = (\omega_r - \omega_{bc})\frac{L}{v_0} \ , \tag{18.46}$$

where v_0 is the atomic velocity.

With n photons in the cavity, this shift becomes

$$\varphi_n = \varphi_0 - n\epsilon \ , \tag{18.47}$$

where ϵ represents the spatial average of the phase shift per photon, that is

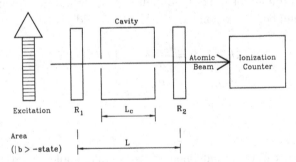

Fig. 18.2. Experimental setup for a QND measurement of photon number in the cavity. The atoms are initially excited into a Rydberg state $\mid e\rangle$ by a laser. They cross the cavity between the two Ramsey fields R_1 and R_2. After that, an ionization detector determines the velocity and state of the atom

$$\varepsilon = \frac{\overline{g(r)^2}}{\delta} \frac{L_c}{v_0} = \frac{g(0)^2 L_c}{2\delta v_0} , \qquad (18.48)$$

where we assumed that g(r) has a sinusoidal dependence that when averaged (g^2) gives the extra $\frac{1}{2}$ factor.

We analyze now the interaction in some detail.

Assume that the cavity contains n photons. The atom-field wavefunction, in the unperturbed representation, can be written as

$$| \psi^{\text{atom}-\text{field}} \rangle = b_b(n,t) \, | \, b, n \rangle + b_c(n,t) \, | \, c, n \rangle , \qquad (18.49)$$

where b_b, b_c are time-dependent functions because of the R_1 and R_2 fields, with initial conditions

$$b_b(n,0) = 1 , \qquad (18.50)$$
$$b_c(n,0) = 0 .$$

During the passage of the atom through R_1, we have the typical couple of differential equations corresponding to a two-level atom interacting with a single mode resonant field. If δL is the length of each zone, the interaction time τ is $\tau = \frac{\delta L}{v}$, and we have

$$\dot{b}_b \, (n,t) = \frac{\Omega_r}{2} b_b(n,t) , \qquad (18.51)$$

$$\dot{b}_c \, (n,t) = -\frac{\Omega_r}{2} b_c(n,t) ,$$

Ω_r being the corresponding Rabi frequency.

The solutions, satisfying the initial conditions are

$$b_b(n,t) = \cos \frac{\Omega_r}{2} \tau , \qquad (18.52)$$

$$b_c(n,t) = -\sin \frac{\Omega_r}{2} \tau .$$

After the atom passed the cavity, b_b suffers a phase shift whereas b_c remains the same.

The initial conditions, before entering the second Ramsey zone are

$$b_b(n) = \exp \left(-in\varepsilon \frac{v_0}{v} \right) \cos \frac{\Omega_r}{2} \tau , \qquad (18.53)$$

$$b_c(n) = -\sin \frac{\Omega_r}{2} \tau, \qquad (18.54)$$

with $\frac{v_0}{v}$ expressing the dependence of the phase shift with time or the inverse velocity.

In the second zone, the differential equations are the same as the first one, except for the added phase $\pm \varphi_0 \frac{v_0}{v}$ that accounts for the shifts between the two Ramsey fields

$$\dot{b}_b(n,t) = \frac{\Omega_r}{2} \exp\left(-i\varphi_0 \frac{v_0}{v}\right) b_b(n,t) \tag{18.55}$$

$$\dot{b}_c(n,t) = -\frac{\Omega_r}{2} \exp\left(i\varphi_0 \frac{v_0}{v}\right) b_c(n,t) .$$

After the second Ramsey zone, the solutions are

$$b_b(n) = \exp\left(-in\varepsilon \frac{v_0}{v}\right) \cos^2 \frac{\Omega_r}{2}\tau - \sin^2 \frac{\Omega_r}{2}\tau \exp\left(-i\varphi_0 \frac{v_0}{v}\right) \tag{18.56}$$

$$b_c(n) = -\frac{1}{2} \sin \Omega_r \tau (1 + \exp i (\varphi_0 - n\varepsilon) \frac{v_0}{v}) .$$

Finally, if we set $\Omega_r \tau = \pi/2$, we get

$$| \psi^{\text{atom}-\text{field}}\rangle_{\text{final}} \tag{18.57}$$

$$= b_b(n, v; \varphi_0, \varepsilon) | b, n\rangle + b_c(n, v; \varphi_0, \varepsilon) | c, n\rangle ,$$

with

$$b_b(n, v; \varphi_0, \varepsilon) = \exp(-i\varphi_0 \frac{v_0}{v}) \left[\cos^2 \frac{\pi v_0}{4v} \exp\left(i\varphi_n \frac{v_0}{v}\right) - \sin^2 \frac{\pi v_0}{4v}\right], \tag{18.58}$$

$$b_c(n, v; \varphi_0, \varepsilon) = -\frac{1}{2} \sin \frac{\pi v_0}{2v} \left[1 + \exp\left(i\varphi_n \frac{v_0}{v}\right)\right] .$$

The atoms are detected when they leave the second Ramsey zone, by a field-ionization detector that determines whether the atom is in state $| b\rangle$ or $| c\rangle$ and also, synchronizing it with that laser excitation, one can determine the velocity v.

The probability P_c of being at c is

$$P_c(n, v; \varphi_0, \varepsilon) = 1 - P_b = | b_c |^2 \tag{18.59}$$

$$= \sin^2 \frac{\pi v_0}{2v} \cos^2 \varphi_n \frac{v_0}{2v} .$$

If instead of a pure $| n\rangle$ state in the cavity, we have a photon distribution $P(n)$, then

$$P_c(p(n), \varphi_0, \varepsilon) = \sum_n p(n) P_c(n, v; \varphi_0, \varepsilon) . \tag{18.60}$$

We may also average over a Maxwellian velocity distribution $D(v)$ thus getting

$$P_c(n, \varphi_0, \varepsilon) = \int D(v) P_c(n, v; \varphi_0, \varepsilon) dv . \tag{18.61}$$

$P_c(n, \varphi_0, \varepsilon)$ for $\varepsilon = 2\pi$ is shown in the Fig. 18.3, for a Fock state (a), a coherent state (b) and a thermal state (c), with $\bar{n} = 3$.

The different fringe patterns allow in principle to distinguish between the various photon statistics. Because these are probabilities, experimentally we should detect, for each φ_0, a large number of atoms and the average.

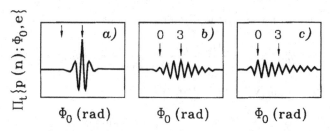

Fig. 18.3. $P_f(p(n), \varphi_0, \epsilon)$ versus φ_0 for a Fock state (**a**), a coherent state (**b**) and a thermal state (**c**). (After [8])

This, of course, implies that the field should be prepared in an initial state with photon statistics p(n), before each atom crosses the cavity, which can be rather cumbersome.

It would be more attractive if we had a scheme involving only one realization of the system, using basically the same experimental setup.

18.3.4 QND Measurement of the Photon Number

We assume a large Q-factor for the cavity and that a bunch of atoms cross it.

At the output, each atom´s velocity and internal state is detected by the counter.

The state of the atom-field system at the entry of the (k+1) atom is given by the density operator

$$\rho_{k+1}^{\text{field+atom}} = \sum_{n,n'} \mid b, n \rangle \langle b, n \mid \rho_{k,n,n'}, \tag{18.62}$$

and after going through the apparatus, will become

$$\rho_{k+1}^{\text{field+atom}} = \sum_{n,n'} (b_b(n, v; \varphi_0, \varepsilon) \mid b, n \rangle + b_c(n, v; \varphi_0, \varepsilon) \mid c, n \rangle) \tag{18.63}$$

$$\times \rho_{k,n,n'} (b_b^*(n', v; \varphi_0, \varepsilon) \langle b, n' \mid + b_c^*(n', v; \varphi_0, \varepsilon) \langle c, n' \mid) \,.$$

After the atomic measurement, the density operator collapses and it is projected giving $\langle \alpha n \mid \rho_{k+1}^{\text{field+atom}} \mid \alpha n \rangle$, so $(\alpha = b, c)$

$$\rho_{k+1,n,n'}^{(\alpha, v)} = \frac{b_\alpha(n, v, \varphi_0, \varepsilon) b_\alpha^*(n', v, \varphi_0, \varepsilon)}{\sum_n \mid b_\alpha(n, v, \varphi_0, \varepsilon) \mid^2} \rho_{k,n,n'}, \tag{18.64}$$

where we introduced the normalizing denominator, as normalization was lost after the state collapsed.

This denominator is nothing but $P_\alpha(p(n), v, \varphi_0, \varepsilon)$, the probability for the atom to be found in the $\mid \alpha \rangle$ state, at a given velocity and for a given photon statistics.

In other words, for an $\mid \alpha \rangle$ and v measurement

$$\rho_{k+1,n,n'}^{(\alpha,v)} = \frac{b_\alpha(n,v,\varphi_0,\varepsilon)b_\alpha^*(n',v,\varphi_0,\varepsilon)}{P_\alpha(p(n),v,\varphi_0,\varepsilon)}\rho_{k,n,n'} . \tag{18.65}$$

If only the atomic velocity is detected

$$\rho_{k+1,n,n'}^{(\alpha=?,v)} = \sum_{\alpha=b,c} b_\alpha(n,v,\varphi_0,\varepsilon)b_\alpha^*(n',v,\varphi_0,\varepsilon)\rho_{k,n,n'} , \tag{18.66}$$

and if the atom is not detected at all

$$\rho_{k+1,n,n'}^{(\alpha=?,v=?)} = \int D(v)dv \sum_{\alpha=b,c} b_\alpha(n,v,\varphi_0,\varepsilon)b_\alpha^*(n',v,\varphi_0,\varepsilon)\rho_{k,n,n'} . \tag{18.67}$$

If one is only interested in the photon number distribution, then only the diagonal part of the field density operator is relevant. Calling $\rho_{nn} = p_n$, we have several cases

a) the atom is detected with velocity v and at the level α

$$p_{k+1}^{(\alpha,v)}(n) = \frac{P_\alpha(n,v,\varphi_0,\varepsilon)}{P_\alpha(p_k(n),v,\varphi_0,\varepsilon)}p_k(n) . \tag{18.68}$$

b) The atomic level is not detected

$$p_{k+1}^{(\alpha=?,v)}(n) = [P_e(n,v,\varphi_0,\varepsilon) + P_f(n,v,\varphi_0,\varepsilon)]\,p_k(n) = p_k(n) , \tag{18.69}$$

and the original photon distribution is not changed at all.

c) If the atom is undetected, then, again

$$p_{k+1}^{(\alpha=?,v=?)}(n) = p_k(n) . \tag{18.70}$$

It is interesting to notice that the unread atom does not alter the photon statistics of the field. As we shall see in the next section, this in general is not true for any type of measurement and it is a signature of the QND nature of this measurement.

Also, the fact that we are getting, through the probe atom, information about the field, modifies the field, although no energy exchange took place.

If we start the field in a pure Fock state $p_k(n) = \delta(n - n_k)$, then

$$P_\alpha(p(n),v,\varphi_0,\varepsilon) = P_\alpha(n,v,\varphi_0,\varepsilon) , \tag{18.71}$$

and no change occurs in the photon statistics

$$p_{k+1}^{(\alpha,v)}(n) = p_k(n) . \tag{18.72}$$

Now, we proceed with the numerical simulation of a continuous QND measurement of the field, initially with a distribution $p_0(n)$.

First, we take randomly a velocity v_1, and compute $P_\alpha(p(n), v, \varphi_0, \varepsilon)$ from (18.60).

Then we decide the result of the measurement of α (energy level) by comparing this probability to a random number between 0 and 1. Next, we multiply $p_0(n)$ by $P_\alpha(n, v, \varphi_0, \varepsilon)$ and normalize, obtaining $p_1(n)$, and so on. This iteration leads to

$$p_k(n) = \frac{\prod_{p=1}^k P_{\alpha_p}(n, v_p, \varphi_0, \varepsilon) p_0(n)}{\sum_{n'} p_0(n') \prod_{p=1}^k P_{\alpha_p}(n', v_p, \varphi_0, \varepsilon)} . \tag{18.73}$$

This simulation can be carried out for different values of φ_0, ε.

Starting from a coherent or thermal distribution, one finds a collapse to a Fock state, as shown in the Fig. 18.4.

As we can see from the figure, a 'decimation' process takes place as we increase the number of detected atoms, until a pure Fock state is reached. This final state is not *a priori* predictable, as the whole process depends on random variables that mimic the measurement process. So, if we repeat this experiment many times, and considering that we are dealing with QND measurements, the statistics of the result will coincide with the initial photon statistics $p_0(n)$.

What we just described in this section is a particular continuous measurement without back-action on the measured observable, in this case, the photon number in a cavity.

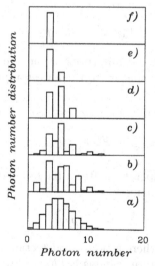

Fig. 18.4. Photon number distribution in a QND sequence. The initial state is coherent with $\bar{n} = 5$ (**a**). The figures **b,c,d,e,f,** correspond to the detection of 1,3,6,10,15 atoms respectively. This run collapses in the Fock state $n = 3$. (After [8])

In general, in a non-QND measurement, the back-action is present and typically a Fock state after repeated measurements becomes a different state. This will be described in detail in the next section.

18.4 Quantum Theory of Continuous Photodetection Process

18.4.1 Introduction

In a series of recent publications, Ueda and co-workers [10, 11], showed that the quantum properties of the field are generally affected by 'yes' and 'no' results, that is a photodetection event or the absence of it, modifying the statistics of the field.

We start our study with a simple model of a two-level atom detector interacting resonantly with a one-mode field, ruled by the Jaynes–Cummings Hamiltonian

$$H = \hbar g(a\sigma^\dagger + \sigma a^\dagger) \,. \tag{18.74}$$

The time evolution of the state vector is

$$|\,\psi(\tau)\rangle = \left[1 - \frac{iH\tau}{\hbar} - \frac{H^2\tau^2}{2\hbar^2}.....\right]|\,\psi(0)\rangle \,. \tag{18.75}$$

Let the initial state of the detector-field system be ($|\,b\rangle$ is the lower state and $|\,f\rangle$ an unspecified field state)

$$|\,\psi(0)\rangle = |\,b\rangle\,|\,f\rangle \,. \tag{18.76}$$

After we measure a photon absorption event, the statevector is projected onto the $|\,a\rangle$ atomic state, that is (to the lowest approximation)

$$\langle a \mid \psi(\tau)\rangle = -\frac{ig\tau}{\hbar}\langle a \mid a\sigma^\dagger + \sigma a^\dagger \mid b\rangle\,|\,f\rangle \tag{18.77}$$

$$= -\frac{ig\tau}{\hbar}a \mid f\rangle \,.$$

So, the field density operator, after a photodetection event, and up to a normalization constant, changes as

$$\rho_f \to a\rho_f a^\dagger \,. \tag{18.78}$$

On the other hand, if there is no photodetection, during the time interval τ, then the final atomic state remains in $|\,b\rangle$, and we have

$$\langle b \mid \psi(\tau)\rangle \tag{18.79}$$

$$= \langle b \mid \left[1 - \frac{iH\tau}{\hbar} - \frac{H^2\tau^2}{2\hbar^2}.....\right]\psi(0)\rangle$$

$$= |f\rangle - \frac{\tau^2 g^2}{2} \langle b | [\sigma^\dagger \sigma a a^\dagger + \sigma \sigma^\dagger a^\dagger a] | b\rangle | f\rangle$$

$$\approx \exp(-\frac{\tau^2 g^2}{2} a^\dagger a) | f\rangle ,$$

where, in the last step, we assumed that the interaction time is short enough as to have

$$\frac{\tau^2 g^2}{2} \langle a^\dagger a\rangle << 1 . \tag{18.80}$$

As we see, up to a normalization constant, the initial field density operator, after a no-absorption event, changes to

$$\rho_f \to \exp(-\frac{\tau^2 g^2}{2} a^\dagger a)\rho_f \exp(-\frac{\tau^2 g^2}{2} a^\dagger a) . \tag{18.81}$$

Now, we assume that the 'yes' events are detected at times $0, t_1, t_2 \ldots, t_n$ and no events in between, with our photodetector being continuously monitored.
(Fig. 18.5)

Each period is subdivided in N small intervals τ, so, for example at $t_1 = N\tau$

$$\rho_f(t_1) = \exp(-\frac{N\tau^2 g^2}{2} a^\dagger a)\rho_f \exp(-\frac{N\tau^2 g^2}{2} a^\dagger a) \tag{18.82}$$
$$= \exp(-2Rt_1 a^\dagger a)\rho_f \exp(-2Rt_1 a^\dagger a) ,$$

with $R = \frac{g^2 \tau}{4}$.

The sequence shown in Fig. 18.5 corresponds to the transformation of the field density matrix operator, as shown

$$\rho_f \to .. \left[\exp(-2R(t_2 - t_1)a^\dagger a)\right] a \left[\exp(-2Rt_1 a^\dagger a)\right] a\rho_f(0) \tag{18.83}$$
$$a^\dagger \left[\exp(-2Rt_1 a^\dagger a)\right] a^\dagger \left[\exp(-2R(t_2 - t_1)a^\dagger a)\right] ..$$

We can group all the a's in one side and the $a^\dagger$'s on the other, giving us numerical factors that can be included in the renormalization constant. The result is

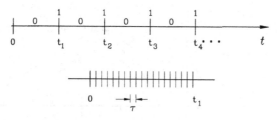

Fig. 18.5. Photodetection sequence in the times $0, t_1, t_2 \ldots$. In between this times, no absorption is detected. The interval between 0 and t_1 is subdivided in smaller τ intervals

$$\rho_f \rightarrow \frac{\left[\exp(-2Rta^\dagger a)\right] a^m \rho_f a^{\dagger m} \left[\exp(-2Rta^\dagger a)\right]}{Tr\left[\exp(-2Rta^\dagger a)\right] a^m \rho_f a^{\dagger m} \left[\exp(-2Rta^\dagger a)\right]} , \qquad (18.84)$$

which corresponds to the modified density operator after m 'yes' counts and t being the total measurement time.

We notice immediately that if the initial state of the field is either a coherent or a Fock state, their nature does not change, but the state does, that is

$$| \alpha\rangle \rightarrow | \alpha\exp(-2Rt)\rangle , \qquad (18.85)$$
$$| N\rangle \rightarrow | N - m\rangle .$$

18.4.2 Continuous Measurement in a Two-Mode System: Phase Narrowing

We consider now the interaction of a three-level Λ system with two-mode radiation field , in a high-Q cavity, as shown in Fig. 18.6.

We assume that the Λ system is initially prepared in a superposition of the split lower states $| b\rangle$ and $| b'\rangle$

$$| \psi_A(0)\rangle = \frac{1}{\sqrt{2}} [| b\rangle + \exp(-i\varphi) | b'\rangle] . \qquad (18.86)$$

Also, we assume that the $| b'\rangle \rightarrow | a\rangle$ transition is resonant with the cavity field (field b) and that the $| b\rangle \rightarrow | a\rangle$ transition is driven by a coherent field $| \alpha\rangle$.

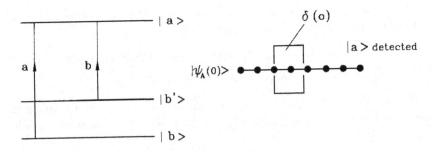

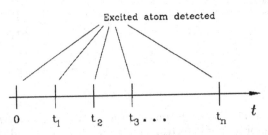

Fig. 18.6. Diagram of the proposed continuous measurement scheme

The relevant Hamiltonian in this system is

$$H_I = \hbar g(a_1 \mid a\rangle\langle b \mid +a_2 \mid a\rangle\langle b' \mid +HC) , \tag{18.87}$$

Outside the high Q cavity, there is a ionization detector that will tell us if the atom is in the excited state or in one of the lower states ($\mid b\rangle$ or $\mid b'\rangle$).

Following a similar argument to the previous section, it is simple to show that after the atom is detected in the excited state $\mid a\rangle$, the field changes as

$$\rho_f(\tau) = \frac{g^2\tau^2}{\hbar^2} A\rho_f(0)A^\dagger , \tag{18.88}$$

with

$$A = \frac{1}{\sqrt{2}} [a_1 + a_2 \exp(-i\varphi)] . \tag{18.89}$$

On the other hand, if the detected atom is not in the excited state, during a measurement period τ, then the field changes as

$$\rho_f(t) = \exp(-2RtA^\dagger A)\rho_f(0) \exp(-2RtA^\dagger A) . \tag{18.90}$$

Once more, if we have n detections in a total measurement time t, then

$$\rho_f^{(n)}(t) = \frac{\exp(-2RtA^\dagger A)A^n\rho_f(0)A^{\dagger n}\exp(-2RtA^\dagger A)}{Tr(\exp(-2RtA^\dagger A)A^n\rho_f(0)A^{\dagger n}\exp(-2RtA^\dagger A))} . \tag{18.91}$$

We notice that in the ordering process we used the relation

$$\exp\left(xA^\dagger A\right)(A)\exp\left(-xA^\dagger A\right) = A\exp(-x) , \tag{18.92}$$

which adds an extra factor that can be included in the normalization of $\rho_f^{(n)}(t)$.

As an example, we assume that initially both fields are coherent

$$\rho(0) = \mid \alpha, \beta\rangle\langle\alpha, \beta \mid , \tag{18.93}$$

so that

$$A \mid \alpha, \beta\rangle = \frac{1}{\sqrt{2}} [\alpha + \beta \exp(-i\varphi)] \mid \alpha, \beta\rangle , \tag{18.94}$$

and the density matrix, after n detections becomes

$$\rho_f^{(n)}(t) = \frac{\exp(-2RtA^\dagger A) \mid \alpha, \beta\rangle\langle\alpha, \beta \mid \exp(-2RtA^\dagger A)}{Tr\left\{\exp(-2RtA^\dagger A) \mid \alpha, \beta\rangle\langle\alpha, \beta \mid \exp(-2RtA^\dagger A)\right\}} . \tag{18.95}$$

We notice that

$$\rho_f^{(n)}(t) = \rho_f^{(0)}(t) , \tag{18.96}$$

because for a coherent state, the A^n factor becomes a numerical one that can be absorbed in the normalization constant. In this case, we may say 'only no counts count'. [12].

Now, we want to simplify the expression given in (18.19).
We define a B operator

$$B = \frac{1}{\sqrt{2}}(a_1 - a_2 \exp -i\varphi) , \qquad (18.97)$$

with

$$[A, B] = [A, B^\dagger] = 0 , \qquad (18.98)$$
$$[B, B^\dagger] = 1 .$$

Now, we look at the effect of the a_1 and a_2 operators on $\exp(-2RtA^\dagger A) \mid \alpha, \beta\rangle$.

$$
\begin{aligned}
a_2 \exp(-2RtA^\dagger A) \mid \alpha, \beta\rangle =& \left(\frac{A - B}{\sqrt{2}}\right) \\
& \times \exp(i\varphi) \exp(-2RtA^\dagger A) \mid \alpha, \beta\rangle \qquad (18.99) \\
=& \exp(-2RtA^\dagger A) \exp(2RtA^\dagger A) \left(\frac{A - B}{\sqrt{2}}\right) \\
& \times \exp(i\varphi) \exp(-2RtA^\dagger A) \mid \alpha, \beta\rangle \\
=& \exp(-2RtA^\dagger A) \left[\frac{A \exp(-2Rt) - B}{\sqrt{2}}\right] \\
& \times \exp(i\varphi) \mid \alpha, \beta\rangle ,
\end{aligned}
$$

where, in the last step, we used the identity given by (18.92).

Now, from (18.99), we observe that $\exp(-2RtA^\dagger A) \mid \alpha, \beta\rangle$ is an eigenstate of a_2 with eigenvalue $\tilde{\beta}$:

$$a_2 \exp(-2RtA^\dagger A) \mid \alpha, \beta\rangle = \tilde{\beta} \exp(-2RtA^\dagger A) \mid \alpha, \beta\rangle , \qquad (18.100)$$
$$\tilde{\beta} = \left[\left(\frac{\alpha \exp i\varphi + \beta}{2}\right) \exp(-2Rt) - \frac{1}{2}(\alpha \exp(i\varphi) - \beta)\right] .$$

Similarly, one can show that

$$a_1 \exp(-2RtA^\dagger A) \mid \alpha, \beta\rangle = \tilde{\alpha} \exp(-2RtA^\dagger A) \mid \alpha, \beta\rangle , \qquad (18.101)$$
$$\tilde{\alpha} = \left\{\left[\frac{\alpha + \beta \exp(-i\varphi)}{2}\right] \exp(-2Rt) - \frac{1}{2}[\alpha - \beta \exp(-i\varphi)]\right\} .$$

So

$$\rho_f^{(n)}(t) = \mid\tilde{\alpha}, \tilde{\beta}\rangle\langle\tilde{\alpha}, \tilde{\beta}\mid . \qquad (18.102)$$

As we can see, as a result of the continuous measurement process, a coherent state, say $\mid \beta\rangle$, has become a new coherent state with a modified amplitude $\mid\tilde{\beta}\rangle$.

Thus, there is a phase noise reduction or phase narrowing if $|\tilde{\beta}|^2 > |\beta|^2$.

On the other hand, if one of the fields is classical (a_1), that is $|\alpha| >> |\beta|$, then approximately

$$|\tilde{\beta}| = \frac{|\alpha|}{2} [1 - \exp(-2Rt)] . \tag{18.103}$$

Writing a phase state

$$|\theta\rangle = \sum_{n=0}^{\infty} \exp(in\theta) |n\rangle , \tag{18.104}$$

one can write an expression for the phase distribution

$$P^n(\theta) = \langle\theta |\tilde{\beta}\rangle\langle\tilde{\beta}| \theta\rangle . \tag{18.105}$$

The above distribution has been numerically evaluated for $\alpha = 10, \beta = 1$, and various times $Rt = 0$ (a), $Rt = 0.5$ (b), $Rt = 1$ (c), $Rt = 1.5$ (d).

In the Fig. 18.7, we observe a striking phase narrowing, that can have interesting applications in small signal detection.

We can also look at the steady state, when one of the fields (say a_2) is in the vacuum, that is $\beta = 0$, then, for $t \to \infty$

$$|0\rangle_{a_2} \to | -\frac{\alpha}{2} \exp(i\varphi)\rangle . \tag{18.106}$$

We may say that we transferred the coherence from the a_1 to the a_2 mode.

Furthermore, this result is independent of the number of 'clicks' (independent of n), that is **the coherence transfer occurs even if we never find an atom in the excited state.** [13].

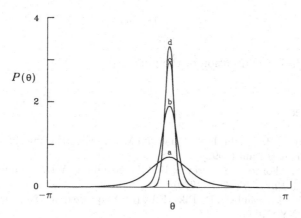

Fig. 18.7. Phase distribution for various times *a)* $Rt = 0$; *b)* $Rt = 0.5$; *c)* $Rt = 1$; *d)* $Rt = 1.5$; The parameters are $\alpha = 10$, $\beta = 1$ (After [12])

Problems

18.1. Verify the solution for the atom-field state given by (18.57, 18.58).

18.2. Prove that the reordering procedure of the creation and annihilation operators give the result from the (18.84).

18.3. Show that

$$\exp(xA)A\exp(-xA) = A\exp(-x) \ .$$

(Appendix A)

18.4. Show that the standard quantum limit for the energy of an oscillator is

$$\Delta E_{\text{SQL}} = \sqrt{\hbar\omega E} \ ,$$

where E is the oscillator's mean energy.

18.5. Suppose that instead of wanting to monitor the position of an oscillator, as we did in the Sect. 18.1.3, we wanted to monitor its energy and derive a condition for the thermal effects.

Prove that for short time measurements

$$\Delta E = \sqrt{\hbar\omega E\frac{\tau}{\tau^*}} \ ,$$

and whether $\Delta E \leqslant \hbar\omega$ implies

$$(n + \frac{1}{2})\frac{k_B T\tau}{\tau^*} \leq \hbar\omega \ .$$

In other words, to monitor the resonator's energy at the level $\hbar\omega$ requires a temperature smaller by a factor $2(n + \frac{1}{2})$ than only set the condition that the oscillator behave quantum mechanically. [1].

18.6. Prove that the standard quantum limit for measurement of an impulsive force, using an oscillator, is

$$F = \frac{1}{\tau_F}\sqrt{\frac{\hbar\omega m}{2}} \ ,$$

where τ_F is the duration of the pulse.

References

1. Braginsky, V.B., Khalil, F.Y.: Quantum Measurements. Cambridge University Press, Great Britain (1992)
2. Caves, C., Thorne, K.S., Drever, R.W.P., Sandberg, V.D., Zimmermann, M.: Rev. Mod. Phys., **52**, 341 (1980)
3. Ozawa, M.: In: Tombesi, P., Pike, E.R. (eds) Squeezed and Nonclassical Light. Plenum, NY (1988)
4. Levenson, M.D., Shelby, R.M., Reid, M., Walls, D.F.: Phys. Rev. Lett., **57**, 2473 (1986)

5. Imoto, N., Watkins, S., Sasaki, Y.: Opt. Comm., **61**, 159 (1987)
6. LaPorta, A., Slusher, R.E., Yurke, B.: Phys. Rev. Lett., **62**, 28 (1989)
7. Grangier, P., Roch, J.F., Roger, G.: Phys. Rev. Lett., **66**, 1418 (1991)
8. Brune, M., Haroche, S., Raimond, J.M., Davidovich, L., Zagury, N.: Phys. Rev. A, **45**, 5193 (1992); also Brune, M., Haroche, S., Lefebre, V., Raimond, J.M., Zagury, N., Phys. Rev. Lett., **65**, 976 (1990)
9. Braginsky, V.B., Vorontsov, Y.I.: Sov. Phys. Usp., **17**, 644 (1975)
10. Ueda, M., Imoto, N., Nagaoka, H., Ogawa, T.: Phys. Rev. A, **46**, 2859 (1992)
11. Ogawa, T., Ueda, M., Imoto, N.: Phys. Rev. Lett., **66**, 1046 (1991); *ibid* Phys. Rev. A, **43**, 6458 (1991)
12. Agarwal, G.S., Scully, M.O., Walther, H.: Physica Scripta, **148**, 128 (1993)
13. Agarwal, G.S., Graf, M., Orszag, M., Scully, M.O., Walther, H.: Phys. Rev. A, **49**, 4077 (1994)

Further Reading

• Braginsky, V.B., VorontsovI, Y.I., Thorne, K.S.: Science, **209**, 5 471 (1980)
• Milburn, G.J., Walls, D.F.: Phys. Rev. A, **28**, 2065 (1983)
• Imoto, N., Haus, H.A., Yamamoto, Y.: Phys. Rev. A, **32**, 2287 (1985)
• Yurke, B.: JOSA B, **2**, 732 (1985)

19. Trapped Ions

In this Chapter, we study the Paul trap for ions. We also analyse the various interactions between the ion's internal and center-of-mass degrees of freedom with light waves.

19.1 Paul Trap [1]

19.1.1 General Properties

This is a trap that confines charged particles in a quadrupole potential, modulated by a radiofrequency field. (Fig. 19.1)

One generates a potential difference U between the two Hyperboloid end caps and a ring of radius r_0, located at z_0 from the upper end cap. (The origin is at the center of the ring.)

We assume a configuration described by the following potential:

$$\phi(x, y, z) = A(x^2 + y^2 - 2z^2) , \tag{19.1}$$

and from the Fig. 19.1

$$\phi(r_0, 0, 0) - \phi(0, 0, z_0) = U . \tag{19.2}$$

From the two equations above, we can calculate A

$$A = \frac{U}{r_0^2 + 2z_0^2} , \tag{19.3}$$

or

$$\phi(x, y, z) = \frac{U}{r_0^2 + 2z_0^2}(x^2 + y^2 - 2z^2) . \tag{19.4}$$

Unfortunately, one cannot confine a charged particle in three dimensions, with a constant electric field, because this electric field would converge from all directions and a net flux would be present, violating the equation $\nabla \cdot \mathbf{E} = 0$.

What we can do, instead, is to modulate the field with a harmonic term proportional to $\cos \Omega t$, thus getting

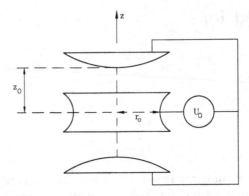

Fig. 19.1. Paul trap to confine charged particles

$$\phi(x, y, z, t) = \frac{V_0}{r_0{}^2 + 2z_0^2}(x^2 + y^2 - 2z^2)\cos \Omega t \,, \qquad (19.5)$$

where we replaced $U \rightarrow V_0 \cos \Omega t$.

If we neglect the B field and

$$\mathbf{E} = -\nabla\phi = A'\cos \Omega t \, (-2x, -2y, 4z) \,, \qquad (19.6)$$

with

$$A' = \frac{V_0}{r_0{}^2 + 2z_0^2} \,, \qquad (19.7)$$

then the classical equations of motion for a charged particle in this field are

$$\ddot{x} = -\frac{2A'q\cos \Omega t}{m}x \,, \qquad (19.8)$$

$$\ddot{y} = -\frac{2A'q\cos \Omega t}{m}y \,, \qquad (19.9)$$

$$\ddot{z} = \frac{4A'q\cos \Omega t}{m}z \,, \qquad (19.10)$$

where m and q are the mass and charge of the particle.

The above equations can be written in a compact form:

$$\ddot{x}_i = C_i x_i \cos \Omega t \,, \qquad (19.11)$$

with

$$C_1 = C_2 = -\frac{2A'q}{m} \qquad (19.12)$$

$$C_3 = \frac{4A'q}{m} \,,$$

and $x_1 = x, x_2 = y, x_3 = z$.

Now we decompose the motion in a fast and a slow time-varying part

$$x_i = \bar{x}_i + \xi_i \, , \tag{19.13}$$

where $\bar{x}_i$ is the slow term, varying with a frequency ω_i, still to be determined, and ξ_i is the fast term changing with frequency $\Omega >> \omega_i$.

So, we can write now

$$\ddot{\bar{x}}_i + \ddot{\xi}_i = C_i\bar{x}_i \cos\Omega t + C_i\xi_i \cos\Omega t \, . \tag{19.14}$$

Now, we want to satisfy

$$| \,\overline{x_i}\, | >> | \,\xi_i\, |, \tag{19.15}$$

$$| \,\ddot{\overline{x_i}}\, | << |\ddot{\xi}_i| \, ,$$

which is possible due to the difference in the frequencies. Then we have

$$\ddot{\xi}_i = C_i\bar{x}_i \cos\Omega t \, . \tag{19.16}$$

For a time scale of the order of $\frac{1}{\Omega}$, $\bar{x}_i$ is basically constant, and we have

$$\xi_i = -\frac{C_i}{\Omega^2}\bar{x}_i \cos\Omega t \, . \tag{19.17}$$

As we can see, the acceleration and position of ξ_i have opposite signs, so that there is a restoring force, similar to a harmonic oscillator, and independent of the sign of C_i, so this is true for all three spatial directions.

If we replace (19.16, 19.17) in (19.14), we get

$$\ddot{\bar{x}}_i + \frac{C_i^2}{\Omega^2} \cos^2\Omega t\,\bar{x}_i = 0 \, . \tag{19.18}$$

Now, if again we average over a time scale $\frac{1}{\Omega}$, the $\cos^2$ averages to $\frac{1}{2}$, and we get a three-dimensional harmonic oscillator

$$\ddot{\bar{x}}_i + \frac{C_i^2}{2\Omega^2}\bar{x}_i = 0 \, , \tag{19.19}$$

with the corresponding frequencies

$$\omega_x = \omega_y = \frac{|\,C_1\,|}{\sqrt{2}\Omega} = \frac{\sqrt{2}A'q}{m\Omega}, \tag{19.20}$$

$$\omega_z = \frac{|\,C_3\,|}{\sqrt{2}\Omega} = \frac{2\sqrt{2}A'q}{m\Omega} \, .$$

Thus, one can write an effective potential energy

$$V_{\text{eff}} = \sum_i \frac{mC_i^2}{4\Omega^2}\bar{x}_i^2 \, . \tag{19.21}$$

As we can see, in the $\Omega >> \omega_i$ limit, the motion of the charged particle is well described by an effective potential corresponding to a three-dimensional harmonic oscillator, superposed by a small amplitude and rapidly varying ξ. This is shown pictorially in the Fig. 19.2.

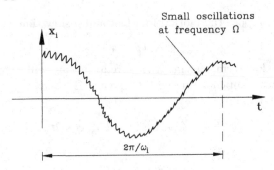

Fig. 19.2. Motion of a charged particle in a Paul trap

19.1.2 Stability Analysis

One can easily verify that if a static field is added to U , that is

$$U = U_0 + V_0 \cos \Omega t , \tag{19.22}$$

or

$$\phi(x,y,z,t) = \frac{U_o + V_0 \cos \Omega t}{r_0{}^2 + 2z_0^2}(x^2 + y^2 - 2z^2) , \tag{19.23}$$

then the effective potential energy becomes

$$V'_{\text{eff}} = \frac{q^2 V_0^2}{m\Omega^2 (r_0{}^2 + 2z_0^2)^2}(\bar{x}^2 + \bar{y}^2 + 4\,\bar{z}^2) \tag{19.24}$$

$$+ \frac{qU_o}{r_0{}^2 + 2z_0^2}(\bar{x}^2 + \bar{y}^2 - 2\,\bar{z}^2) ,$$

which becomes isotropic if

$$U_0 = \frac{qV_0^2}{m\Omega^2 (r_0{}^2 + 2z_0^2)} \cdot \tag{19.25}$$

One can also have essentially a one-dimensional harmonic oscillator, along the z-axis, if one chooses U_0 to be large and negative.

In general, for any value of U_0 and V_0, the equations of motion for x and z are

$$\ddot{x} = \frac{-2q}{m(r_0^2 + 2z_0^2)}(U_0 + V_0 \cos \Omega t)\, x , \tag{19.26}$$

$$\ddot{z} = \frac{4q}{m(r_0^2 + 2z_0^2)}(U_0 + V_0 \cos \Omega t)\, z .$$

We perform, now, a change of variables

$$\Omega t = 2\tau, \tag{19.27}$$

$$x_1 = x, x_2 = y, x_3 = z,$$

$$a_3 \equiv \frac{-16qU_0}{m\Omega^2(r_0^2 + 2z_0^2)} = -2a_1 = -2a_2,$$

$$q_3 \equiv \frac{8qV_0}{m\Omega^2(r_0^2 + 2z_0^2)} = -2q_1 = -2q_2 .$$

With the above definitions, the equations of motion now read

$$\frac{\mathrm{d}^2 x_i}{\mathrm{d}^2\tau} + (a_i - 2q_i \cos 2\tau) x_i = 0 . \tag{19.28}$$

$$i = 1, 2, 3 .$$

Equation (19.28) is Mathieu's equation, with a periodic coefficient, with period π.

Thus, if $x(\tau)$ is a solution of Mathieu's Equation, then $x(\pi + \tau)$ is also a solution.

Now, we explore the possibility of finding solutions of Mathieu's Equation, where

$$x(\pi + \tau) = \mu x(\tau) , \tag{19.29}$$

μ being a complex number to be determined later. (Floquet's Theorem)

Now, let $h(\tau)$ and $g(\tau)$ be two independent solutions of Mathieu's Equation. As this equation is linear, then

$$x(t) = Ag(\tau) + Bh(\tau) , \tag{19.30}$$

is also a solution.

Furthermore,

$$g(\tau + \pi) = \alpha_1 g(\tau) + \alpha_2 h(\tau) , \tag{19.31}$$
$$h(\tau + \pi) = \beta_1 g(\tau) + \beta_2 h(\tau) .$$

Now, replacing 19.30 and 19.31 in 19.29, we get

$$x(\pi + \tau) = Ag(t + \tau) + Bh(t + \tau) \tag{19.32}$$
$$= A[\alpha_1 g(\tau) + \alpha_2 h(\tau)]$$
$$+ B[\beta_1 g(\tau) + \beta_2 h(\tau)]$$

$$= (A\alpha_1 + B\beta_1)g(\tau) + (A\alpha_2 + B\beta_2)h(\tau)$$
$$= \mu Ag + \mu Bh ,$$

or

$$(\alpha_1 - \mu)A + \beta_1 B = 0 , \tag{19.33}$$

$$\alpha_2 A + (\beta_2 - \mu)B = 0 .$$

For non-trivial solutions, the determinant of the coefficients must vanish, giving us two possible values for μ :

$$x_1(\pi + \tau) = \mu_1 x_1(\tau), \tag{19.34}$$

$$x_2(\pi + \tau) = \mu_2 x_2(\tau) .$$

If we define two quantities

$$\exp \sigma_i \pi \equiv \mu_i \tag{19.35}$$

$$F_i(\tau) \equiv x_i(\tau) \exp(-\sigma_i \tau) ,$$

then it is simple to show that $F_i(\tau)$ is π-periodic

$$\begin{aligned} F_i(\tau + \pi) &= x_i(\tau + \pi) \exp(-\sigma_i \tau) \exp(-\sigma_i \pi) \tag{19.36} \\ &= x_i(\tau) \exp(\sigma_i \pi) \exp(-\sigma_i \tau) \exp(-\sigma_i \pi) \\ &= x_i(\tau) \exp(-\sigma_i \tau) = F_i(\tau) . \end{aligned}$$

From the last line of (19.36), we write

$$x_i(\tau) = (\mu_i)^{\frac{\tau}{\pi}} F_i(\tau) , \tag{19.37}$$

with $F_i(\tau)$ being a periodic function, with period π.

Some properties of μ_1 and μ_2

If x_1 and x_2 are solutions of the Mathieu Equation, one can easily prove that

$$\ddot{x}_1 x_2 - \ddot{x}_2 x_1 = 0,$$

or

$$\frac{\mathrm{d}(\dot{x}_1 x_2 - \dot{x}_2 x_1)}{\mathrm{d}\tau} = 0 , \tag{19.38}$$

leading to

$$\dot{x}_1 x_2 - \dot{x}_2 x_1 = c . \tag{19.39}$$

But, since

$$x_1(\pi + \tau) = \mu_1 x_1(\tau);$$

$$x_2(\pi + \tau) = \mu_2 x_2(\tau) ,$$

then

$$\dot{x}_1 \, (\pi + \tau) x_2(\pi + \tau) - \dot{x}_2 \, (\pi + \tau) x_1(\pi + \tau) = c$$
$$= \mu_1 \mu_2 (\dot{x}_1 \, x_2 - \dot{x}_2 \, x_1)$$
$$= (\dot{x}_1 \, x_2 - \dot{x}_2 \, x_1) \,,$$

thus

$$\mu_1 \mu_2 = 1 \,. \tag{19.40}$$

We can clearly differentiate two cases

a) μ_1 and μ_2 are reals; then

$$\mu_1 = (\mu_2)^{-1} = \exp \sigma_1 \pi \,, \tag{19.41}$$

and because σ is real, we get a stable and an unstable solution

$$x_1(\tau) = \exp(\sigma_1 \tau) F_1(\tau),$$
$$x_2(\tau) = \exp(-\sigma_1 \tau) F_2(\tau) \,.$$

So, in this case, one always has an unstable solution.

b) $\mu_1 = \mu_2^*$, $\mid \mu_{1,2} \mid^2 = 1$.

So, if $\mu_1 = \exp i\beta\pi$, $\mu_2 = \exp -i\beta\pi$, then

$$x_1(\tau) = \exp(i\beta\tau) F_1(\tau), \tag{19.42}$$
$$x_2(\tau) = \exp(-i\beta\tau) F_2(\tau) \,,$$

which are both stable solutions .

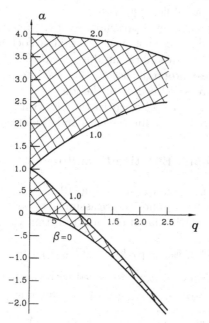

Fig. 19.3. Parameter space of Mathieu's Equation separating the stable region (with lines) from the unstable one

Boundary between stable and unstable region.

$\mu = \exp i\beta\pi$ becomes real for

$\beta = 0 \rightarrow \mu_1 = \mu_2 = 1,$

$\beta = 1 \rightarrow \mu_1 = \mu_2 = -1,$

in which case, the solutions are periodic functions of τ, with period $\pi(\beta = 0)$ or $2\pi(\beta = 1)$.

One can plot [2] the values of a and q that admit periodic solutions, with period 2π. This curve separates the stable from the unstable regions in the parameter space (Fig. 19.3).

19.2 Trapped Ions

19.2.1 Introduction

Recently, single quantum systems have been investigated, based on the advancement on low temperature [3] and confinement techniques. For example, in the case of single trapped ions, much work has been devoted to the coupling of a few internal electronic levels with the center of mass motion of the ion.

These couplings, in the case of the ion traps, can be achieved by direct transition [4, 5], or non-resonant Raman transition. In the latter case, this coupling is via two optical fields [6, 7, 8].

Several proposals have been presented for experiments leading to observation of non-classical states such as generation of Fock and squeezed vibrational states and also vibrational Schrodinger cats [9].

In most cases, these proposals use an effective interaction, between internal and external degrees of freedom of the ion.

Single trapped ions have also led to the observation of quantum jumps [5], antibunching in resonance fluorescence and Quantum Zeno effects. [10]

19.2.2 The Model and Effective Hamiltonian

The basic level scheme is shown in the Fig. 19.4.

The electronic levels $\mid a\rangle$ and $\mid b\rangle$ are assumed to be metastable, separated by $\hbar\omega_0$ and coupled by stimulated Raman transition via two classical optical fields

$$\mathbf{E}_i = \mathbf{E}_{0i} \left[\exp i(\mathbf{k}_i\mathbf{x} - \omega_i t + \phi_i) + cc\right], \tag{19.43}$$

where $\mathbf{x}$ is the position operator associated with the center of mass motion, and $\omega_1 - \omega_2 = (k_1 - k_2)c = \omega_0 - \delta$, δ being of the order of ω, the vibrational frequency of the ion.

Both fields 1 and 2 are detuned from the b–c and a–c transition by Δ and $\Delta - \delta$, respectively.

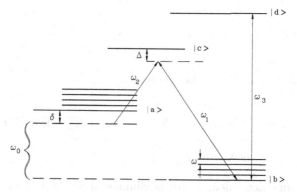

Fig. 19.4. Energy level diagram. A Raman-stimulated transition is induced between the levels $| a \rangle$ and $| b \rangle$ by laser beams 1 and 2. Detection of the electronic state is provided by the scattered photons resulting from the cycling transition $| b \rangle$ and $| d \rangle$, produced by a resonant pulse 3

A fourth level $| d \rangle$ is introduced for detecting electronic states and precooling, as it will be shown later.

We assumed that the ion is trapped in a harmonic potential. The centre of mass position operator can be written as

$$\mathbf{x}_i = \sqrt{\frac{\hbar}{2m\omega_i}}(a_i + a_i^\dagger), i = 1, 2, 3 , \tag{19.44}$$

where ω_i is the oscillatory frequency along the i direction.

The Hamiltonian that describes the system between detections (so the level $| d \rangle$ does not participate in the dynamics) can be written as

$$H = H_0 + H_1 , \tag{19.45}$$

with

$$H_0 = \sum_{i=x,y,z} \hbar\omega_i a_i^\dagger a_i + \hbar\omega_a \mid a \rangle\langle a \mid +\hbar\omega_b \mid b \rangle\langle b \mid +\hbar\omega_c \mid c \rangle\langle c \mid , \tag{19.46}$$

and

$$
\begin{aligned}
H_1 = &\hbar g_1 \exp\left[-i(\mathbf{k}_1 \cdot \mathbf{x} - \omega_1 t + \phi_1)\right] \mid b \rangle\langle c \mid \\
&+\hbar g_2 \exp\left[-i(\mathbf{k}_2 \cdot \mathbf{x} - \omega_2 t + \phi_2)\right] \mid a \rangle\langle c \mid +hc .
\end{aligned} \tag{19.47}
$$

For shorthand notational purposes, if we set

$$g_1 \exp\left[i(\mathbf{k}_1 \cdot \mathbf{x} + \phi_1)\right] \rightarrow g_1,$$

$$g_2 \exp\left[i(\mathbf{k}_2 \cdot \mathbf{x} + \phi_2)\right] \rightarrow g_2,$$

and

$$\omega_{cb} - \omega_1 = \Delta \, , \tag{19.48}$$

$$\omega_{ca} - \omega_2 = \Delta - \delta \, , \tag{19.49}$$

then the Hamiltonian H_0 can be written as

$$H_0 = \sum_{i=x,y,z} \hbar\omega_i a_i^\dagger a_i + \hbar(\omega_b + \Delta) \mid b\rangle\langle b \mid + \hbar\left[\omega_a + (\Delta - \delta)\right] \mid a\rangle\langle a \mid + \hbar\omega_c \mid c\rangle\langle c \mid$$
$$- \hbar\Delta \mid b\rangle\langle b \mid - \hbar(\Delta - \delta) \mid a\rangle\langle a \mid \, , \tag{19.50}$$

so that the time-dependent factor is eliminated in the interaction picture, because

$$\omega_c - (\omega_b + \Delta) = \omega_1 \, , \tag{19.51}$$

$$\omega_c - (\omega_a + \Delta - \delta) = \omega_2 \, . \tag{19.52}$$

The Hamiltonian in the interaction picture now reads

$$H' = -\hbar\Delta \mid b\rangle\langle b \mid - \hbar(\Delta - \delta) \mid a\rangle\langle a \mid$$
$$+ \hbar(g_1 \mid c\rangle\langle b \mid + g_1^* \mid b\rangle\langle c \mid)$$
$$+ \hbar(g_2 \mid c\rangle\langle a \mid + g_2^* \mid a\rangle\langle c \mid) \, . \tag{19.53}$$

Now we proceed to calculate the effective Hamiltonian or the a and b levels only, by adiabatic elimination of the c level [11].

Starting from the Liouville equation

$$\dot{\rho} = -\frac{i}{\hbar} [H', \rho] \, , \tag{19.54}$$

we can write

$$\dot{\rho} = i\Delta \left[\mid b\rangle\langle b \mid, \rho\right] + i(\Delta - \delta) \left[\mid a\rangle\langle a \mid, \rho\right]$$
$$- i \left[g_1 \mid c\rangle\langle b \mid + g_1^* \mid b\rangle\langle c \mid, \rho\right]$$
$$- i \left[g_2 \mid c\rangle\langle a \mid + g_2^* \mid a\rangle\langle c \mid, \rho\right] \, . \tag{19.55}$$

Writing Liouville equation by its matrix elements, we get

$$\dot{\rho}_{cc} = -ig_1\rho_{bc} + ig_1^*\rho_{cb} - ig_2\rho_{ac} + ig_2^*\rho_{ca} \, , \tag{19.56}$$

$$\dot{\rho}_{bb} = ig_1\rho_{bc} - ig_1^*\rho_{cb} \, , \tag{19.57}$$

$$\dot{\rho}_{aa} = ig_2\rho_{ac} - ig_2^*\rho_{ca} \, , \tag{19.58}$$

$$\dot{\rho}_{ca} = -i(\Delta - \delta)\rho_{ca} - ig_2\rho_{aa} + ig_2\rho_{cc} - ig_1\rho_{ba} \, , \tag{19.59}$$

$$\dot{\rho}_{cb} = -i\Delta\rho_{cb} - ig_1\rho_{bb} + ig_1\rho_{cc} - ig_2\rho_{ab} \, , \tag{19.60}$$

$$\dot{\rho}_{ab} = -i\Delta\rho_{ab} + i(\Delta - \delta)\rho_{ab} + ig_1\rho_{ac} - ig_2^*\rho_{cb} \, . \tag{19.61}$$

Now, we take $\Delta \gg \delta$ and set ρ_{ca}, ρ_{cb} to steady state

$$\rho_{cb} = \frac{1}{i\Delta} \left[ig_1(\rho_{cc} - \rho_{bb}) - ig_2\rho_{ab} \right] , \tag{19.62}$$

$$\rho_{ca} = \frac{1}{i\Delta} \left[ig_2(\rho_{cc} - \rho_{aa}) - ig_1\rho_{ba} \right] , \tag{19.63}$$

and replacing (19.62, 19.63) in (19.56) we get

$$\dot{\rho}_{cc} = 0,$$

so, if $\rho_{cc}(0) = 0$, then $\rho_{cc}(t) = 0$.

The rest of the equations become

$$\dot{\rho}_{bb} = \frac{ig_1^* g_2}{\Delta}\rho_{ab} - \frac{ig_2^* g_1}{\Delta}\rho_{ba} , \tag{19.64}$$

$$\dot{\rho}_{aa} = \frac{ig_2^* g_1}{\Delta}\rho_{ba} - \frac{ig_1^* g_2}{\Delta}\rho_{ab}, \tag{19.65}$$

$$\dot{\rho}_{ab} = -i\left(\delta'\right)\rho_{ab} - \frac{ig_2^* g_1}{\Delta}(\rho_{aa} - \rho_{bb}) , \tag{19.66}$$

with

$$\delta' = \left(\delta + \frac{|g_1|^2}{\Delta} - \frac{|g_2|^2}{\Delta}\right) . \tag{19.67}$$

We notice that δ' differs from δ by the Stark shifts $\frac{|g_1|^2}{\Delta}$ and $\frac{|g_2|^2}{\Delta}$.
Now, if we write the effective Hamiltonian in the following form

$$H = \hbar\alpha\frac{|a\rangle\langle a| - |b\rangle\langle b|}{2} + \hbar\beta |a\rangle\langle b| + \hbar\beta^* |b\rangle\langle a| , \tag{19.68}$$

then the Liouville equation gives us

$$\dot{\rho}_{bb} = -i\beta^*\rho_{ab} + i\beta\rho_{ba} , \tag{19.69}$$

$$\dot{\rho}_{aa} = -i\beta\rho_{ba} + i\beta^*\rho_{ab} , \tag{19.70}$$

$$\dot{\rho}_{ab} = -i\alpha\rho_{ab} + i\beta(\rho_{aa} - \rho_{bb}) . \tag{19.71}$$

By direct comparison of the above equations with (19.64, 19.65, 19.66), we get

$$\alpha = \delta', \tag{19.72}$$

$$\beta = -\frac{g_2^* g_1}{\Delta} .$$

The final effective Hamiltonian, is

$$H = \sum_{i=x,y,z} \hbar\omega_i a_i^\dagger a_i + \hbar\delta\sigma_3 - \frac{\hbar\Omega_0}{2} \left[| a\rangle\langle b | \exp\left[i(k_1 - k_2)x + i\phi\right] + hc\right] ,$$

(19.73)

with $\Omega_0 \equiv \left(\frac{2|g_1 g_2|}{\Delta}\right)$, $\sigma_3 \equiv \left(\frac{|a\rangle\langle a| - |b\rangle\langle b|}{2}\right)$.

In terms of the phonon raising and lowering operators, we can also write it as

$$H = \sum_{i=x,y,z} \hbar\omega_i a_i^\dagger a_i + \hbar\delta\sigma_3 - \frac{\hbar\Omega_0}{2} \left\{ | a\rangle\langle b | \exp\left[i \sum_i \eta_i(a_i + a_i^\dagger) + i\phi\right] + hc \right\} ,$$

(19.74)

where η_i is the Lamb–Dicke parameter, defined as

$$\eta_i \equiv \delta k_i \sqrt{\frac{\hbar}{2m\omega_i}},$$

(19.75)

$$\delta k_i \equiv (k_1 - k_2)_i .$$

(19.76)

We notice that according to the definition of η_i, η_i^2 is the ratio between the recoil energy and the quantum vibrational energy, both taken in the i-th direction.

In the rest of the chapter, we are going to assume, for the sake of simplicity, that only vibrations in a given i direction are excited, which could be the case if ω_i is much larger than the other frequencies or if $\delta\mathbf{k}$ is in the i direction. Then the Hamiltonian is

$$H = \hbar\omega a^\dagger a + \hbar\delta\sigma_3 - \frac{\hbar\Omega_0}{2} \left\{ | a\rangle\langle b | \exp\left[i\eta(a + a^\dagger) + i\phi\right] + hc \right\} .$$ (19.77)

Expanding the exponential and using the B.C.H. identity (see appendix A), we get

$$\exp i\eta(a + a^\dagger) = \exp\left(-\frac{\eta^2}{2}\right) \sum_{l,l'} \frac{(i\eta)^{l+l'}}{l!l'!} a^{\dagger l} a^{l'} ,$$

(19.78)

and

$$H = \hbar\omega a^\dagger a + \hbar\delta\sigma_3 - \frac{\hbar\Omega_0}{2} \left[| a\rangle\langle b | \exp\left(-\frac{\eta^2}{2}\right) \sum_{l,l'} \frac{(i\eta)^{l+l'}}{l!l'!} a^{\dagger l} a^{l'} \exp(i\phi) + hc \right] .$$

(19.79)

In the interaction picture, the Hamiltonian becomes

$$H_I = -\frac{\hbar\Omega_0}{2} \left(| a\rangle\langle b | \exp\left(-\frac{\eta^2}{2} + i\phi\right) \sum_{l,l'} \frac{(i\eta)^{l+l'}}{l!l'!} a^{\dagger l} a^{l'} \exp\left\{it\left[(l - l')\omega + \delta\right]\right\} + hc \right) .$$

(19.80)

Let $k = l' - l$, so that

$$\Delta_k \equiv \delta' - k\omega \, , \qquad (19.81)$$

and we define

$$\Omega_k \equiv \frac{\Omega_0}{2} \exp\left(-\frac{\eta^2}{2} + i\phi\right) (i\eta)^k \sum_l \frac{(i\eta)^{2l}}{l!(l+k)!} a^{\dagger l} a^l \, . \qquad (19.82)$$

Considering now a near resonant condition, where for a particular value of k, $\Delta_k \ll k\omega$, and neglecting the fast rotating terms, one can write approximately

$$H_I = -\hbar \begin{pmatrix} 0 & \Omega_k \exp(i\Delta_k t) a^k \\ a^{\dagger k} \Omega_k^{\dagger} \exp(-i\Delta_k t) & 0 \end{pmatrix} \, . \qquad (19.83)$$

19.2.3 The Lamb–Dicke Expansion and Raman Cooling

In the Lamb–Dicke limit $\eta\sqrt{n} \ll 1$, we mention several interesting cases (in this section, we will neglect the Stark shifts thus δ' reduces to δ). We concentrate on the time-independent case, that is for $\Delta_k = 0$, or $\delta' = k\omega$.

'Carrier' Transition

This case corresponds to $\delta = 0, k = 0$, and the Hamiltonian lowest order term in η is

$$H_I = -\frac{\hbar\Omega_0}{2} \left[\exp(i\phi)\sigma_+ + \exp(-i\phi)\sigma_-\right] \, , \qquad (19.84)$$

Hamiltonian that produces Rabi oscillations between vibrational sublevels of the same degree of excitation, as shown in the Fig. 19.5.

First 'Red' Sideband

This case corresponds to

$$\delta = \omega \, , \qquad (19.85)$$

$$k = 1$$

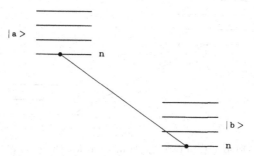

Fig. 19.5. Rabi oscillation between two vibrational sublevels of the same degree of excitation. This case corresponds to $k = 0 = \delta$

and we get, to the lowest order in η,

$$H_I = -i\eta\hbar\frac{\Omega_0}{2}(a\sigma_+ \exp i\phi - a^\dagger\sigma \exp -i\phi) , \qquad (19.86)$$

which is the Jaynes–Cummings Hamiltonian. The vibrational transitions described by this Hamiltonian is shown in the Fig. 19.6

Similarly, there is a first blue sideband, corresponding to a 'anti-Jaynes–Cummings' model and higher order sidebands corresponding to non-linear optical models.

For example, if $\delta = 2\omega$ or $k = 2$ corresponds to the two-photon Jaynes–Cummings model

$$H_I = \frac{\Omega_0}{2}\frac{\eta^2\hbar}{2}(a^2\sigma_+ \exp i\phi + a^{\dagger 2}\sigma \exp -i\phi) , \qquad (19.87)$$

and so on.

Recent experimental results [8] reported that the first red sideband was used for cooling purposes. However, if they just tuned the $\delta = \omega$ transition, one can go from $n \to n - 1$, but the reverse process is also possible and on the average, no cooling occurred [12].

To achieve resolved sideband stimulated Raman cooling , they needed the following sequence.

First, a π red sideband pulse, producing $| b \rangle | n \rangle \to | a \rangle | n - 1 \rangle$ transition. Second, an additional 'repumper' pulse populates the $| c \rangle$ level, followed by spontaneous emission, to the same $| n - 1 \rangle$ level, but now associated to the $| b \rangle$ electronic state. This is showed in the Fig. 19.7.

To have a quantitative understanding of the above effects, we require the study of the dynamics of our system [11, 13].

19.2.4 The Dynamical Evolution

An exact solution for the time evolution operator can be derived, in some simple cases, like the Hamiltonian given by (19.83).

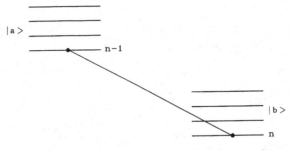

Fig. 19.6. First sideband vibrational transition for the case $\delta = \omega$, $k = 1$, that corresponds to the Jaynes–Cummings dynamics. The upwards transition ($b \to a$) produces a vibrational cooling from $n \to n - 1$

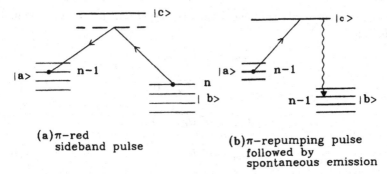

**(a)π−red
sideband pulse**

**(b)π−repumping pulse
followed by
spontaneous emission**

Fig. 19.7. Resolved sideband stimulated Raman-stimulated Raman cooling. First, a red sideband π pulse produces the $|b\rangle\,|n\rangle \to |a\rangle\,|n-1\rangle$ transition. A second π repumping pulse followed by spontaneous emission produces the $|a\rangle\,|n-1\rangle \to |b\rangle\,|n-1\rangle$ transition having a net cooling effect

We show here the detailed calculation to illustrate the method. We have to solve the equation

$$i\hbar\frac{dU}{dt} = HU , \tag{19.88}$$

or

$$i\begin{pmatrix} \dot{U}_{aa} & \dot{U}_{ab} \\ \dot{U}_{ba} & \dot{U}_{bb} \end{pmatrix} = \begin{pmatrix} 0 & -\Omega_k \exp(i\Delta_k t)a^k \\ -a^{\dagger k}\Omega_k^{\dagger}\exp(-i\Delta_k t) & 0 \end{pmatrix}\begin{pmatrix} U_{aa} & U_{ab} \\ U_{ba} & U_{bb} \end{pmatrix} , \tag{19.89}$$

giving us the set of equations

$$i\dot{U}_{aa} = -\Omega_k\exp(i\Delta_k t)a^k U_{ba} , \tag{19.90}$$

$$i\dot{U}_{ba} = -a^{\dagger k}\Omega_k^{\dagger}\exp(-i\Delta_k t)U_{aa} , \tag{19.91}$$

$$i\dot{U}_{ab} = -\Omega_k\exp(i\Delta_k t)a^k U_{bb} , \tag{19.92}$$

$$i\dot{U}_{bb} = -a^{\dagger k}\Omega_k^{\dagger}\exp(-i\Delta_k t)U_{ab} . \tag{19.93}$$

We can easily eliminate the time-dependent factors with the transformation

$$\bar{U}_{ba} = \exp i\frac{\Delta_k t}{2}U_{ba} , \tag{19.94}$$

$$\bar{U}_{aa} = \exp -i\frac{\Delta_k t}{2}U_{aa} , \tag{19.95}$$

$$\bar{U}_{ab} = \exp -i\frac{\Delta_k t}{2}U_{ab} , \tag{19.96}$$

$$\bar{U}_{bb} = \exp i\frac{\Delta_k t}{2}U_{bb} . \tag{19.97}$$

It is simple to verify that the differential equations for the various matrix elements are

$$\ddot{\overline{U_{bb}}} + \bar{\mu}^2\,\overline{U_{bb}} = 0\;, \tag{19.98}$$

$$\ddot{\overline{U_{ba}}} + \bar{\mu}^2\,\overline{U_{ba}} = 0\;, \tag{19.99}$$

$$\ddot{\overline{U_{aa}}} + \mu^2\,\overline{U_{aa}} = 0\;, \tag{19.100}$$

$$\ddot{\overline{U_{ab}}} + \mu^2\,\overline{U_{ab}} = 0\;, \tag{19.101}$$

where

$$\mu^2 = \frac{\Delta_k^2}{4} + \Omega_k a^k a^{\dagger k}\Omega_k^\dagger\;, \tag{19.102}$$

$$\bar{\mu}^2 = \frac{\Delta_k^2}{4} + a^{\dagger k}\Omega_k^\dagger\Omega_k a^k\;. \tag{19.103}$$

The reader may verify that the solutions of the above equations for the initial conditions $U_{bb}(0) = U_{aa}(0) = 1$ and $U_{ab}(0) = U_{ba}(0) = 0$ are [14]:

$$U_{aa}(t) = \exp i\frac{\Delta_k t}{2}\left(\cos\mu t - i\frac{\Delta_k}{2\mu}\sin\mu t\right)\;, \tag{19.104}$$

$$U_{ab}(t) = \exp\left(i\frac{\Delta_k t}{2}\right)i\frac{\sin\mu t}{\mu}\Omega_k a^k\;, \tag{19.105}$$

$$U_{ba}(t) = \exp\left(-i\frac{\Delta_k t}{2}\right)i\frac{\sin\bar{\mu}\,t}{\bar{\mu}}a^{\dagger k}\Omega_k^\dagger\;, \tag{19.106}$$

$$U_{bb}(t) = \exp -i\frac{\Delta_k t}{2}\left(\cos\bar{\mu}\,t + i\frac{\Delta_k}{2\,\bar{\mu}}\sin\bar{\mu}\,t\right)\;. \tag{19.107}$$

For the resonant case ($\Delta_k = 0$), we can expand these results in η (Lamb–Dicke regime), getting

$$\Omega_k \approx \frac{\Omega_0}{2}\exp i\left(\phi + \frac{k\pi}{2}\right)\eta^k\left\{1 - \eta^2\left[\frac{a^\dagger a}{(k+1)!} + \frac{1}{2} + ...\right]\right\}\;, \tag{19.108}$$

$$\mu^2 \approx \frac{\Omega_0^2}{4}\eta^{2k}a^k a^{k\dagger}\left\{1 - 2\eta^2[\frac{a^\dagger a}{(k+1)!} + \frac{1}{2}]...\right\}\;, \tag{19.109}$$

$$\bar{\mu}^2 \approx \frac{\Omega_0^2}{4}\eta^{2k}a^{k\dagger}a^k\left\{1 - 2\eta^2[\frac{a^\dagger a}{(k+1)!} + \frac{1}{2}]...\right\}\;, \tag{19.110}$$

$$U_{aa}(t) = \cos\mu t,\; U_{bb}(t) = \cos\bar{\mu}\,t, \tag{19.111}$$

$$U_{ab}(t) = i\frac{\sin\mu t}{\mu}\Omega_k a^k,\; U_{ba}(t) = i\frac{\sin\bar{\mu}\,t}{\bar{\mu}}a^{\dagger k}\Omega_k^\dagger\;.$$

For the Jaynes–Cummings model, applied for cooling, $k = 1$, and initially with the ion in the state $| n \rangle \, | b \rangle$, after some time t

$$| \psi(t) \rangle = \begin{pmatrix} \cos \mu t & i \frac{\sin \mu t}{\mu} \Omega_1 a \\ i \frac{\sin \bar{\mu} t}{\bar{\mu}} a^\dagger \Omega_1^\dagger & \cos \overline{\mu} t \end{pmatrix} \begin{pmatrix} 0 \\ 1 \end{pmatrix} | n \rangle , \tag{19.112}$$

$$| \psi(t) \rangle = \left[\left(i \frac{\sin \mu t}{\mu} \Omega_1 a \right) | a \rangle + \cos \bar{\mu} \, t \, | b \rangle \right] | n \rangle . \tag{19.113}$$

In the lowest order, $\bar{\mu} \approx \mu \approx (\frac{\Omega_0}{2}) \eta \sqrt{n}$. A π pulse corresponds to $\Omega_0 \eta \sqrt{n} t = \pi$, and the net effect of such a pulse is to change the state

$$| n \rangle \, | b \rangle \rightarrow i \frac{\Omega_1 \sqrt{n}}{\mu} \, | n - 1 \rangle \, | a \rangle , \tag{19.114}$$

that corresponds to the experimental situation described in the previous section.

19.2.5 QND Measurements of Vibrational States

We assume [15, 16], in this case , that $\delta = k = 0$, and if

$$| \psi(0) \rangle = \sum_n C_n \, | n \rangle \, | b \rangle , \tag{19.115}$$

then at a later time t

$$| \psi(t) \rangle_b = \sum_n C_n \left[\left(i \exp i\phi \sin \frac{\Omega_n}{2} t \right) | a \rangle + \cos \frac{\Omega_n}{2} t \, | b \rangle \right] | n \rangle , \tag{19.116}$$

where $\Omega_n = 2 \mu_{nn} = 2 \, \bar{\mu}_{nn} = \Omega_0 \left[1 - \eta^2 (n + \frac{1}{2}) \right] ...$ is the usual Rabi frequency.

In the case the ion is initially in the $| a \rangle$ electronic state, at time t it will be at

$$| \psi(t) \rangle_a = \sum_n C_n \left[\left(i \exp -i\phi \sin \frac{\Omega_n}{2} t \right) | b \rangle + \cos \frac{\Omega_n}{2} t \, | a \rangle \right] | n \rangle . \tag{19.117}$$

The linear dependence of Ω_n with n is the basis for the QND measurement of the vibrational population for the trapped ion. We proceed as follows:

The ion is submitted to a Raman pulse of duration τ, resonant with the electronic transition, so that the vibrational occupation number is not changed. This time τ is assumed to be much smaller than the lifetimes of the electronic states or the vibrational states.

Right after the pulse, the electronic state of the ion is detected, by resonant excitation of the $| b \rangle \rightarrow | d \rangle$ transition with circularly polarized light. This is the way the experimental detection was reported in the reference [9].

We assume that the area of this pulse is sufficiently large, so that in this stage, a large amount of photons are scattered by the ion, thus generating a detection efficiency close to one. This determines the electronic state of the ion. A fluorescent signal implies that the ion is in the $| \, b \rangle$ state, and therefore, as a consequence of the measurement, the ion is projected into this state, while the absence of fluorescence projects the ion into the $| \, a \rangle$ state.

We notice that in the Lamb–Dicke regime, each photon scattering leads to a negligible recoil ($\eta \propto \Delta k$).

However, for many photons, appreciable heating is possible, spoiling our QND procedure, based on the fact that approximately no energy exchange takes place during the measurement.

This puts an upper limit on the photon number scattered by the ion.

A rough estimate of this limitation goes as follows. Let ΔE be the recoil energy of a single scattering process, then we must assume, for a near QND measurement, that

$$N \Delta E << \hbar \omega \ , \tag{19.118}$$

N being the number of scattered photons, and because $\eta^2 = \frac{\Delta E}{\hbar \omega}$, the condition becomes

$$N << \eta^{-2} \ . \tag{19.119}$$

For a typical experimental value, $\eta \sim 0.1$ so $N < \, < 100$.

On the other hand, for a saturating cycling process, $N \approx \frac{\Gamma T}{2}$, where Γ is the width of the $| \, d \rangle$ level and T the duration of the pulse, so one must have $T << \frac{200}{\Gamma}$ and for $\frac{\Gamma}{2\pi} = 20\mathrm{MHz}$, we get

$$T < 2\mu s \ . \tag{19.120}$$

If the above conditions are satisfied, then very little heating takes place, which can be neglected, and, if $| \, \psi_\varepsilon^{(1)} \rangle$ is the state of the ion after the first measurement, where $\varepsilon = 1$ corresponds to the case in which the detected electronic state coincides with the initial one, and $\varepsilon = 0$ otherwise, then

$$| \, \psi_\varepsilon^{(1)} \rangle = \frac{\sum_n \exp\left[i \left(\frac{\pi}{2} \pm \phi\right) (1 - \varepsilon)\right] C_n \sin\left(\frac{\Omega_n \tau}{2} + \frac{\varepsilon \pi}{2}\right)}{\sqrt{\sum_{n=0}^{\infty} | \, C_n \, |^2 \sin^2\left(\frac{\Omega_n \tau}{2} + \frac{\varepsilon \pi}{2}\right)}} | \, n \rangle \ . \tag{19.121}$$

Equation (19.121) shows that the original vibrational distribution $P(n) = | \, C_n \, |^2$, after the measurement is modified

$$P_\varepsilon^{(1)}(n) = \frac{P(n) \sin^2\left(\frac{\Omega_n \tau}{2} + \frac{\varepsilon \pi}{2}\right)}{\sum_{n'=0}^{\infty} P(n') \sin^2\left(\frac{\Omega_n \tau}{2} + \frac{\varepsilon \pi}{2}\right)} \ . \tag{19.122}$$

This implies a decimation of the population, depending on the phase $\theta(\tau) = \eta^2 \tau \frac{\Omega_0}{2}$.

To enhance the n dependence of Ω_n, we choose a pulse duration τ such that $\theta(\tau) = \pi$.

On the other hand, the term $\frac{\Omega_0 \tau}{2}$ is a large number, but independent of both n and η and produces an irrelevant phase shift.

After the first sequence of Raman pulse and detection, a new cycle is initiated with a different pulse area $\Omega_0 \tau$, thus multiplying the original $P(n)$ distribution by sines and cosines. After the i-th cycle

$$P_\varepsilon^{(i)}(n) = \frac{P^{(i-1)}(n) \sin^2 \left(\frac{\Omega_n \tau_i}{2} + \frac{\varepsilon \pi}{2}\right)}{\sum_{n'=0}^{\infty} P^{(i-1)}(n') \sin^2 \left(\frac{\Omega_n \tau_i}{2} + \frac{\varepsilon \pi}{2}\right)}, \qquad (19.123)$$

where τ_i is the duration of the i-th Raman cycle, and we defined $P_\varepsilon^{(0)} = P(n)$, and $P^{(i-1)}(n)$ is the probability distribution of vibronic excitations of the previous cycle.

The numerical simulation of this procedure shows that this may result in a decimation of more and more population, until a Fock state is reached, and then, the state does not change anymore.

If this procedure is done in all three directions, a three-dimensional Fock state is formed.

As this process depends on the random nature of the detected state, which Fock state is obtained is something that cannot be predicted a priori.

This experiment can be turned into a 'numerical experiment', by feeding the computer with data about the successive state detection and pulse duration.

We notice here an interesting point. Doing the experiment many times gives us every time a different Fock state, thus building up a distribution of vibrational states. As the QND measurement did not change the vibrational states and adding all possible measurements implies no measurement at all, this vibrational distribution should be identical to the original distribution $P(n)$.

Also, the probability of detecting an ion in the $\mid a \rangle$ or $\mid b \rangle$ state, at the end of the i-th cycle is

$$P_\varepsilon^{(i)}(n) = \sum_{n'=0}^{\infty} P^{(i-1)}(n') \sin^2 \left(\frac{\Omega_{n'} \tau_i}{2} + \frac{\varepsilon \pi}{2}\right), \qquad (19.124)$$

where, as above, $\varepsilon = 1$ or 0 depending on whether this state coincides or not with the electronic state in the beginning of the cycle.

In order to do the numerical experiment, we choose a random number with a flat distribution, between 0 and 1, and according to the above probabilities of being in the $\mid a \rangle$ or $\mid b \rangle$ state, we decide every time the outcome of the measurement.

The result is shown in the Fig. 19.8.

19.2.6 Generation of Non-Classical Vibrational States

Generation and detection of vibrational Fock , coherent and squeezed states was recently achieved, with a single $^9Be^+$ ion, confined in a rf Paul trap, with a frequency $\frac{\omega}{2\pi} = 11.2\,\mathrm{MHz}$, along the x-axis and $\eta = 0.2$.

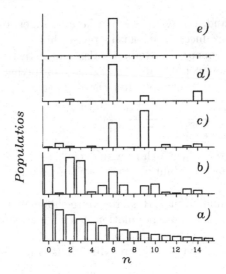

Fig. 19.8. Population distribution as a function of the vibronic excitation number, for $\eta = 0.1$, and (**a**) Initial state, which was chosen to be a thermal distribution with $\langle n \rangle = 5$. (**b**) After the first cycle. (**c**) After the fifth cycle. (**d**) After the tenth cycle. (**e**) After the thirteenth cycle. θ was chosen at random and which level was detected was chosen according to the probability distribution

Once the ion is prepared in the $| b \rangle | 0 \rangle$ state, a high n-Fock state can be created by simply applying a sequence of π-pulses of laser radiation on the first blue sideband ($k = -1$), the first red sideband ($k = 1$) and carrier ($k = 0$).

For example we want to generate the $| a \rangle | 2 \rangle$ state. This can be achieved in the following sequence

$$| b \rangle | 0 \rangle \quad \overset{\pi\text{-}pulse}{\underset{\substack{blue \\ sideband}}{\longrightarrow}} \quad | a \rangle | 1 \rangle \quad \overset{\pi\text{-}pulse}{\underset{\substack{red \\ sideband}}{\longrightarrow}} \quad | b \rangle | 2 \rangle \quad \overset{\pi\text{-}pulse}{\underset{carrier}{\longrightarrow}} | a \rangle | 2 \rangle.$$

This can be also seen in the Fig. 19.9.

Coherent states can be produced by a classical driving field, oscillating at the trap frequency.

We can also apply a two-photon Raman coupling within one atomic state, with $k = 1$ ($\delta = \omega$)

Starting from $n = 0$, the vibronic excitation number diffuses upwards. It can be shown that applying these sequence of pulses is equivalent to apply the displacement operator

$$| 0 \rangle \rightarrow | \alpha \rangle,$$

with $\alpha = \eta \Omega_0 \tau$.

Finally, **Zoller** et al. [17] proposed to prepare squeezed states of motion in an ion trap with a combination of standing and traveling wave light fields.

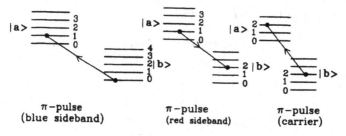

π-pulse
(blue sideband)

π-pulse
(red sideband)

π-pulse
(carrier)

Fig. 19.9. $\mid n\rangle$ state created by applying a sequence of Rabi π-pulses, first on the blue sideband ($k = -1$), next on the red sideband ($k = 1$) and finally on the carrier frequency ($k = 0$)

Also, one can irradiate the ion in the $\mid n = 0\rangle$ state with two Raman beams that differ by 2ω, driving transitions between even-n levels and creating a squeezed vacuum state. The interested reader can check the references for more details [18].

Furthermore, there has been recently interesting advances in creating arbitrary superpositions of coherent states [22], entangled states with two trapped ions [23, 24] and laser cooling with two trapped ions [25].

Problems

19.1. Consider a single two-level ion moving in a one-dimensional Paul trap and interacting with a classical laser field [19].

The corresponding Hamiltonian is

$$H(t) = H_{cm} + H_a + H_{In},$$

where H_{cm} represents the center of mass motion, H_a is the ion's internal energy and H_{In} is the interaction term. The three parts of the Hamiltonian can be written as

$$H_{cm} = \frac{p^2}{2m} + \frac{1}{2}\omega(t)^2 x^2,$$

$$H_a = \frac{\hbar\omega_a}{2}\sigma_z,$$

$$H_{In} = \hbar g[\sigma^+ \exp -i(\omega_L t - kx) + hc],$$

where

$$\omega^2(t) = \frac{\Omega^2}{4}[a + 2q \cos \Omega t].$$

That parameters a and q are proportional to the applied DC and AC fields of the trap and Ω is the frequency of the AC field.

Show that in the Interaction picture

$$H_{\text{INT}} = \hbar g \{\sigma^+ \exp(-i\Delta t) D\left[\alpha(t)\right] + hc\},$$

where

$$\Delta = \omega_L - \omega_a,$$
$$\alpha(t) = i\eta\varepsilon(t),$$
$$\eta = k\sqrt{\frac{\hbar}{2m\Omega}},$$

and $\varepsilon(t)$ satisfies the Mathieu's differential [21]

$$\ddot{\varepsilon} + \omega(t)^2 \varepsilon = 0.$$

19.2. In the problem 19.1, the interaction involves multiphoton transitions. To make this more evident, prove that

$$H_{\text{INT}} = \sum_{n=0}^{\infty} \sum_{s=-n}^{\infty} \hbar \Omega^{(n,n+s)}(t) \sigma^+ \mid n \rangle \langle n+s \mid + hc,$$

where the generalized Rabi frequency is defined as

$$\Omega^{(n,n+s)}(t) = g\sqrt{\frac{n!}{(n+s)!}} \exp(-i\Delta t) \left[i\eta\varepsilon^*(t)\right]^s$$

$$\exp\left[-\frac{\eta^2}{2}\varepsilon(t)^2\right] L_n^s(\eta \mid \varepsilon(t) \mid^2)$$

for $s \geq 0$, and

$$\Omega^{(n,n+s)}(t) = g\sqrt{\frac{(n+s)!}{n!}} \exp(-i\Delta t) \left[i\eta\varepsilon(t)\right]^{-s}$$

$$\exp\left[-\frac{\eta^2}{2}\varepsilon(t)^2\right] L_{n+s}^{-s}(\eta \mid \varepsilon(t) \mid^2)$$

for $-n \leq s < 0$.
Hint: use the relations [20]

$$\langle n \mid D(\alpha(t)) \mid m \rangle = \sqrt{\frac{m!}{n!}} \exp\left(-\frac{1}{2} \mid \alpha \mid^2\right) \alpha^{n-m} L_m^{n-m}(\mid \alpha \mid^2), m \leq n,$$

$$\langle n \mid D(\alpha(t)) \mid m \rangle = \sqrt{\frac{n!}{m!}} \exp\left(-\frac{1}{2} \mid \alpha \mid^2\right) (-\alpha^*)^{m-n} L_n^{m-n}(\mid \alpha \mid^2), m \geq n.$$

References

1. Cohen-Tannoudji, C., Cours de Physique Atomique et Moleculaire (notes, 1986)
2. Ince, E.L.: Ordinary Differential Equation Dover, NewYork (1956)
3. Javanainen, J.: JOSA.B, **5**, 73 (1988)
4. Dietrich, F., Bergquist, J.C., Itano, W.M., Wineland, D.J.: Phys. Rev. Lett., **62**, 403 (1989)
5. Nagourney, W., Sandberg, J., Dhemelt, H.G.: Phys. Rev. Lett., **56**, 2 797 (1986); Sauter, T.H., Neuhauser, W., Blatt, R., Toschek, P.: Phys. Rev. Lett., **57**, 1696 (1986); Berquist, J.C., Hulet, R.G., Itano, I.W.M., Wineland, D.J.: Phys. Rev. Lett., **57**, 1699 (1986)
6. Kasevich, M., Chu, S.: Phys. Rev. Lett., **69**, 1741 (1992)
7. Heinzen, D.J., Wineland, D.J.: Phys. Rev. A, **42**, 2977 (1990)
8. Monroe, C., Meekhof, D.M., King, B.E., Jefferts, S.R., Itano, W.M., Wineland, J.M., Gould, P., Phys. Rev. Lett., **75**, 4011 (1995); Monroe, C., Meekhof, D.M., King, B.E., Itano, W.M., Wineland, J.M.: Phys. Rev. Lett., **75**,4 714 (1995)
9. Monroe, C., Meekhof, D.M., King, B.E., Wineland, J.M.: Science, **272**, 1131 (1996)
10. Itano, W.M., Heinzen, D.J., Bollinger, J.J., Wineland, J.M.: Phys. Rev. A, **41**, 2295 (1990)
11. Retamal, J.C.: Notes and private communication.
12. Monroe, C.: Lecture notes, Swieca School, Rio de Janeiro (1996)
13. Marzoli, I., Cirac, J.I., Blatt, R., Zoller, P.: Phys. Rev. A, **49**, 2771 (1994)
14. Retamal, J.C., Zagury, N.: Phys. Rev. A, **55** , 2387 (1997)
15. Davidovich, L., Orszag, M., N. Zagury, Phys. Rev. A, **54**, 5118 (1996)
16. Vogel, W., deMathos Filho, R.L.: Phys. Rev. Lett., **22**, 4608 (1996)
17. Cirac, J.I., Parkins, A.S., Blatt, R., Zoller, P.: Phys. Rev. Lett., **70**, 556 (1993)
18. Meekhof, D.M., Monroe, C., King, B.E., Itano, W.M., Wineland, J.M., Gould, P.: Phys. Rev. Lett., **76**, 1796 (1996)
19. Bardroff, P.J., Leichtle, C., Schrade, G., Schleich, W.P.: Acta. Phys. Slov., **46**, 231 (1996)
20. Cahill, K.E., Glauber, R.J.: Phys. Rev., **177**, 1857 (1969)
21. Cirac, J.I., Garay, L.J., Blatt, R., Parkins, A.S., Zoller, P.: Phys. Rev. A, **49**, 421 (1994)
22. Moya-Cessa, H., Wallentowitz, S., Vogel, V.W.: Phys. Rev. A, **59**, 2920 (1999)
23. Turchette, Q.A., Wood, C.S., King, B.E., Myat, C.J., Leibfied, D., Itano, W.M., Monroe, C., Wineland, D.J.: Phys. Rev. Lett., **81**, 3631 (1998)
24. Solano, E., deMatos Filho, R.L., Zagury, N.: Phys. Rev. A, **59**, R2539 (1999)
25. Morigi, G., Eschner, J., Cirac, J.I., Zoller, P.: Phys. Rev. A, **59**, 3797(1999)

Further Reading

Observation of Quantum Jumps on Single Ions

- Nagourney, W., Sandberg, J., Dhemelt, H.: Phys. Rev. Lett., **56**, 2797 (1986)
- Sauter, Th., Neuhauser, W., Blatt, R., Toschek, P.E.: Phys. Rev. Lett., **57**, 1696 (1986)

- Bregquist, J.C., Hulet, R.G., Itano, W.M., Wineland D.J.: Phys. Rev. Lett., **57**, 1699 (1986)

Photon Antibunching in Single Ions

- Dietrich, F., Walther, H.: Phys. Rev. Lett., **58**, 203 (1987)

20. Decoherence

In this chapter, we study the general concept of decoherence, the dynamics of the correlations, an example where we calculate the decoherence time and how to avoid these effects, via the Decoherence Free Subspaces (DFS). The conditions for DFS are given as well as some examples at the end of the chapter.

Quantum Mechanics is a very successful theory that explains a huge number of physical phenomena.

Quantum states evolve according to Schrödinger's equation, which is linear. As such, the superposition principle plays a major role.

An important factor is that macroscopic systems are coupled to the environment, and therefore, we are dealing, in general, with open systems where the Schrödinger equation is no longer applicable, or, to put it in a different way, the **coherence leaks out of the system into the environment, and, as a result, we have decoherence** [1, 2, 3].

Niels Bohr [4] proposed that, according to the Copenhagen interpretation of Quantum Mechanics, a classical apparatus was necessary to carry out the measurements, thus implying **a sharp borderline between the Classical and the Quantum world.**

Traditionally, the Classical Systems are associated to the macroscopic and Quantum to the microscopic [1], but this distinction is actually not very adequate considering recently studied effects of macroscopic systems that behave quantum mechanically. We also have the non-classical squeezed states with large number of photons, etc.

As opposed to **Bohr, Von Neumann** [5] considered quantum measurements .

Let us assume that we have a system with states $\mid a \rangle$ and $\mid b \rangle$ and a metre that can be in the states $\mid d_a \rangle$ and $\mid d_b \rangle$.

If the detector is initially in the $\mid d_b \rangle$ state, we assume that it switches when the system is in the $\mid a \rangle$ state and does not change if the system is in the $\mid b \rangle$ state, that is

$$\mid a \rangle \mid d_b \rangle \rightarrow \mid a \rangle \mid d_a \rangle , \qquad (20.1)$$
$$\mid b \rangle \mid d_b \rangle \rightarrow \mid b \rangle \mid d_b \rangle .$$

If, on the other hand, we assume that the system is in a superposition state

$$| \psi_{\text{initial}} \rangle = \alpha \, | \, a \rangle + \beta \, | \, b \rangle \, , \qquad (20.2)$$

with

$$| \, \alpha \, |^2 + | \, \beta \, |^2 = 1 \, , \qquad (20.3)$$

then

$$| \psi_{\text{initial}} \rangle = (\alpha \, | \, a \rangle + \beta \, | \, b \rangle) \, | \, d_b \rangle \overset{after}{\underset{meas}{\Longrightarrow}} \qquad (20.4)$$
$$\alpha \, | \, a \rangle \, | \, d_a \rangle + \beta \, | \, b \rangle \, | \, d_b \rangle \equiv | \, \Psi^c \rangle \, ,$$

where the state $| \, \Psi^c \rangle$ is a correlated one, and this process can be achieved, as we will see soon, just with Schrödinger's equation, with an appropriate interaction.

Thus, if the detector is in the $| \, d_a \rangle$ state, one can be certain that the system is in the $| \, a \rangle$ state.

However, we are ignorant about the quantum state of the system, and it is more realistic to approach the system in a statistical way, with the density matrix.

According to Von Neumann, besides the unitary evolution that rules the dynamics of the of the quantum phenomena, there is also a **non-unitary reduction of the wavefunction** $| \, \Psi^c \rangle$ that takes the pure state density matrix $| \, \Psi^c \rangle \langle \Psi^c \, |$ and converts it into a mixed state, by eliminating the off-diagonal elements

$$\rho^c = | \, \Psi^c \rangle \langle \Psi^c \, | \qquad (20.5)$$
$$= | \, \alpha \, |^2 | \, a \rangle \langle a \, || \, d_a \rangle \langle d_a \, | + | \, \beta \, |^2 | \, b \rangle \langle b \, || \, d_b \rangle \langle d_b \, |$$
$$+ \alpha^* \beta \, | \, b \rangle \langle a \, || \, d_b \rangle \langle d_a \, | + \alpha \beta^* \, | \, a \rangle \langle b \, || \, d_a \rangle \langle d_b \, |$$

$$\overset{Non}{\underset{unitary}{\Longrightarrow}} \rho^r = | \, \alpha \, |^2 | \, a \rangle \langle a \, || \, d_a \rangle \langle d_a \, | + | \, \beta \, |^2 | \, b \rangle \langle b \, || \, d_b \rangle \langle d_b \, | \, .$$

The difference between the original ρ^c and the 'after the measurement' reduced density matrix ρ^r, is that because in the latter case, the off-diagonal elements are missing, one could safely describe the system with alternative states ruled by classical probabilities $| \, \alpha \, |^2$ and $| \, \beta \, |^2$.

On the other hand, in the quantum case (ρ^c), things are more complicated, because we may use a different basis, say

$$| \, c \rangle = \frac{1}{\sqrt{2}} (| \, a \rangle + | \, b \rangle) \, , \qquad (20.6)$$

$$| \, d \rangle = \frac{1}{\sqrt{2}} (| \, a \rangle - | \, b \rangle) \, ,$$

and choosing $\alpha = -\beta = \frac{1}{\sqrt{2}}$, we write

$$| \Psi^c \rangle = \frac{1}{\sqrt{2}} (| a \rangle | d_a \rangle - | b \rangle | d_b \rangle) \tag{20.7}$$

$$= \frac{1}{\sqrt{2}} \left[\frac{1}{\sqrt{2}} (| c \rangle + | d \rangle) | d_a \rangle - \frac{1}{\sqrt{2}} (| c \rangle - | d \rangle) | d_b \rangle \right]$$

$$= \frac{1}{\sqrt{2}} [| c \rangle | d_c \rangle + | d \rangle | d_d \rangle] \, ,$$

where

$$| d_c \rangle = \frac{1}{\sqrt{2}} (| d_a \rangle - | d_b \rangle) \, , \tag{20.8}$$

$$| d_d \rangle = \frac{1}{\sqrt{2}} (| d_a \rangle + | d_b \rangle) \, .$$

We see that the diagonal elements of ρ^c give us different alternatives. In the first basis

$$(\rho^c)_{\text{diag}} = \frac{1}{2} | a \rangle \langle a \, || \, d_a \rangle \langle d_a | + \frac{1}{2} | b \rangle \langle b \, || \, d_b \rangle \langle d_b | \, , \tag{20.9}$$

while in the second basis

$$(\rho^c)_{\text{diag}} = \frac{1}{2} | c \rangle \langle c \, || \, d_c \rangle \langle d_c | + \frac{1}{2} | d \rangle \langle d \, || \, d_d \rangle \langle d_d | \, . \tag{20.10}$$

The problem, once more, is that we do not know the quantum state of the system.

Now, as we mentioned before, the first step of the measurement is to obtain the correlated wavefunction $| \Psi^c \rangle$, which can be achieved via a unitary operator.

The second step, however, was the **Von Neumann** non-unitary reduction.

Can this step be achieved in a different way? Perhaps, by another unitary operator?

The answer to this question is yes [1], and the way to do it is by coupling the system–detector pair to the environment, to dispose of the extra information.

We call the environment states $| \varepsilon \rangle$. Then

$$| \Psi^c \rangle | \varepsilon_0 \rangle = (\alpha | a \rangle | d_a \rangle + \beta | b \rangle | d_b \rangle) | \varepsilon_0 \rangle \tag{20.11}$$

$$\to (\alpha | a \rangle | d_a \rangle | \varepsilon_a \rangle + \beta | b \rangle | d_b \rangle | \varepsilon_b \rangle) = | \psi \rangle \, ,$$

where the correlation has been extended from the system–detector to system–detector–environment, getting a 'chain of states'.

If the environment states $| \varepsilon_a \rangle$ and $| \varepsilon_b \rangle$, corresponding to the detector states $| d_a \rangle$ and $| d_b \rangle$ respectively, are orthogonal, then we can trace (average) over the environment variables

$$\rho_{SD} = Tr_\varepsilon \mid \psi \rangle \langle \psi \mid = \sum_i \langle \varepsilon_i \mid \psi \rangle \langle \psi \mid \varepsilon_i \rangle = \rho^r , \qquad (20.12)$$

getting precisely the **Von Neumann** reduced density matrix, but this time by only unitary transformations, without ad hoc assumptions.

20.1 Dynamics of the Correlations

Here we discuss in more detail [6] the **'chain of states'** mentioned in the previous section.

We assume that the system is coupled to the environment by a Hamiltonian of the following form:

$$H_{\text{int}} = \hbar \sum_n \mid n \rangle \langle n \mid \mathbf{A}_n , \qquad (20.13)$$

where $\mathbf{A}_n$ are n-dependent operators acting on the Hilbert space of the environment and $\mid n \rangle$ is an eigenstate of a system observable to be measured.

The environment acquires the information about the state $\mid n \rangle$, in the sense that changes according to

$$\mid n \rangle \mid \phi_0 \rangle \xrightarrow{t} \exp\left(-i\frac{H_{\text{int}}}{\hbar}t\right) \mid n \rangle \mid \phi_0 \rangle = \mid n \rangle \exp(-iA_n t) \mid \phi_0 \rangle (20.14)$$

$$= \mid n \rangle \mid \phi_n(t) \rangle .$$

We notice that here, the 'measurement' is made not in the sense of Von Neumann but rather as a dynamical evolution of the joint system, according to Schrödinger's equation, with the appropriate coupling.

The resulting environment states $\mid \phi_n(t) \rangle$ are called **'pointer states'**. In case the environment is interpreted as the measuring apparatus, they would correspond to particular apparatus states.

From the linearity of the Schrödinger's equation, one can also write

$$\sum_n C_n \mid n \rangle \mid \phi_0 \rangle \rightarrow \sum_n C_n \mid n \rangle \mid \phi_n(t) \rangle , \qquad (20.15)$$

that is, we get a correlated state.

The density matrix of the system changes according to

$$\rho_S = Tr_{\text{envir}} \sum_{n,m} C_n C_m^* \mid n \rangle \langle m \mid\mid \phi_n(t) \rangle \langle \phi_m(t) \mid \qquad (20.16)$$

$$= \sum_{n,m} C_n C_m^* \mid n \rangle \langle m \mid \langle \phi_n(t) \mid \phi_m(t) \rangle ,$$

and for orthogonal states

$$\langle \phi_n(t) \mid \phi_m(t) \rangle = \delta_{nm} , \qquad (20.17)$$

the system density matrix becomes diagonal

$$\rho_S \rightarrow \sum_n \mid C_n \mid^2 \mid n \rangle \langle n \mid . \qquad (20.18)$$

During this evolution, the interference was destroyed and the system appears to be classical with respect to the quantum number n.

If the above evolution is viewed as a model of system–apparatus coupling, unfortunately, the apparatus, being macroscopic, will invariably interact with the environment ε.

By the same mechanism, the information about the measurement is rapidly transferred to the environment

$$\sum_n C_n \mid n \rangle \mid \phi_n \rangle \mid \varepsilon_0 \rangle \rightarrow \sum_n C_n \mid n \rangle \mid \phi_n \rangle \mid \varepsilon_n \rangle , \qquad (20.19)$$

and if the environment states are orthogonal, then

$$\rho_{\text{system}-\text{apparatus}} = \sum_n \mid C_n \mid^2 \mid n \rangle \langle n \mid\mid \phi_n \rangle \langle \phi_n \mid . \qquad (20.20)$$

Once more, we have defined the interaction of the apparatus with the environment by a Hamiltonian of the form given by (20.13), defining in this way, the pointer states $\mid \phi_n \rangle$.

20.2 How Long Does It Take to Decohere?

As discussed in the two previous sections, both measurements and coupling the system to an environment has, as a net effect, the loss of coherence, that is, the off-diagonal elements of the density matrix of our system vanish.

In this section, we want to find out, in some particular example, the damped harmonic oscillator, how long it takes for the coherence to vanish.

We start with an initial condition for the oscillator that consists in a superposition of two coherent states, and we study its evolution to discover that actually two very different time scales are present. One is the time it takes for the coherent amplitude to decay, γ^{-1}, while the off-diagonal elements of the density matrix, or the quantum coherence, decays much faster, with a characteristic time (in the case the initial superposition is $\propto [\mid \alpha \rangle + \mid -\alpha \rangle]$):

$$t_c = \frac{\gamma^{-1}}{2 \mid \alpha \mid^2} .$$

We first study the harmonic oscillator without losses, with an initial state [7]:

$$| \psi(0) \rangle = N (| \alpha_1 \rangle + | \alpha_2 \rangle) , \qquad (20.21)$$

where N is just a normalization factor.

As the Hamiltonian is $H = \hbar \omega a^\dagger a$, at time t it will evolve to

$$| \psi(t) \rangle = U(t) | \psi(0) \rangle = \exp(-i\omega a^\dagger a t) N (| \alpha_1 \rangle + | \alpha_2 \rangle) \qquad (20.22)$$
$$= N [| \alpha_1 \exp(-i\omega t) \rangle + | \alpha_2 \exp(-i\omega t) \rangle] .$$

To derive the above relation (last line), we used the property

$$\exp(-i\omega a^\dagger a t) \, | \, \alpha \rangle = \exp \left(-\frac{| \alpha |^2}{2} \right) \sum_n \frac{\alpha^n}{\sqrt{n!}} \exp(-i\omega t n) \, | \, n \rangle \qquad (20.23)$$

$$= \exp \left(-\frac{| \alpha |^2}{2} \right) \sum_n \frac{(\alpha \exp(-i\omega t))^n}{\sqrt{n!}} \, | \, n \rangle = | \, \alpha \exp(-i\omega t) \rangle .$$

Thus,

$$\rho(t) = N^2 \sum_{i,j=1}^{2} | \, \alpha_i(t) \rangle \langle \alpha_j(t) \, | , \qquad (20.24)$$

with

$$\alpha_i(t) = \alpha_i \exp(-i\omega t) . \qquad (20.25)$$

On the other hand, we look at the position representation of the coherent states

$$\langle q' | \alpha \rangle = \left(\frac{\omega}{\pi \hbar} \right)^{\frac{1}{4}} \exp \left[-\frac{\omega}{2\hbar} q'^2 + \sqrt{\frac{2\omega}{\hbar}} \alpha q' - \frac{\alpha^2 + | \alpha |^2}{2} \right] , \qquad (20.26)$$

so that

$$\langle q' | \rho(t) | q' \rangle = N^2 \left[\begin{array}{c} | \langle q' | \alpha_1(t) \rangle |^2 + | \langle q' | \alpha_2(t) \rangle |^2 \\ + 2 \operatorname{Re} \langle q' | \alpha_1(t) \rangle \langle \alpha_2(t) | q' \rangle \end{array} \right] . \qquad (20.27)$$

As $\alpha_i(t)$ is a complex number, one can separate the real and imaginary parts of $\langle q' | \alpha \rangle$

$$\langle q' | \alpha \rangle = \left(\frac{\omega}{\pi \hbar} \right)^{\frac{1}{4}} \exp \left[-\left(\sqrt{\frac{\omega}{2\hbar}} q' - \alpha \cos \omega t \right)^2 \right] \qquad (20.28)$$

$$\exp i \left[-\sqrt{\frac{2\omega}{\hbar}} \alpha q' \sin \omega t + \frac{\alpha^2}{2} \sin 2\omega t \right] ,$$

so, we can write, in the case $\alpha_1 = \alpha, \alpha_2 = -\alpha$

$$| \langle q' | \alpha_{1,2} \rangle |^2 = \left(\frac{\omega}{\pi \hbar} \right)^{\frac{1}{2}} \exp \left[- \left(\sqrt{\frac{\omega}{\hbar}} q' \pm \sqrt{2} \alpha \cos \omega t \right)^2 \right] \equiv I_{1,2}^2 , \qquad (20.29)$$

and

$$2 \, \mathrm{Re} \langle q' | \alpha_1(t) \rangle \langle \alpha_2(t) | q' \rangle = 2 I_1 I_2 \cos \theta(t) , \qquad (20.30)$$

so finally

$$\langle q' | \rho(t) | q' \rangle = N^2 \left[I_1^2 + I_2^2 + 2 I_1 I_2 \cos \theta(t) \right] , \qquad (20.31)$$

with

$$\theta(t) = 2 \sqrt{\frac{2\omega}{\hbar}} \alpha q' \sin \omega t . \qquad (20.32)$$

As we can see, the quantum interference term $2 I_1 I_2 \cos \theta(t)$ is present.

Now, it is interesting to study the effects of damping. This will give us information about the characteristic decoherence time due to the interaction with the environment.

The Master equation for the damped harmonic oscillator, at zero temperature, is given by

$$\frac{d\rho}{dt} = \frac{\gamma}{2} (2 a \rho a^\dagger - a^\dagger a \rho - \rho a^\dagger a) . \qquad (20.33)$$

The normally ordered characteristic function is defined as

$$X_N(\eta, t) = Tr(\rho(t) \exp(\eta a^\dagger) \exp(-\eta^* a)) . \qquad (20.34)$$

One can write [8]

$$\frac{\partial X_N(\eta, t)}{\partial t} = Tr \left[\frac{d\rho}{dt} \exp(\eta a^\dagger) \exp(-\eta^* a) \right]$$

$$= \frac{\gamma}{2} Tr \left[(2 a \rho a^\dagger - a^\dagger a \rho - \rho a^\dagger a) \exp(\eta a^\dagger) \exp(-\eta^* a) \right]$$

$$= \frac{\gamma}{2} Tr[2 \rho a^\dagger \exp(\eta a^\dagger) \exp(-\eta^* a) a - \rho \exp(\eta a^\dagger) \exp(-\eta^* a) a^\dagger a -$$

$$\rho a^\dagger a \exp(\eta a^\dagger) \exp(-\eta^* a)]$$

$$= -\frac{\gamma}{2} \{ \eta Tr[\rho a^\dagger \exp(\eta a^\dagger) \exp(-\eta^* a)]$$

$$- \eta^* Tr[\rho \exp(\eta a^\dagger) \exp(-\eta^* a) a] \}$$

$$\frac{\partial X_N(\eta, t)}{\partial t} = -\frac{\gamma}{2} \left[\eta \frac{\partial X_N(\eta, t)}{\partial \eta} + \eta^* \frac{\partial X_N(\eta, t)}{\partial \eta^*} \right] . \qquad (20.35)$$

In the last steps, we used the following properties

$$[a, f(a, a^\dagger)] = \frac{\partial f(a, a^\dagger)}{\partial a^\dagger} , \qquad (20.36)$$

$$[a^\dagger, f(a, a^\dagger)] = -\frac{\partial f(a, a^\dagger)}{\partial a} , \tag{20.37}$$

$$[a^\dagger a, \exp(\eta a^\dagger) \exp(-\eta^* a)] = \eta a^\dagger \exp(\eta a^\dagger) \exp(-\eta^* a)$$
$$+\eta^* \exp(\eta a^\dagger) \exp(-\eta^* a) a . \tag{20.38}$$

One can show that the solution to (20.35) is

$$X_N(\eta, t) = X_N \left(\eta \exp\left(-\frac{\gamma t}{2}\right), 0 \right) = X_N(\eta(t), 0) . \tag{20.39}$$

We can check the above result as follows (considering η as a complex number)

$$\frac{\partial X_N(\eta, t)}{\partial t} = \frac{\partial X_N(\eta, t)}{\partial \eta(t)} \frac{\partial \eta(t)}{\partial t} + \frac{\partial X_N(\eta, t)}{\partial \eta^*(t)} \frac{\partial \eta^*(t)}{\partial t}$$
$$= -\frac{\gamma}{2} \left[\eta \frac{\partial X_N(\eta, t)}{\partial \eta} + \eta^* \frac{\partial X_N(\eta, t)}{\partial \eta^*} \right] .$$

Now, the initial condition is

$$X_N(\eta, 0) = Tr(\rho(0) \exp(\eta a^\dagger) \exp(-\eta^* a)) , \tag{20.40}$$
$$= N^2 Tr \sum_{i,j} [| \alpha_i \rangle \langle \alpha_j | \exp(\eta a^\dagger) \exp(-\eta^* a)]$$
$$= N^2 \sum_{i,j} [\langle \alpha_j | \exp(\eta a^\dagger) \exp(-\eta^* a) | \alpha_i \rangle]$$
$$= N^2 \sum_{i,j} [\langle \alpha_j | \alpha_i \rangle \exp(\eta \alpha_j^* - \eta^* \alpha_i)] ,$$

So

$$X_N(\eta, t) = N^2 \sum_{i,j} \left[\langle \alpha_j | \alpha_i \rangle \exp(\eta \alpha_j^* - \eta^* \alpha_i) \exp\left(-\frac{\gamma t}{2}\right) \right] . \tag{20.41}$$

It is not difficult to show that the corresponding density matrix ρ is given by

$$\rho = N^2 \sum_{i,j=1}^{2} \langle \alpha_i | \alpha_j \rangle^{(1-\exp(-\gamma t))} | \alpha_j \exp\left(-\frac{\gamma t}{2}\right) \rangle \langle \alpha_i \exp\left(-\frac{\gamma t}{2}\right) | . \tag{20.42}$$

In the particular case $\alpha_1 = \alpha; \alpha_2 = -\alpha$, we have

$$\langle \alpha | -\alpha \rangle = \exp(-2 | \alpha |^2) , \tag{20.43}$$

and

$$\rho = N^2 \left[| \alpha \exp\left(-\frac{\gamma t}{2}\right) \rangle \langle \alpha \exp\left(-\frac{\gamma t}{2}\right) | \right.$$
$$\left. + | -\alpha \exp\left(-\frac{\gamma t}{2}\right) \rangle \langle -\alpha \exp\left(-\frac{\gamma t}{2}\right) | \right] \tag{20.44}$$

$$+ N^2 \exp\left[-2 | \alpha |^2 (1 - \exp(-\gamma t)] \{ | \alpha \exp\left(-\frac{\gamma t}{2}\right) \rangle \langle -\alpha \exp\left(-\frac{\gamma t}{2}\right) |$$
$$+ | -\alpha \exp\left(-\frac{\gamma t}{2}\right) \rangle \langle \alpha \exp\left(-\frac{\gamma t}{2}\right) | \}$$

If $\gamma t \ll 1$, then the relevant exponential factor multiplying the crossed terms becomes

$$\exp(-2 | \alpha |^2 \gamma t) \equiv \exp\left(-\frac{t}{t_c}\right) , \tag{20.45}$$

where

$$t_c = \frac{\gamma^{-1}}{2 | \alpha |^2} , \tag{20.46}$$

and

$$\langle q | \rho(t) | q \rangle = N^2 \left[I_1^2 + I_2^2 + 2 I_1 I_2 \cos\theta(t) \exp\left(-\frac{t}{t_c}\right) \right] , \tag{20.47}$$

where all the definitions are the same as before, except that $\alpha \to \alpha \exp(-\frac{\gamma t}{2})$, so

$$I_{1,2}^2 = \left(\frac{\omega}{\pi\hbar}\right)^{\frac{1}{2}} \exp\left\{ -\left[\sqrt{\frac{\omega}{\hbar}} q \pm \sqrt{2}\alpha \exp\left(-\frac{\gamma t}{2}\right) \cos\omega t \right]^2 \right\} , \tag{20.48}$$

which is a Gaussian distribution, whose centre is oscillating, the amplitude of which decreases in a time scale γ^{-1}. On the other hand, the quantum interference term $2 I_1 I_2 \cos\theta(t) \exp\left(-\frac{t}{t_c}\right)$ will vanish, for $| \alpha |^2 \gg 1$, in a much shorter time t_c. This may explain the difficulties in observing the quantum coherence in a macroscopic situation.

It has been recently shown [9] that the decay time of the quantum coherence , in a phase sensitive reservoir, for an initial superposition of $| \alpha \rangle$ and $| -\alpha \rangle$ is given by

$$t_c^{(sq)} = \frac{\gamma^{-1}}{2 \left[N + 2\alpha^2(N - M + \frac{1}{2}) \right]} , \tag{20.49}$$

where the notation is the same as in Chap. 9.

In the vacuum reservoir, $M = N = 0$, the result coincides with (20.46).

On the other hand, it is interesting to notice that for an ideally squeezed reservoir $[| M |^2 = N(N + 1)]$, with $M > 0$, the decay rate of the quantum coherence is significantly suppressed, and for large N is independent of α, namely

$$t_c^{(sq)} = \frac{\gamma^{-1}}{2\,[N]}\,, \tag{20.50}$$

which means that the decay rate of the quantum coherence (off-diagonal terms in the density matrix) is of the same order of magnitude as the decay rate of the energy (diagonal part of the density matrix).

Also, if $M < 0$, the decay rate of the coherence increases.

As we can see, one could in principle control the decay rate of the quantum coherence, by monitoring the phase of the squeezing parameter of the reservoir [9, 10, 11, 12], which may have interesting applications in quantum computing.

Finally, there are some recent publications on decoherence in the non-classical motion of trapped ions [13, 14, 15].

20.3 Decoherence Free Subspaces

As we have seen in the previous sections, decoherence is a consequence of the inevitable coupling of any quantum system to its environment, causing information loss from the system to the environment. In other words, we consider the decoherence as a non-unitary dynamics that is a consequence of the system–environment coupling.

This includes both dissipative and dephasing contributions.

Dissipation is a process by which the populations of the quantum states are modified by the interactions with the environment, while dephasing is a process of randomization of the relative phases of the quantum states.

Both effects are caused by the entanglement of the system with the environment degrees of freedom, leading to the non-unitary dynamics of our system.

Lidar et al. [16] introduced the term 'decoherence-free subspaces', referring to robust states against perturbations, in the context of Markovian Master equations.

One uses the symmetry of the system–environment coupling to find a 'quiet corner' in the Hilbert Space not experiencing this interaction.

A more formal definition of the DFS is as follows:

A system with a Hilbert space H is said to have a decoherence free subspace $\widetilde{H} \subset H$, if the evolution inside $\widetilde{H}$ is purely unitary.

20.3.1 Simple Example: Collective dephasing [17]

Consider a system of F two-level systems coupled to a common bath, whose effect is dephasing.

We define a qubit as a two-level system that in the basis $| \, 0 \rangle$ and $| \, 1 \rangle$ can be written as

$$| \, \psi \rangle = a \, | \, 0 \rangle + b \, | \, 1 \rangle \, . \tag{20.51}$$

The effect of the dephasing bath over these basis states is the following:

$$| \, 0 \rangle_j \rightarrow | \, 0 \rangle_j \tag{20.52}$$
$$| \, 1 \rangle_j \rightarrow \exp(i\phi) \, | \, 1 \rangle_j$$

where ϕ is a random phase.

This transformation can be written as a matrix

$$R_z(\phi) = \begin{bmatrix} 1 & 0 \\ 0 & \exp(i\phi) \end{bmatrix} \, , \tag{20.53}$$

acting on the $\{| \, 0 \rangle, \, | \, 1 \rangle\}$ basis.

We assume in this particular example that this transformation is collective, implying that the phase ϕ is the same for all two-level systems.

Now, we study the effect of the bath on an initial qubit $| \, \psi \rangle_j = a \, | \, 0 \rangle_j + b \, | \, 1 \rangle_j$.

We can write now the average density matrix over all possible phases, distributed with a probability density $p(\phi)$ as:

$$\rho_j = \int_{-\infty}^{\infty} R_z(\phi) \, | \, \psi \rangle_j \langle \psi \, | \, R_z(\phi)^{\mathsf{T}} p(\phi) d\phi \, , \tag{20.54}$$

and we assume all the qubits in the same state.

To be more specific, if we take a Gaussian distribution for the phase

$$p(\phi) = \frac{1}{\sqrt{4\pi\gamma}} \exp\left(-\frac{\phi^2}{4\gamma}\right) \, , \tag{20.55}$$

then it is simple to show that

$$\rho_j = \begin{bmatrix} | \, a \, |^2 & ab^* \exp(-\gamma) \\ a^* b \exp(-\gamma) & | \, b \, |^2 \end{bmatrix} \, , \tag{20.56}$$

basically showing the decoherence as the exponential decay of the non-diagonal elements of the density matrix.

Now, let us go and hunt for decoherence free subspaces, considering, for example, two and three particles.

Two particles

In this case we have four basis states, and the effect of the bath is the following

$$| \, 0 \rangle_1 \otimes | \, 0 \rangle_2 = | \, 00 \rangle \rightarrow | \, 00 \rangle \, ,$$

$$| \, 0 \rangle_1 \otimes | \, 1 \rangle_2 = | \, 01 \rangle \rightarrow \exp(i\phi) \, | \, 0 \rangle_1 \otimes | \, 1 \rangle_2 = \exp(i\phi) \, | \, 01 \rangle \, ,$$

$$| 1\rangle_1 \otimes | 0\rangle_2 = | 10\rangle \rightarrow \exp(i\phi) | 1\rangle_1 \otimes | 0\rangle_2 = \exp(i\phi) | 10\rangle \,,$$

$$| 1\rangle_1 \otimes | 1\rangle_2 = | 11\rangle \rightarrow \exp(2i\phi) | 1\rangle_1 \otimes | 1\rangle_2 = \exp(2i\phi) | 11\rangle \,. \qquad (20.57)$$

We notice that the states $| 10\rangle$ and $| 01\rangle$ transform with the same phase factor $\exp(i\phi)$; so any combination of these states will have only a global phase

$$| \chi\rangle = \alpha | 10\rangle + \beta | 01\rangle \rightarrow \exp(i\phi)(\alpha | 10\rangle + \beta | 01\rangle) = \exp(i\phi) | \chi\rangle, \quad (20.58)$$

thus, we have in this example a DFS of dimension 2

$$DFS_2 = \{| 10\rangle, | 01\rangle\}. \qquad (20.59)$$

We also have a couple of trivial DFS of dimension 1, such as $\{| 00\rangle\}$ and $\{| 11\rangle\}$. However, as the global phases of the various DFS are different, there is decoherence between them.

Three particles

We have two three-dimensional DFS, namely

$$\{| 001\rangle, | 010\rangle, | 100\rangle\} = DFS_3^{(1)}, \{| 110\rangle, | 101\rangle, | 011\rangle\} = DFS_3^{(2)}, \quad (20.60)$$

plus the trivial one-dimensional subspaces $\{| 000\rangle\}$, $\{| 111\rangle\}$.

There has been experimental verification of DFS_2 using trapped ions [18].

20.3.2 General Treatment [19]

Consider the Hamiltonian of a system (living in a Hilbert space H) interacting with a bath

$$H = H_S \otimes I_B + I_S \otimes H_B + H_I \,, \qquad (20.61)$$

where H_S, H_B and H_I are, respectively, the system, bath and system–bath interaction Hamiltonians, and I is the identity operator.

The time evolution of the whole system is given by

$$\rho_{SB}(t) = U \rho_{SB}(0) U^\dagger \,, \qquad (20.62)$$

where ρ_{SB} is the combined system–bath density operator, and U is the usual time evolution operator $U = \exp\left(-\frac{i}{\hbar} H t\right)$.

At this point, we assume that initially the system and the bath are decoupled, so $\rho_{SB}(t) = U \rho_S(0) \otimes \rho_B(0) U^\dagger$.

The interaction term can be written, quite generally as

$$H_I = \sum_\alpha S_\alpha \otimes B \,, \qquad (20.63)$$

where S_α and B are system and bath operators, respectively. Next, we trace over the bath variables to get

$$\rho = Tr_B \left[U\rho(0) \otimes \rho_B(0)U^\dagger \right] . \tag{20.64}$$

This reduced density matrix represents the system alone. If we diagonalize the density matrix of the bath, $\rho_B(0) = \sum_\nu \lambda_\nu \mid \nu \rangle\langle \nu \mid$, by performing the trace over the bath variables, we obtain

$$\rho = \sum \langle \mu \mid U(t)(\rho(0) \otimes \sum_\nu \lambda_\nu \mid \nu \rangle\langle \nu \mid)U^\dagger(t) \mid \mu \rangle \tag{20.65}$$

$$= \sum_a A_a \rho(0) A_a^\dagger ,$$

with the 'Kraus operators' defined as

$$A_a = \sqrt{\lambda_\nu} \langle \mu \mid U(t) \mid \nu \rangle, \ a = \mu, \nu . \tag{20.66}$$

As the density matrix is normalized, one can write

$$\sum_a A_a^\dagger A_a = I_S . \tag{20.67}$$

20.3.3 Condition for DFS: Hamiltonian Approach

As we said before, the decoherence is the direct consequence of the system–bath entanglement, caused in the present model, by the Hamiltonian H_I. Of course, if $H_I = 0$, then there is no decoherence and the dynamics follows a unitary evolution. Unfortunately, in practice, one cannot switch off the interaction with the reservoir. We have to look for alternatives, such as a particular subspace which is free of decoherence.

Zanardi et al.[20], [21] has shown that there exists a set of states $\{ \mid \widetilde{k} \rangle \}$ of eigenvectors of S_α such that

$$S_\alpha \mid \widetilde{k} \rangle = c_\alpha \mid \widetilde{k} \rangle, \ \forall \alpha, \mid \widetilde{k} \rangle . \tag{20.68}$$

These are degenerate eigenvectors of the system operators whose eigenvalues depend only on the index α, but not on the state index k.

If H_S leaves the Hilbert space $\widetilde{H} = span \left\{ \mid \widetilde{k} \rangle \right\}$ invariant, and if we start within $\widetilde{H}$, then the evolution of the system will be decoherence free. We let the reader check this proof.

20.3.4 Condition for DFS: Lindblad Approach

Lindblad has shown that the most general evolution of a system density matrix ρ_s is governed by the Master equation

$$\frac{d\rho}{dt} = -\frac{i}{\hbar} [H_s, \rho] + L_D(\rho) \tag{20.69}$$

with

$$L_D(\rho(t)) = \frac{1}{2} \sum_{\alpha,\beta=1}^{M} d_{\alpha,\beta} \left(\left[F_\alpha, \rho F_\beta^\dagger \right] + \left[F_\alpha \rho, F_\beta^\dagger \right] \right) \tag{20.70}$$

where H_s is the system Hamiltonian, the F_α is a family of the 'Lindblad' operators in an M-dimensional space and $d_{\alpha,\beta}$ are elements of a positive Hermitian matrix (some cases are discussed in Chap. 16 in the context of quantum trajectories).

All the non-unitary, decohering dynamics is accounted for by L_D. That is, in the Lindblad form, there is a clear separation between the unitary and decohering dynamics.

Let $\left\{ | \widetilde{k} \rangle_{k=1}^{N} \right\}$ be a basis for an N-dimensional subspace

$$\widetilde{H}(DFS) \subseteq H(TOTAL\ SYSTEM\ HILBERT\ SPACE) . \tag{20.71}$$

In this basis, we may express the density matrix as

$$\rho = \sum_{k,j=1}^{N} \rho_{kj} | \widetilde{k} \rangle \langle \widetilde{j} | . \tag{20.72}$$

Now, we consider the action of the Lindblad operators F_α on $| \widetilde{k} \rangle$,

$$F_\alpha | \widetilde{k} \rangle = \sum_{j=1}^{N} C_{kj}^\alpha | \widetilde{j} \rangle .$$

Substituting in (20.70), we find

$$L_D(\rho) = \frac{1}{2} \sum_{\alpha,\beta=1}^{M} d_{\alpha,\beta} \sum_{k,j,m,n=1}^{N} \rho_{kj} (2 C_{jm}^{\beta*} C_{kn}^{\alpha} | \widetilde{n} \rangle \langle \widetilde{m} | - C_{mn}^{\beta*} C_{kn}^{\alpha} | \widetilde{m} \rangle \langle \widetilde{j} |$$

$$\tag{20.73}$$

$$- C_{jm}^{\beta*} C_{nm}^{\alpha} | \widetilde{k} \rangle \langle \widetilde{n} |) = 0 .$$

Notice that we have used the condition $L_D(\rho) = 0$, which is precisely the definition of the DFS.

The coefficients $d_{\alpha,\beta}$ represent information about the bath, which we assume is uncontrollable. So we require that each term in the α, β sum vanishes separately.

Furthermore, we expect no dependence on the initial conditions, i.e., no dependence on ρ_{kj}, which implies that each term in the parenthesis vanishes separately.

This can be done if all the projectors are the same, requirement that is satisfied if

$$C_{kn}^\alpha = C_n^\alpha \delta_{k,n},$$

so the (20.73) becomes

$$\sum_{k,j=1}^{N} \rho_{kj} \mid \tilde{k}\rangle\langle \tilde{j} \mid (2C_j^{\beta*}C_k^{\alpha} - C_k^{\beta*}C_k^{\alpha} - C_j^{\beta*}C_j^{\alpha}) = 0 . \tag{20.74}$$

Assuming that the $C_k^{\alpha} \neq 0$, this yields

$$(2C_j^{\beta*}C_k^{\alpha} - C_k^{\beta*}C_k^{\alpha} - C_j^{\beta*}C_j^{\alpha}) = 0 , \tag{20.75}$$

or

$$2 = Z^* + Z^{-1} , \tag{20.76}$$

with $Z \equiv \frac{C_j^{\alpha}}{C_k^{\alpha}}$.

The (20.76) has the unique solution $Z = 1$. Thus, $C_j^{\alpha} = C_k^{\alpha} \equiv C^{\alpha}$

Thus, we proved the following Theorem.

The necessary and sufficient condition for a subspace $\tilde{H}(DFS) = \left\{\mid \tilde{k}\rangle_{k=1}^{N}\right\}$ *to be decoherence free is that the basis states* $\mid \tilde{k}\rangle$ *are degenerate eigenstates of all Lindblad operators* $F\alpha$,

$$F_{\alpha} \mid \tilde{k}\rangle = C^{\alpha} \mid \tilde{k}\rangle \; for \; \forall \, \alpha, k . \tag{20.77}$$

The above condition can be also written as

$$[F_{\alpha}, F_{\beta}] \mid \tilde{k}\rangle = 0 . \tag{20.78}$$

If one can write

$$[F_{\alpha}, F_{\beta}] = \sum_{\gamma=1}^{M} f_{\alpha,\beta}^{\gamma} F_{\gamma} ,$$

with $f_{\alpha,\beta}^{\gamma} \neq 0$ and linearly independent (the Fs forming a 'semi-simple Lie algebra'). In this case, the condition (20.78) can be written as

$$\sum_{\gamma=1}^{M} f_{\alpha,\beta}^{\gamma} C^{\gamma} = 0 \tag{20.79}$$

that can only be satisfied for $C^{\gamma} = 0$.

Thus, the condition for a DFS , for the semisimple case, is that the set of states should be degenerate eigenstates of all Lindblad operators with zero eigenvalue,

$$F_{\alpha} \mid \tilde{k}\rangle = 0, \; for \; \forall \, \alpha, k . \tag{20.80}$$

We consider, in the next section, some simple examples.

20.3.5 Example: N Spins in Boson Bath [19]

Consider the following Hamiltonian

$$H_I = \sum_{i=1}^{N} \sum_k \left[g_{ik}^+ \sigma_i^\dagger b_k + g_{ik}^- \sigma_i b_k^\dagger + g_{ik}^z \sigma_i^z (b_k + b_k^\dagger) \right] ,$$

where $\sigma_i^\dagger, \sigma_i$ σ_i^z are the Pauli matrices acting on the i-th spin and b_k ,$b_k^\dagger$ are the boson operators for the k-th mode, and g_{ik}^α are the coupling constants.

This Hamiltonian describes a collection of spins (two-level systems) exchanging energy via the terms $g_{ik}^+ \sigma_i^\dagger b_k$, $g_{ik}^- \sigma_i b_k^\dagger$ and exchanging phase via $g_{ik}^z \sigma_i^z (b_k + b_k^\dagger)$.

As the couplings are different for each spin, there are $3N$ S_α operators that are N triple operators $\{\sigma_i^\dagger, \sigma_i, \sigma_i^z\}$. It is simple to see that there are no DFS in this case.

On the other hand, if we assume a collective interaction with the bath, that is if the coupling are independent of the spin index i,

$$g_{ik}^\alpha \equiv g_k^\alpha , \alpha = +, -, z,$$

then one defines three collective operators $S_\alpha = \sum_{i=1}^{N} \sigma_i^\alpha$, $\alpha = +, -, z$ and the interaction Hamiltonian becomes

$$H_I = \sum_{\alpha=+,-,z} S_\alpha \otimes B_\alpha,$$

where

$$B_+ = \sum_k g_k^+ b_k, \ B_- = \sum_k g_k^- b_k^\dagger, \ B_z = \sum_k g_k^z (b_k + b_k^\dagger) .$$

As the angular momentum algebra is semi-simple, one can write the DFS condition as

$$S_\alpha \mid \widetilde{k} \rangle = 0, \ \forall \alpha.$$

For **two spins**, the singlet state will satisfy the above condition, so we get a one-dimension DFS

$$\mid \widetilde{k} \rangle_{12} = \frac{1}{\sqrt{2}} (\mid 01 \rangle - \mid 10 \rangle) = \mid J = 0, m_j = 0 \rangle ,$$

where the indices 1,2 refer to the singlet state corresponding to qubits 1 and 2.

If we take **four spins**, by adding two angular momentum $j_1 = 1$ and $j_2 = 1$, a possible result is $\mid J = 0, m_j = 0 \rangle$, which is just a combination of the triplet states, giving

$$\mid J = 0, m_j = 0 \rangle_1$$

$$= \frac{1}{\sqrt{3}} \left[|\, 00 \rangle_{12} \,|\, 11 \rangle_{34} + |\, 11 \rangle_{12} \,|\, 00 \rangle_{34} - \frac{1}{2} \left(|\, 01 \rangle_{12} + |\, 10 \rangle_{12} \right) \left(|\, 01 \rangle_{34} + |\, 10 \rangle_{34} \right) \right].$$

The second state in the DFS is the combination of singlet states, namely

$$|\, J = 0, m_j = 0 \rangle_2 = |\, \tilde{k} \rangle_{12} \,|\, \tilde{k} \rangle_{34}.$$

The dimension of the DFS is given by $DIM(DFS) = \frac{N!}{(\frac{N}{2})!(\frac{N}{2}+1)!}$, which is equal 2 for $N = 4$. There is no DFS for N odd.

Problems

20.1. Prove (20.42).
20.2. Show that for a phase sensitive reservoir, one has

$$t_c^{(\mathrm{sq})} = \frac{\gamma^{-1}}{2 \left[N + 2\alpha^2 \left(N - M + \frac{1}{2} \right) \right]}.$$

Hint: See [9].

References

1. Zurek, W.H.: Physics. Today, **44**, 36 (1991)
2. Zurek, W.H.: Phys. Rev. A, **24**, 1516 (1981)
3. Zurek, W.H.: Phys. Rev. A, **26**, 1862 (1982)
4. Bohr, N.: Nature, **121**, 580 (1928)
5. Von Neumann, J.: Matematische Grundlagen der Quanten mechanik. Springer, Berlin (1932)
6. Giulini, D., Joos, E., Kiejer, C., Kupsch, J., Stamatescu, I.O., Zeh, H.D.: Decoherence and the Appearence of Classical World in Quantum Theory. Springer, Berlin (1997)
7. Walls, D.F., Milburn, G.J.: Phys. Rev. A, **31**, 2403 (1985)
8. Walls, D.F., Milburn, G.J.: Quantum Optics. Springer, Berlin (1994)
9. Buzek, V., Knight P.L.: In: Wolf, E. (ed.) Progress in Optics. vol XXXIV,1 Elsevier, Amsterdam (1995)
10. Kim, M.S., Buzek, V.: Phys. Rev. A, **46**, 4239 (1992)
11. Kim, M.S., Buzek, V.: J. Mod. Opt., **39**, 1609 (1992)
12. Kim, M.S., Buzek, V.: Phys. Rev. A, **47**, 610 (1993)
13. Schneider, S., Milburn, G.J.: Phys. Rev. A, **57**, 3748 (1998)
14. Munrao, M., Knight, P.L.: Phys. Rev. A, **58**, 663 (1998)
15. Schneider, S., Milburn, G.J.: Phys. Rev. A, **59**, 3766 (1999)
16. Lidar, D.A., Chuang, I.L., Whaley, K.B.: Phys. Rev. Lett., **81**, 2594 (1998)
17. Palma, G.M., Suominen K.A., Ekert, A.K.: Proc. Roy. Soc., London Ser, A, 452 (1996)
18. Kielpinski, D., Meyer, V., Rowe, M.A., Sackett, C.A., Itano, W.M., Monroe, C., Wineland, D.J.: Science, **291**, 1013 (2001)

19. Lidar, D.A., Whaley, K.B.: quant-phys/0301032
20. Zanardi, P., Rasetti, M.: Mod. Phys. Lett. B, **11**, 1085(1997)
21. Zanardi, P., Rasetti, M.: Phys. Rev. Lett., **79**, 3306 (1997)

Further Reading

Quantum Coherence in Cavity QED

- Davidovich, L., Brune, M., Raimond, J.M., Haroche, S.: Phys. Rev. A, **53**, 1295 (1996)

21. Quantum Bits, Entanglement and Applications

In this chapter, we deal with a general introduction to quantum computation and describe the bipartite entanglement of pure and mixed states. We also present quantum teleportation as an application.

21.1 Qubits and Quantum Gates

A *quantum bit* or qubit is a quantum system in which the classical Boolean states 0 and 1 are replaced by a pair of mutually orthogonal quantum states labeled by $\{|\,0\rangle, |\,1\rangle\}$. These two states form a computational basis for a pure state, a qubit, written as [1]

$$|\,\psi\rangle = \alpha\,|\,0\rangle + \beta\,|\,1\rangle, \tag{21.1}$$

with $|\,\alpha\,|^2 + |\,\beta\,|^2 = 1$.

Physically speaking, a qubit corresponds typically to the two levels of some microscopic system such as a polarized photon, a trapped ion, a nuclear spin etc.

A *Quantum Logic Gate* is a device that performs fixed a unitary operation during a fixed period of time.

One of the most common single qubit gate is the Hadamard Gate, defined by the matrix

$$H = \frac{1}{\sqrt{2}} \begin{bmatrix} 1 & 1 \\ 1 & -1 \end{bmatrix}, \tag{21.2}$$

written on the $\{|\,0\rangle, |\,1\rangle\}$ computational basis. If we represent $|\,0\rangle = \begin{pmatrix} 1 \\ 0 \end{pmatrix}$ and $|\,1\rangle = \begin{pmatrix} 0 \\ 1 \end{pmatrix}$, then, the effect of the Hadamard Gate on the basis vectors is the following:

$$H\,|\,0\rangle = \frac{1}{\sqrt{2}} \begin{bmatrix} 1 & 1 \\ 1 & -1 \end{bmatrix} \begin{pmatrix} 1 \\ 0 \end{pmatrix} = \frac{1}{\sqrt{2}} \begin{pmatrix} 1 \\ 1 \end{pmatrix} \tag{21.3}$$

$$= \frac{1}{\sqrt{2}}(|\,0\rangle + |\,1\rangle), \tag{21.4}$$

$$H \mid 1\rangle = \frac{1}{\sqrt{2}} \begin{bmatrix} 1 & 1 \\ 1 & -1 \end{bmatrix} \begin{pmatrix} 0 \\ 1 \end{pmatrix} = \frac{1}{\sqrt{2}} \begin{pmatrix} 1 \\ -1 \end{pmatrix}$$
$$= \frac{1}{\sqrt{2}} (\mid 0\rangle - \mid 1\rangle). \tag{21.5}$$

A *Quantum Network* is a device that consists in a collection of Quantum Logic Gates , synchronized in some way.

A network of size two, for example, is a Hadamard Transform on two qubits. Assuming that we start with the state $\mid 00\rangle_{12} \equiv \mid 0\rangle_1 \otimes \mid 0\rangle_2$, then (for simplicity we will skip the sub-indices)

$$\mid 0\rangle \rightarrow [H] \rightarrow \frac{\mid 0\rangle + \mid 1\rangle}{\sqrt{2}},$$

$$\mid 0\rangle \rightarrow [H] \rightarrow \frac{\mid 0\rangle + \mid 1\rangle}{\sqrt{2}}, \tag{21.6}$$

so that acting on $\mid 00\rangle$ gives

$$\mid 00\rangle \rightarrow [H][H] \rightarrow$$

$$\frac{1}{2} \{\mid 00\rangle + \mid 01\rangle + \mid 10\rangle + \mid 11\rangle\}. \tag{21.7}$$

Notice that in the above parenthesis, we have the digital representation of 0, 1, 2, 3, so in a 'decimal' basis we can write it as

$$\mid 00\rangle \rightarrow [H][H] \rightarrow$$

$$\frac{1}{2} \{\mid 0\rangle + \mid 1\rangle + \mid 2\rangle + \mid 3\rangle\}. \tag{21.8}$$

So, if the initial state is $\mid 00\rangle$, then the output is the superposition of the states $\mid 0\rangle$ to $\mid 3\rangle$.

On the other hand, if the initial state is $\mid 01\rangle$ then the output is

$$\mid 01\rangle \rightarrow [H][H] \rightarrow$$

$$\frac{1}{2} \{\mid 0\rangle - \mid 1\rangle + \mid 2\rangle - \mid 3\rangle\}. \tag{21.9}$$

Another single qubit gate is the phase gate, defined in the computational basis as

$$P = \begin{pmatrix} 1 & 0 \\ 0 & \exp i\phi \end{pmatrix}, \tag{21.10}$$

whose net effect is

$$\mid 0\rangle \rightarrow \mid 0\rangle,$$
$$\mid 1\rangle \rightarrow \exp i\phi \mid 1\rangle. \tag{21.11}$$

Fig. 21.1. Phase gate

Symbolically, this is written as Fig. 21.1 for $j = 0, 1$.

We can combine the Hadammard and phase gates, to get *the most general single qubit unitary transformation*, starting from $| 0 \rangle$ (Fig. 21.2).

It is a simple to show that the above operation yields

$$\exp \left(i \frac{\theta}{2} \right) \left[| 0 \rangle \cos \left(\frac{\theta}{2} \right) + | 1 \rangle \exp(i\phi) \sin \left(\frac{\theta}{2} \right) \right]. \tag{21.12}$$

As we can see from the above result, a combination of a Hadamard and Phase gates is sufficient to construct any unitary operation on a single qubit.

Pictorially, a qubit can be represented as a unit vector in the Bloch sphere, as shown in Fig. 21.3.

An example of a two-qubit gate is a CONTROLLED-NOT or C-NOT gate, represented by the following 4×4 matrix

$$C = \begin{bmatrix} 1 & 0 & 0 & 0 \\ 0 & 1 & 0 & 0 \\ 0 & 0 & 0 & 1 \\ 0 & 0 & 1 & 0 \end{bmatrix}, \tag{21.13}$$

and normally, the symbol for it is (Fig. 21.4).

This gate flips the second qubit (called 'target', k in Fig 21.4) if the first qubit is $| 1 \rangle$ (called control, j in Fig 21.4), and it does nothing if the control is $| 0 \rangle$, for example

$$C \, | j \rangle \, | 0 \rangle \rightarrow | j \rangle \, | j \rangle, \text{ for } j = 0,1. \tag{21.14}$$

If the control is in a superposition state $| \psi \rangle = \alpha \, | 0 \rangle + \beta \, | 1 \rangle$, and the target is in $| 0 \rangle$, then the C-NOT gate generates a state that is not separable, an *entangled state*

$$C \left[(\alpha \, | 0 \rangle + \beta \, | 1 \rangle) \, | 0 \rangle \right] \rightarrow \alpha \, | 00 \rangle + \beta \, | 11 \rangle,$$

$$\neq | \psi_1 \rangle \otimes | \psi_2 \rangle. \tag{21.15}$$

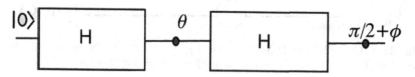

Fig. 21.2. General single qubit transformation

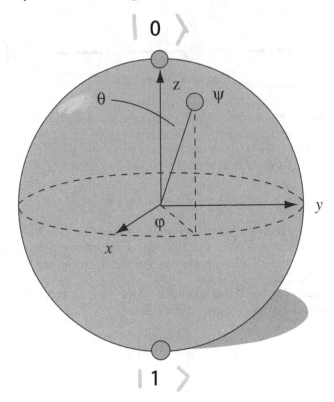

Fig. 21.3. Pictorial representation of a qubit in the Bloch sphere

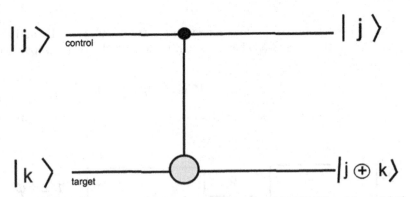

Fig. 21.4. CONTROLLED-NOT GATE. $|j\rangle$ is the CONTROL qubit and $|k\rangle$ is the TARGET. $|j\rangle$ has to be in the state $|j=1\rangle$ to change the target

For gates with a higher number of qubits, we have for example the Toffoli Gate, which acts on three qubits, two of them (j and k in Fig. 21.5) acting as controls and the third one (m) as the target, so that the target is only changed when both controls are in the $|1\rangle$ state.

This gate is shown in Fig (21.5)

For example

$$T \mid 000\rangle \rightarrow \mid 000\rangle,$$

$$T \mid 110\rangle \rightarrow \mid 111\rangle,$$

$$T \mid 111\rangle \rightarrow \mid 110\rangle, \text{ etc.} \tag{21.16}$$

21.2 Entanglement

21.2.1 Pure States

Consider a quantum system composed of two subsystems A and B whose states are in the Hilbert spaces H_A and H_B of a finite dimension $d_{A,B} \equiv \dim(H_{A,B})$ respectively. The complete space is $H = H_A \otimes H_B$.

Let us consider now states of the whole system, in which the subsystem is in $\mid i\rangle_A$ and the other one in $\mid \mu\rangle_B$. For clarity, we use Latin letters for the states in A and Greek for the states in B

The whole system is denoted by $\mid i\rangle_A \otimes \mid \mu\rangle_B \in H$. For example, we could have product states such as $\mid 01\rangle, \mid 10\rangle, \mid 00\rangle, \mid 11\rangle$.

However, as any superposition in Hilbert space is possible, we could also have

$$\mid \psi\rangle = \alpha \mid 00\rangle + \beta \mid 11\rangle \in H. \tag{21.17}$$

As we mentioned earlier, this state cannot be written in a factorized form, and we refer to as an '*entangled state*'.

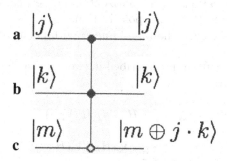

Fig. 21.5. TOFFOLI GATE. $\mid j\rangle$ and $\mid k\rangle$ are the CONTROL qubits and $\mid m\rangle$ is the TARGET. To change the target, the controls have to be in the state $\mid j = 1\rangle$ $\mid k = 1\rangle$, otherwise nothing happens

These states play a very important role in quantum communications, teleportation, quantum cryptography, etc.

The more formal definition of two-partite entanglement is the following one:

Consider two systems A and B.

$| \psi \rangle \in H_A \otimes H_B$ is a product state if there exists $| \phi_{1,2} \rangle_{A,B} \in H_{A,B}$ such that $| \psi \rangle = | \phi_1 \rangle \otimes | \phi_2 \rangle$. Otherwise, we say that $| \psi \rangle$ is an entangled state.

Example 1

The *Bell* states

$$| \phi^{\pm} \rangle = \frac{1}{\sqrt{2}}(| 00 \rangle \pm | 11 \rangle),$$

$$| \psi^{\pm} \rangle = \frac{1}{\sqrt{2}}(| 01 \rangle \pm | 10 \rangle), \tag{21.18}$$

are entangled states

The most important property of the entangled states is that they carry correlations, that is, a measurement of an observable in A and B are correlated.

For example, defining the Pauli matrices in the computational basis

$$\sigma_x = | 1 \rangle \langle 0 | + | 0 \rangle \langle 1 |,$$
$$\sigma_y = i(| 1 \rangle \langle 0 | - | 0 \rangle \langle 1 |),$$
$$\sigma_z = - | 1 \rangle \langle 1 | + | 0 \rangle \langle 0 |, \tag{21.19}$$

and if the system is in $| \psi^- \rangle$, then a measurement of the observable σ_z will always give the opposite results in A and B.

Correlations can be rather subtle, as we show in example 2.

Example 2

Suppose that we have a two-photon source providing pairs of photons with opposite polarizations in the horizontal and vertical basis, without us knowing which photon has which polarization.

Let us say that we have photons A and B with polarizations H (horizontal) and V (vertical).

The above situation can be described in two ways, a classical and a quantum mechanical one:

$$\rho_{\text{class}} = \frac{1}{2}(| H_A V_B \rangle \langle H_A V_B |)$$
$$+ \frac{1}{2}(| V_A H_B \rangle \langle V_A H_B |), \tag{21.20}$$

$$\rho_{QM} = | \psi^+ \rangle \langle \psi^+ |, \tag{21.21}$$

with

$$| \psi^+ \rangle = \frac{1}{\sqrt{2}} (| H_A V_B \rangle + | V_A H_B \rangle). \qquad (21.22)$$

If we observe, for example, an H polarization for photon A, expressed as

$$\frac{Tr_A (\rho \, | \, H_A \rangle \langle H_A \,|)}{Tr (\rho \, | \, H_A \rangle \langle H_A \,|)} = | V_B \rangle \langle V_B \,|, \qquad (21.23)$$

this gives certain V-polarization of the photon B in *both cases* .

As we shall see very soon, ρ_{class} is not entangled while ρ_{QM} is.

To appreciate the difference between the two cases, we write these two density matrices as rows and columns ordered as HH, HV, VH, VV. The result is:

$$\rho_{\text{class}} = \frac{1}{2} \begin{pmatrix} 0\,0\,0\,0 \\ 0\,1\,0\,0 \\ 0\,0\,1\,0 \\ 0\,0\,0\,0 \end{pmatrix}, \qquad (21.24)$$

$$\rho_{QM} = \frac{1}{2} \begin{pmatrix} 0\,0\,0\,0 \\ 0\,1\,1\,0 \\ 0\,1\,1\,0 \\ 0\,0\,0\,0 \end{pmatrix}. \qquad (21.25)$$

We notice that the non-zero off-diagonal elements or 'coherences', are only present in ρ_{QM}, in the mixed positions HV-VH and VH-HV.

In both cases, the reduced density matrix, say ρ_A yields

$$(\rho_{\text{class}})_A = (\rho_{QM})_A = \frac{1}{2} I, \qquad (21.26)$$

I being the identity matrix [2].

As we can see, the off-diagonal elements of ρ_{QM} does not have an impact in individual particles, and therefore, we get the same reduced density matrix. The difference between the two cases shows up when doing violations of Bells inequalities, which are purely quantum effects. [3, 4].

The Schmidt decomposition

For two-partite entanglement, one can always write the states in terms of the basis states of the parts A and B [4]

$$| \psi_{AB} \rangle = \sum_{i=1}^{d_A} \sum_{\mu=1}^{d_B} C(i, \mu) \, | \, i \rangle \otimes | \, \mu \rangle, \qquad (21.27)$$

where $C(i, \mu)$ are the matrix elements of

$$C = \sum_{i=1}^{d_A} \sum_{\mu=1}^{d_B} C(i, \mu) \, | \, i \rangle \langle \mu \, |. \qquad (21.28)$$

One can easily check that the reduced density matrices ρ_A and ρ_B can be given by

$$\rho_A = CC^\dagger, \quad \rho_B = C^\dagger C. \tag{21.29}$$

Now, we write the eigenvalue equation

$$\rho_A \mid f^i \rangle = CC^\dagger \mid f^i \rangle = \lambda_i \mid f^i \rangle, \tag{21.30}$$

thus getting a basis for the A space

$$\langle f^i \mid f^j \rangle = \delta_{i,j},$$
$$\sum_j \langle i \mid f^j \rangle \langle f^j \mid j \rangle = \delta_{i,j}. \tag{21.31}$$

Now, multiplying (21.30) by $C^\dagger$, we get

$$C^\dagger \rho_A \mid f^i \rangle = C^\dagger C(C^\dagger \mid f^i \rangle) = \lambda_i(C^\dagger \mid f^i \rangle). \tag{21.32}$$

Equation (21.32) shows that $C^\dagger \mid f^i \rangle$ is an eigenstate of $\rho_B = C^\dagger C$ with the same eigenvalue λ_i,

$$\rho_B C^\dagger \mid f^i \rangle = \lambda_i C^\dagger \mid f^i \rangle.$$

The new orthonormal set of states, which are the eigenstates of ρ_B, we denote by $\mid \phi^i \rangle$, defined as

$$\mid \phi^i \rangle = \frac{1}{\sqrt{\lambda_i}} C^\dagger \mid f^i \rangle. \tag{21.33}$$

To get an explicit expression for $C(i,\mu)$, we multiply (21.33) by C and then by $\langle j \mid$, getting

$$\sqrt{\lambda_i}\langle j \mid f^i \rangle = \sum_\mu C(j,\mu)\langle \mu \mid \phi^i \rangle. \tag{21.34}$$

Defining $f^i_j = \langle j \mid f^i \rangle$ and $\phi^i_\mu = \langle \mu \mid \phi^i \rangle$, we get by multiplying by $\langle \phi^i \mid \mu \rangle$, summing over i and using the orthonormality of the $\mid \phi^i \rangle$ states

$$C(j,\mu) = \sum_i \sqrt{\lambda_i} f^i_j (\phi^i_\mu)^*.$$

Now, we are ready to write $\mid \psi_{AB} \rangle$ in the so-called Schmidt form

$$\mid \psi_{AB} \rangle = \sum_{i=1}^{d_A} \sum_{\mu=1}^{d_B} C(i,\mu) \mid i \rangle \otimes \mid \mu \rangle$$
$$= \sum_{i=1}^{d_A} \sum_{\mu=1}^{d_B} \left[\sum_s \sqrt{\lambda_s} f^s_j (\phi^s_\mu)^* \mid i \rangle \otimes \mid \mu \rangle \right]$$

$$= \sum_s \sqrt{\lambda_s} \left[\sum_{i=1}^{d_A} (f_j^s) \mid i \rangle \right] \otimes \left[\sum_{\mu=1}^{d_B} (\phi_\mu^s)^* \mid \mu \rangle \right]$$

$$\equiv \sum_s \sqrt{\lambda_s} \mid F^s \rangle \mid \Phi^s \rangle. \tag{21.35}$$

We notice that the original $\mid \psi_{AB} \rangle$ was written in terms of a double sum while the present 'diagonal form' has been reduced to a single one.

In terms of the density operator, one can write

$$\rho = \sum_{i,k} \sqrt{\lambda_i \lambda_k} \mid F^i \rangle \langle F^k \mid \otimes \mid \Phi^i \rangle \langle \Phi^k \mid . \tag{21.36}$$

It is simple to see that the reduced density matrices are now diagonal:

$$\rho_F = \sum_k \lambda_k \mid F^k \rangle \langle F^k \mid ,$$

$$\rho_\Phi = \sum_k \lambda_k \mid \Phi^k \rangle \langle \Phi^k \mid . \tag{21.37}$$

As the λ_s are eigenvalues of density matrices, they obey the following conditions:

$$0 \leq \lambda_s \leq 1,$$

$$\sum_s \lambda_s = 1. \tag{21.38}$$

In the case of a product state, there is only one Schmidt term different from zero, and the corresponding eigenvalue is 1.

Conversely, if we have a state with only one Schmidt coefficient, it must be a product state.

Thus, $\mid \psi_{AB} \rangle$ is a product state if and only if the corresponding reduced density matrices correspond to pure states.

This implies that if we have an entangled state, the corresponding reduced density operators must correspond to a mixed state, with more than one Schmidt eigenvalue different from zero.

We see that entanglement of a state is directly related to the mixedness of the reduced density operator.

A simple way of measuring the degree of entanglement in a two-partite pure state is via the 'Schmidt number', defined as the reciprocal of the purity of the reduced density matrix [5].

$$K = \frac{1}{Tr_A \rho_A^2} = \frac{1}{Tr_B \rho_B^2} = \frac{1}{\sum_n \lambda_n^2}. \tag{21.39}$$

If only one eigenvalue is 1, and the rest are zero, then $K = 1$ and we have a product state.

On the other hand, if all the λ_s are equal, implying that all the N terms are equally important in the Schmidt decomposition, then $\lambda_s = \frac{1}{N}$ and $K = N$.

So, if D is the dimension of the space, then

$$1 \leq K \leq D. \tag{21.40}$$

Example 3

In the case of the two-photon source (Example 2), we had $\rho_{QM} = |\psi^+\rangle\langle\psi^+|$ and the reduced density matrices $(\rho_{QM})_A = (\rho_{QM})_B = \frac{1}{2}I$, so $\lambda_1 = \lambda_2 = \frac{1}{2}$ and $K = 2$, which is the maximum possible value since $D = 2$.

Example 4 Two-mode squeezed states. [6]

In the previous example, we had the Schmidt decomposition of two particles.

A more complex situation is the case of optical entanglement of two field modes, in particular, a highly entangled state called two-mode squeezed state. In Chap. 5, we described with details the squeezed states and mentioned that one of its quadratures has quantum fluctuations below than that of a perfectly coherent field. Of course this example has interest in low-noise quantum measurement and noise reduction in optical systems, in general.

We are going to analyse the two-mode squeezed vacuum state . Its generation is achieved in an Optical Parametric Amplifier . The experiment is as follows: Laser light at frequency Ω acting as the pump is focussed onto a non-linear crystal (with a large second-order non-linearity), which absorbs a pump photon and emits two photons into the signal and idler modes of the amplifier, with frequencies ω_u and ω_v , such that

$$\Omega = \omega_u + \omega_v.$$

The effective Hamiltonian for this interaction is

$$H = i\hbar\lambda(c^\dagger a_u a_v - c a_u^\dagger a_v^\dagger). \tag{21.41}$$

If the laser field is intense (with an amplitude E), then it can be considered as classical, and we may write a simplified Hamiltonian

$$H = i\hbar\lambda(E^* a_u a_v - E a_u^\dagger a_v^\dagger).$$

The time evolution operator corresponding to this Hamiltonian is

$$U(t - t_0) = \exp\left[\eta^* a_u a_v - \eta a_u^\dagger a_v^\dagger\right], \tag{21.42}$$

where $\eta = \lambda E(t - t_0)$, which is the generator of two-mode squeezed states.

The above operator can be written as [7]

$$\exp\left(\eta^* a_u a_v - \eta a_u^\dagger a_v^\dagger\right)$$

$$= \frac{1}{\cosh(r)} \exp\left[-a_u^\dagger a_v^\dagger \tanh(r)\exp(i\theta)\right]$$

$$\exp\left[-(a_u^\dagger a_u + a_v^\dagger a_v)\ln(\cosh(r))\right]$$

$$\exp\left[-a_u a_v \tanh(r)\exp(-i\theta)\right]. \tag{21.43}$$

If the initial state is the two-mode vacuum, then

$$|\psi\rangle = \exp\left[-\eta^* a_u a_v + \eta a_u^\dagger a_v^\dagger\right] |0\rangle_u \otimes |0\rangle_v$$

$$= \frac{1}{\cosh(r)} \exp\left[-a_u^\dagger a_v^\dagger \tanh(r)\exp(i\theta)\right] |0\rangle_u \otimes |0\rangle_v. \tag{21.44}$$

Expanding in power series the r.h.s. of 21.44, we get

$$|\psi\rangle = \frac{1}{\cosh(r)} \sum_n (\tanh r)^n \exp(in\theta) |n\rangle_u \otimes |n\rangle_v.$$

This state is already in the Schmidt form written in the two-mode biorthogonal Fock basis.

Therefore, according to the above discussion, the reduced density matrices are diagonal and have the form

$$\rho_u = \sum_n \frac{1}{[\cosh(r)]^2} (\tanh r)^{2n} |n\rangle_{uu}\langle n|, \tag{21.45}$$

$$\rho_v = \sum_n \frac{1}{[\cosh(r)]^2} (\tanh r)^{2n} |n\rangle_{vv}\langle n|. \tag{21.46}$$

Notice that these reduced density matrices describe a thermal state, with $\langle n\rangle = \sinh^2 r$. In both cases, one gets $\rho_{u,v} = \sum_n \frac{1}{1+\langle n\rangle}\left(\frac{\langle n\rangle}{1+\langle n\rangle}\right)^n |n\rangle_{u,v}\langle n|$, which is precisely a thermal state.

21.2.2 Mixed states

For mixed states, the situation is more complicated. There are several criteria for separability [8].

A mixed state is called separable, if it can be prepared by the two-parties (popularly called 'Alice' and 'Bob') in a 'classical way', which means agreeing by direct communication on the local preparation of states. The corresponding Density Matrix of a *separable* state should have only classical correlations or mathematically should be of the form [9]

$$\rho = \sum_i p_i |a_i\rangle\langle a_i| \otimes |b_i\rangle\langle b_i|, \tag{21.47}$$

otherwise, it is *entangled*.

Here the coefficients p_i are probabilities with $0 \leq p_i \leq 1$ and $\sum_i p_i = 1$. This decomposition is not unique.

Example 5

An example of a mixed state that has classical but not quantum correlations is $\rho = \frac{1}{2}(| \ 00 \ \rangle\langle 00 \ | \ + \ | \ 11 \rangle\langle 11 \ |)$. Another example is the state $\rho = \frac{1}{2}(| \ \phi^+ \rangle\langle \phi^+ \ | \ + | \ \phi^- \rangle\langle \phi^- \ |)$, which is separable because it can be written as the previous example, as it can be seen from the definitions of the Bell states.

However, we should warn the reader that a criteria for separability like 21.47 is not easy to use. Or, in other words, finding for ρ a form like that or proving that it does not exist is not a simple task. Therefore, we must find a simpler criteria to detect entanglement.

Peres–Horodecki criteria (positive partial transpose) [10, 11]

The partial transpose of a composite density matrix is given by transposing only one of the two subsystems. For example, the partial transposition with respect to Alice is

$$(\rho^{T_A})_{m\mu,n\nu} = (\rho)_{n\mu,m\nu}, \tag{21.48}$$

where again, we are using Latin subindices for the Alice subsystem and Greek for Bobs.

Thus, for any separable state, one can write the partial transpose as:

$$\rho_{\text{sep}}^{T_A} = \sum_i p_i (| \ a_i \rangle\langle a_i \ |)^T \otimes | \ b_i \rangle\langle b_i \ | \ . \tag{21.49}$$

As the $(| \ a_i \rangle\langle a_i \ |)^T$ are again valid density matrices for Alice, one immediately finds that $\rho_{\text{sep}}^{T_A} \geq 0$ implying non-negative eigenvalues.

The same holds for the partial transposition with respect to Bob. In summary, the partial transpose of a separable density matrix is positive. This means that it has only positive non-vanishing eigenvalues (or equivalently, a positive operator has a positive or zero expectation value *with any state*).

The converse, that is, if $\rho^{T_A} \geq 0$, then ρ is separable is true only for low dimensional systems, namely for composite systems 2×2 and 2×3. In these cases, the positivity of the partial transpose (PPT) is a necessary and sufficient condition for separability. For higher dimensions, the PPT condition is only necessary.

The partial transposition criterion for detecting entanglement is simple: given a bipartite state ρ_{AB}, find the eigenvalues of any of its partial transpositions. A negative eigenvalue immediately implies that the state is entangled. Examples of such states include the singlet state.

The partial transposition criterion allows to detect in a simple way, all entangled states that are NPT (negative partial transpose density matrices with at least a negative eigenvalue), which is a large class of states. However, in higher dimensions, there are PPT states that are not separable, called 'bound entangled states' [12, 13].

21.2.3 Bell Inequalities

(see Fig. 21.6)

As we have seen in the previous section, entanglement is a purely quantum mechanical effect. Schrödinger was the first one to deal with this problem, in connection with the famous Schrödinger cat, or entanglement on a macroscopic scale.

Later, Einstein, Podolsky and Rosen thought about a 'Gedanken' (thought) Experiment as a criticism of Quantum Mechanics. They suggest that quantum mechanics cannot be a complete theory but should be supplemented by additional variables (hidden variables) , to be able to restore reality and locality.

In particular, the requirement of locality , which creates the main difficulty, says that the measurement on one system should be unaffected by

Fig. 21.6. John Bell (1972)

operations on a distant system with which the first system interacted in the past.

As far as reality is concerned, I quote Einstein [14] *"In a complete theory there is an element corresponding to each element of reality. A sufficient condition for the reality of a physical quantity is the possibility of predicting it with certainty, without disturbing the system. In quantum mechanics, in the case of two physical quantities described by non-commuting operators, the knowledge of one precludes the knowledge of the other. Then either (1) the description of the wavefunction in quantum mechanics is not complete or (2) these two quantities cannot have simultaneous reality.... One is thus led to conclude that the description of reality as given by a wavefunction is not complete"(See Fig. 21.7, 1940)*

Bell made the assumptions of local realism, to derive an inequality that must be satisfied by any physical theory of nature, that is local and realistic, concepts that will be more clear in an example. He then goes on to show that there are states in Quantum Mechanics that violate this inequality, both theoretically and experimentally (as proven later on).

Fig. 21.7. Albert Einstein 1940

We will derive here the inequality and give examples of the its violation [15, 16].

We consider two spin-$\frac{1}{2}$ particles, far apart, called A and B. Let us perform Stern–Gerlach measurements in the directions $\mathbf{a}$ and $\mathbf{b}$ on A and B. The outcomes of the measurements performed on the particles A and B are A_a and B_b. The results of these measurements $A_a(B_b)$ whose values can be ± 1 may depend on the directions $a(b)$ and some uncontrollable parameter λ. We assume that $A_a(\lambda)$ and $B_b(\lambda)$ have a definite pre-measurement value and that the measurement merely discovers this value. This is the *reality*. λ is usually called the hidden variable. Furthermore, the result of a measurement at A does not depend on measurements performed on B and viceversa. This is Einstein's locality assumption.

Now we assume that the parameter λ has a probability distribution $P(\lambda)$ such that

$$\int P(\lambda)\mathrm{d}\lambda = 1, P(\lambda) \geq 0. \tag{21.50}$$

The correlation function of two spin-$\frac{1}{2}$ for a measurement in a direction $\mathbf{a}$ for particle A and a direction $\mathbf{b}$ for particle B, and assuming the existence of hidden variables, is given by:

$$E(a, b) = \int A_a(\lambda)B_b(\lambda)P(\lambda)\mathrm{d}\lambda. \tag{21.51}$$

Let us now assume that observations are made at the particle A in the directions $\mathbf{a}$ and $\mathbf{a'}$ and at the particle B in the directions $\mathbf{b}$ and $\mathbf{b}$. If the outcome of those measurements are $A_a, A_{a'}$ and $B_b, B_{b'}$, respectively. Then

$$E(a, b) + E(a, b') + E(a', b) - E(a', b')$$
$$= \int \{A_a(\lambda)[B_b(\lambda) + B_{b'}(\lambda)] + A_{a'}(\lambda)[B_b(\lambda) - B_{b'}(\lambda)]\}P(\lambda)\mathrm{d}\lambda. \tag{21.52}$$

now, $[B_b(\lambda)+B_{b'}(\lambda)]$ and $[B_b(\lambda)-B_{b'}(\lambda)]$ can be $\pm 2, 0$ and $0, \pm 2$, respectively, so

$$| E(a, b) + E(a, b') + E(a', b) - E(a', b') | \leq 2. \tag{21.53}$$

This is actually a generalization of Bell's inequality called Clauser, Horne, Shimony and Holt (BHSH) derived in 1969 [17].

Example 6

An interesting example is the use of the singlet state to detect entanglement via Bell inequality. For the singlet state, one can prove that

$$E(a, b) = \langle \psi^- | \sigma_a \cdot \sigma_b | \psi^- \rangle = -\cos(\theta_{ab}), \tag{21.54}$$

where $\sigma_a = \overrightarrow{\sigma} \cdot \mathbf{a}$ and $\sigma_b = \overrightarrow{\sigma} \cdot \mathbf{b}$, with $\overrightarrow{\sigma} = (\sigma_x, \sigma_y, \sigma_z)$. Also θ_{ab} is the angle between the two measurement directions $\mathbf{a}$ and $\mathbf{b}$.

So, for the singlet state, we can define

$$B = E(a, b) + E(a, b') + E(a', b) - E(a', b')$$
$$= -\cos(\theta_{ab}) - \cos(\theta_{ab'}) - \cos(\theta_{a'b}) + \cos(\theta_{a'b'}).$$

The maximum value of this function is obtained when

a a′ b b′

are contained in a plane. In that case $B = 2\sqrt{2}$ thus violating the BHSH inequality.

Thus, a singlet state, a state allowed by quantum mechanics (an entangled state) violates the BHSH inequality, or in other words, it does not obey a theory, which is local and realistic.

To put it in a different way, quantum mechanics is not local realistic. This is precisely the message of Bell's Theorem.

21.3 Quantum Teleportation

The dream of Teleportation is to travel by simply reappearing in some distant location (see Fig. 21.8).

Classically, one can characterize the system by measurements and make a copy at the distant location. One does not need parts and pieces of the original. We only have to send the information that can be used to reconstruct the object.

However, what happens if these parts and pieces are electrons, photons, atoms, etc.? What happens with their quantum properties, considering that we cannot have measurements with absolute precision, according to Heisenberg's uncertainty principle?

Bennett et al. [18] suggested that is possible to transfer the quantum state of a particle to another particle – process baptized as *quantum teleportation* – provided one does not acquire any information about the state to be teleported.

This can be achieved via a fundamental quantum property, *entanglement*, that describes, as we have seen in the previous sections, a correlation much stronger than the classical one.

Bob and Alice could, separated at a certain distance, implement a *Teleportation* procedure as follows:

They share a maximally entangled pair of quantum two-level systems Such a state could be, for example the Bell state $| \psi_{AB} \rangle = \frac{1}{\sqrt{2}}(| 0 \rangle_A | 0 \rangle_B + | 1 \rangle_A | 1 \rangle_B)$ where A stands for Alice and B for Bob.

On the other hand, Alice receives an unknown qubit $| \Phi \rangle = (a | 0 \rangle + b | 1 \rangle)$. She wants to Teleport this state to Bob. This state is unknown to Alice, otherwise she can inform Bob classically (phone call, fax, etc.). If Alice does not know the state, she requires a very large number of measurements to determine a and b. But she has another possibility. She can use the state

Fig. 21.8. Movie fiction (Star Trek)

shared with Bob $\mid \psi_{AB}\rangle$. The total state of three qubits is now (the first two qubits belong to Alice and the third one to Bob)

$$\mid \Phi_{AB}\rangle =\mid \Phi\rangle \mid \psi_{AB}\rangle = (a\mid 0\rangle + b\mid 1\rangle)\frac{1}{\sqrt{2}}(\mid 00\rangle +\mid 11\rangle),$$

which can be written as

$$\mid \Phi_{AB}\rangle = \frac{1}{\sqrt{2}}(a\mid 000\rangle + a\mid 011\rangle + b\mid 100\rangle + b\mid 111\rangle),$$

that can be conveniently written in terms of the Bell basis:

$$| \Phi_{AB} \rangle = \frac{1}{2} \left[| \phi^+ \rangle (a \,|\, 0 \rangle + b \,|\, 1 \rangle) + | \phi^- \rangle (a \,|\, 0 \rangle - b \,|\, 1 \rangle) \right]$$
$$+ \frac{1}{2} \left[| \psi^+ \rangle (a \,|\, 0 \rangle + b \,|\, 1 \rangle) + | \psi^- \rangle (a \,|\, 0 \rangle - b \,|\, 1 \rangle) \right] .$$

Now the Teleportation proceeds as follows:

1. Alice performs a joint measurement on the Bell basis. Thus, she will get one of the four Bell states randomly with equal probability.
2. Suppose that she gets $| \phi^- \rangle$. Then, the three-qubit state collapses to

$$| \phi^- \rangle (a \,|\, 0 \rangle - b \,|\, 1 \rangle).$$

Alice now communicates this result to Bob over a classical channel how this state differs from the original one. In this particular example, the unitary transformation to go from $(a \,|\, 0 \rangle - b \,|\, 1 \rangle) \to a \,|\, 0 \rangle + b \,|\, 1 \rangle$. It is simple to see that the transformation is σ_z. For all four possibilities, we have

Alice finds $| \phi^+ \rangle \to$ (0) Bob does nothing
Alice finds $| \phi^- \rangle \to$ (1) Bob performs σ_z
Alice finds $| \psi^+ \rangle \to$ (2) Bob performs σ_x
Alice finds $| \psi^- \rangle \to$ (3) Bob performs $\sigma_z \sigma_x$.

So the classical information that Alice sends to Bob about her joint measurement can be encoded, for example '1', meaning he has to perform σ_z. Once he has applied this transformation on his particle, he can be certain that he has reproduced the original state. Notice that this process is different from cloning, because the original is destroyed by the measurement. Also, the operations performed are local, in either location of Alice

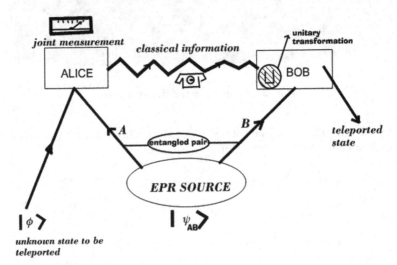

Fig. 21.9. Quantum teleportation protocol

and Bob. But we never performed a global transformation involving both. A pictorial description of the Quantum teleportation Protocol in described in Fig. 21.9.

Problems

21.1. Show that for a two-mode squeezed state, $K = 1 + 2\langle n \rangle$, where $\langle n \rangle = \sinh^2 r$. Notice that for $r = 0$, $K = 1$ that correspond to a factorized state, while for $r \to \infty$, $K \to \infty$, approaching the maximum value of K for an ∞ dimensional system (maximally entangled).

21.2. Show that the state $\rho = |\phi^+\rangle\langle\phi^+|$ is entangled, using the NPT criterion.

21.3. Show that the state with the corresponding density matrix [19]

$$\rho = \frac{1}{20}\begin{bmatrix} 7 & 1 & 2 & 2 \\ 1 & 3 & 2 & 2 \\ 2 & 2 & 3 & 1 \\ 2 & 2 & 1 & 7 \end{bmatrix},$$

is separable.

21.4. Prove that $E(a, b) = \langle\psi^- | \sigma_a \cdot \sigma_b | \psi^-\rangle = -\cos(\theta_{ab})$.

References

1. Nielsem, M.A., Chuang, I.L.: Quantum Computation and Quantum Information. Cambridge University Press UK (2000)
2. Eberly, J.H.: quant-ph/0508019
3. Freedman, S.J., Clauser, J.F.: Phys. Rev. Lett., **28**, 938 (1972)
4. Eberly, J.H.: Am. J. Phys., **70**, 276 (2002)
5. Grobe, R., Rzazewski, K., Eberly, J.H.: J. Phys. B, **27**, L503 (1994)
6. Ekert, A., Knight, P.L.: Am. J. Phys., **63**, 415 (1995)
7. Perelmov, A.: Generalized Coherent States and Applications. Springer, Berlin (1986)
8. Bruss, D.: J. Math. Phys., **43**, 4237 (2002)
9. Werner, R.: Phys. Rev. A, **40**, 4277 (1989)
10. Peres, A.: Phys. Rev. Lett., **77**, 1413 (1996)
11. Horodecki, M., Horodecki, P., Horodecki, R.: Phys. Lett. A, **223**, 1 (1996)
12. Horodecki, P.: Phys. Lett. A **232**, 333 (1997)
13. Sen, A., Sen, U., Lewenstein, M., Sampera, A.: LANL Report No quant-ph\0508032
14. Einstein, A., Podolsky, B., Rosen, N.: Phys. Rev., **57**, 777 (1935)
15. Bell, J.S.: Physics, **1**, 195 (1964)
16. Sen, A., Sen, U., Lewestein, M., Sampera, A.: quant-phys/0508032

17. Clauser, J.F., Horne, M.A., Shimony, A., Holt, R.A.: Phys. Rev. Lett., **23**, 880 (1969); See also Bell, J.S.: Speakable and Unspeakable in Quantum Mechanics. Cambridge Univ. Press (1987)
18. Bennet, C.H., Brassard, G., Crepeau, C., Jozsa, R., Peres, A., Wooters, W.K.: Phys. Rev. Lett., **70**, 1895 (1993)
19. Beige, A., Huelga, S.F., Knight, P.L., Plenio, M.B., Thompson, R.C.: J. Mod. Opt., Cambridge Univ. Press **47**, 401 (2000)

22. Quantum Cloning and Processing

22.1 The No-Cloning Theorem

The no-cloning theorem, derived in 1982 by Wootters and Zurek ([1]), showed that it is not possible to construct a device that will produce an exact copy of an arbitrary quantum state.

This theorem is an unexpected quantum effect, because of the linear superposition of quantum states, as opposed to the classical physics case, where the copying process presents no difficulties, and represents the most significant difference between classical and quantum mechanics.

Let us assume that the unitary operator U_{ab} acts on the two-qubit space, such that (C-NOT Gate):

$$U_{ab} \mid 0\rangle_a \mid 0\rangle_b \rightarrow \mid 0\rangle_a \mid 0\rangle_b \tag{22.1}$$
$$U_{ab} \mid 1\rangle_a \mid 0\rangle_b \mid \rightarrow \mid 1\rangle_a \mid 1\rangle_b .$$

The question is, Is it possible, in general, to have a copying machine that will perform the following operation?

$$U_{ab} \mid \psi\rangle_a \mid \varphi\rangle_b \rightarrow \mid \psi\rangle_a \mid \psi\rangle_b , \tag{22.2}$$

for an arbitrary $\mid \psi\rangle = \alpha \mid 0\rangle + \beta \mid 1\rangle$, with α and β being complex coefficients satisfying $\mid \alpha \mid^2 + \mid \beta \mid^2 = 1$. Acting with the unitary operator on $\mid \psi\rangle$, we should get, according to (22.2)

$$U_{ab}(\alpha \mid 0\rangle + \beta \mid 1\rangle)_a \mid 0\rangle_b \rightarrow (\alpha \mid 0\rangle + \beta \mid 1\rangle)_a \otimes (\alpha \mid 0\rangle + \beta \mid 1\rangle)_b \tag{22.3}$$
$$= \mid \alpha \mid^2 \mid 00\rangle_{ab} + \alpha\beta \mid 01\rangle_{ab} + \beta\alpha \mid 10\rangle_{ab} + \mid \beta \mid^2 \mid 11\rangle_{ab} .$$

On the other hand, the operator U is linear, so applying the rule of (22.3), we get

$$U_{ab}(\alpha \mid 0\rangle + \beta \mid 1\rangle)_a \mid 0\rangle_b = U_{ab}\alpha \mid 0\rangle_a \mid 0\rangle_b + U_{ab}\beta \mid 1\rangle_a \mid 0\rangle_b \tag{22.4}$$
$$\rightarrow \alpha \mid 00\rangle_{ab} + \beta \mid 11\rangle_{ab}$$
$$\neq \mid \psi\rangle_a \mid \psi\rangle_b .$$

As we can see, the operation 'Cloning' of (22.2) is not possible, because the coefficients are different and the crossed terms are missing. (For a more general proof, see Appendix G.)

22.2 The Universal Quantum Copying Machine (UQCM)

According to the no-cloning theorem , we cannot copy the quantum information about an arbitrary state exactly. However, we can still have imperfect copies. This problem was dealt with by V. Buzek and M. Hillery ([2]) in 1996. In term of notation, the ideal copying would be achieved via the following transformation

$$| \psi \rangle_a | \varphi \rangle_b | Q \rangle_x \rightarrow | \psi \rangle_a | \psi \rangle_b | \widetilde{Q} \rangle_x, \qquad (22.5)$$

where $| \psi \rangle_a$ is the original state to be copied, $| \varphi \rangle_b$ is the 'blank' copy while $| Q \rangle_x$ is the state of the copying machine, whose output is described by $| \widetilde{Q} \rangle_x$

We now detail the requirements for the UQCM:

1) In terms of the density matrix, we require that for the final outputs

$$\rho_a^{out} = \rho_b^{out}. \qquad (22.6)$$

2) To determine the 'quality' of the copy, we define a distance between two operators as

$$D(\rho_1, \rho_2) = Tr \left[(\rho_1 - \rho_2)^2 \right]. \qquad (22.7)$$

Also, Schumacker ([3]) defined a fidelity between two density matrices as

$$F = Tr \left[\rho_1^{\frac{1}{2}} \rho_2 \rho_1^{\frac{1}{2}} \right]^{\frac{1}{2}}. \qquad (22.8)$$

which ranges from 0 to 1. A fidelity of one means that the two density matrices are equal.

In the particular case of the copying machine, we want to compare the output density operators $\rho_{a_j}^{(out)}$ for $j = 1, 2$ and $a_1 = a, a_2 = b$ with the ideal output $\rho_{a_j}^{(id)}$ that corresponds to the original input density operator. If we measure the distance between these two operators , it will depend in general on the input parameters, so the copying procedure will be better for some input states than others.

To avoid this, we want

$$D \left(\rho_{a_j}^{(out)} - \rho_{a_j}^{(id)} \right) = \text{constant}, j = 1, 2. \qquad (22.9)$$

3) The next condition is that the copies should be as close as possible to the ideal case, that is

$$D \left(\rho_{a_j}^{(out)} - \rho_{a_j}^{(id)} \right) = \min D \left(\rho_{a_j}^{(out)} - \rho_{a_j}^{(id)} \right), j = 1, 2 \qquad (22.10)$$

4) The two copies of an ideal copying machine should be independent, condition that can be written as

$$\rho_{ab}^{(id)} = \rho_a^{(id)} \otimes \rho_b^{(id)}. \tag{22.11}$$

However, the copies will have a non-vanishing correlation.

The UQCM was thought so as to satisfy all three requirements, when applied to an arbitrary qubit $| \psi \rangle_{a_1} = \alpha | 0 \rangle_{a_1} + \beta | 1 \rangle_{a_1}$. The final result is (see Appendix H for details)

$$\rho_{a_j}^{(out)} = \frac{5}{6} | \psi \rangle_{a_j} \langle \psi | + \frac{1}{6} | \psi_\perp \rangle_{a_j} \langle \psi_\perp |, \tag{22.12}$$
$$j = 1,2$$

where $| \psi_\perp \rangle_{a_j}$ is the state orthogonal to $| \psi \rangle_{a_j}$. From (22.12) we see that the copy presents $\left(\frac{5}{6}\right)$ of the desired state and $\left(\frac{1}{6}\right)$ of the undesired orthogonal state.

For a general review, the reader is referred to [4].

22.3 Quantum Copying Machine Implemented by a Circuit

In this section, we are going to implement the UQCM via a quantum circuit. [5, 6].

As we mention in the previous section, the original state to be copied is $| \psi \rangle_{a_1}^{(in)} = \alpha | 0 \rangle_{a_1} + \beta | 1 \rangle_{a_1}$ where α and β are complex numbers that can be written as $\alpha = \sin\theta \exp i\varphi$ and $\beta = \cos\theta$ satisfying $| \alpha |^2 + | \beta |^2 = $. The blank state over which a copy is made was called $| 0 \rangle_{a_2}$ and the state of the machine $| Q \rangle_x$. However, as this last state (machine) is also a qubit of dimension two, it could be also used as a second qubit to make copies.

In other words, we can have three input qubits in a circuit, with three output qubits, two of which are going to be the two copies. The state of the third qubit, instead of calling it x, we will refer to it as a_3.

The circuit of the Fig. 22.1 is divided into two parts.

The dynamics of the present circuit can be described as:

$$| \psi \rangle_{a_1}^{(in)} | 0 \rangle_{a_2} | 0 \rangle_{a_3} \rightarrow | \psi \rangle_{a_1}^{(in)} | \psi \rangle_{a_2,a_3}^{(prep)} \rightarrow | \psi \rangle_{a_1,a_2,a_3}^{(out)}. \tag{22.13}$$

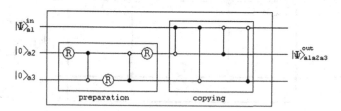

Fig. 22.1. Circuit for a universal quantum copying machine

22.3.1 Preparation Stage

The first section is the 'preparation' of the qubits a_2 and a_3, in an arbitrary state $| \psi \rangle_{a_2,a_3}^{(\mathrm{prep})}$. (see Fig. 22.2). This stage of the circuit consists in three rotations (denoted by **R**) and two C-NOT gates.

A rotation has the following effect on a qubit:

$$R_j(\theta) \, | \, 0 \rangle_j = \cos(\theta) \, | \, 0 \rangle_j + \sin(\theta) \, | \, 1 \rangle_j, \qquad (22.14)$$
$$R_j(\theta) \, | \, 1 \rangle_j = - \sin(\theta) \, | \, 0 \rangle_j + \cos(\theta) \, | \, 1 \rangle_j.$$

The C-NOT gates, mentioned already in Chap. 21, can be written in terms of an operator P_{kl}, such that

$$P_{kl} \, | \, 0 \rangle_k \, | \, 0 \rangle_l = | \, 0 \rangle_k \, | \, 0 \rangle_l, \qquad (22.15)$$
$$P_{kl} \, | \, 0 \rangle_k \, | \, 1 \rangle_l = | \, 0 \rangle_k \, | \, 1 \rangle_l,$$
$$P_{kl} \, | \, 1 \rangle_k \, | \, 0 \rangle_l = | \, 1 \rangle_k \, | \, 1 \rangle_l,$$
$$P_{kl} \, | \, 1 \rangle_k \, | \, 1 \rangle_l = | \, 1 \rangle_k \, | \, 0 \rangle_l,$$

where k is the control and l the target .

If we assume that the two preparation qubits are initially in the state $| \psi \rangle_{a_2,a_3}^{(\mathrm{in})} = | \, 0 \rangle_{a_2} \, | \, 0 \rangle_{a_3}$, the state after going through the preparation stage will be of the form

$$| \psi \rangle_{a_2,a_3}^{(\mathrm{prep})} = c_1 \, | \, 00 \rangle_{a_2,a_3} + c_2 \, | \, 01 \rangle_{a_2,a_3} + c_3 \, | \, 10 \rangle_{a_2,a_3} + c_4 \, | \, 11 \rangle_{a_2,a_3}. \quad (22.16)$$

According to the above figure, the preparation state is given by

$$| \psi \rangle_{a_2,a_3}^{(\mathrm{prep})} = R_2(\theta_3) P_{32} R_3(\theta_2) P_{23} R_2(\theta_1) \, | \, 00 \rangle_{a_2,a_3}. \qquad (22.17)$$

By combining (22.14–22.17), we get

$$\cos(\theta_1) \cos(\theta_2) \cos(\theta_3) + \sin(\theta_1) \sin(\theta_2) \sin(\theta_3) = c_1,$$
$$- \cos(\theta_1) \sin(\theta_2) \sin(\theta_3) + \sin(\theta_1) \cos(\theta_2) \cos(\theta_3) = c_2,$$

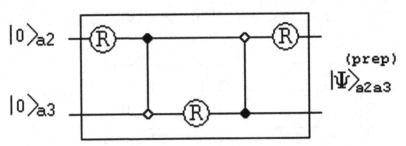

Fig. 22.2. Preparation stage of the universal quantum copying machine

$$\cos(\theta_1)\cos(\theta_2)\sin(\theta_3) - \sin(\theta_1)\sin(\theta_2)\cos(\theta_3) = c_3,$$
$$\cos(\theta_1)\sin(\theta_2)\cos(\theta_3) + \sin(\theta_1)\cos(\theta_2)\sin(\theta_3) = c_4.$$

As we can see, the rotation angles completely specify the c_i parameters, which are chosen according to the design of the copying machine.

22.3.2 Copying Stage and Output

The second section corresponds to the "copying", where the information of the original qubit (a_1)is redistributed among the three qubits (see Fig 22.3).

For an arbitrary input and given the preparation state, the output is given by

$$|\psi\rangle^{(out)}_{a_1,a_3,a_3} = P_{31}P_{21}P_{13}P_{12}\,|\psi\rangle^{(in)}_{a_1}\,|\psi\rangle^{(prep)}_{a_2,a_3} \qquad (22.18)$$

$$
\begin{aligned}
= \ & \alpha c_1\,|000\rangle_{a_1,a_2,a_3} + \alpha c_2\,|101\rangle_{a_1,a_2,a_3}\\
& + \alpha c_3\,|110\rangle_{a_1,a_2,a_3} + \alpha c_4\,|011\rangle_{a_1,a_2,a_3}\\
& + \beta c_1\,|111\rangle_{a_1,a_2,a_3} + \beta c_2\,|010\rangle_{a_1,a_2,a_3}\\
& + \beta c_3\,|001\rangle_{a_1,a_2,a_3} + \beta c_4\,|100\rangle_{a_1,a_2,a_3}\ .
\end{aligned}
$$

Now, because we need the density operator for further analysis, we calculate the total and reduced operators, written as:

$$\rho^{(out)}_{a_1,a_2,a_3} = |\psi\rangle^{(out)}_{a_1,a_2,a_3}\langle\psi|, \qquad (22.19)$$

$$\rho^{(out)}_{a_1} = Tr_{a_2,a_3}\left[\rho^{(out)}_{a_1,a_2,a_3}\right],$$

$$\rho^{(out)}_{a_2} = Tr_{a_1,a_3}\left[\rho^{(out)}_{a_1,a_2,a_3}\right],$$

$$\rho^{(out)}_{a_3} = Tr_{a_1,a_2}\left[\rho^{(out)}_{a_1,a_2,a_3}\right].$$

In a simple calculation, one finds that

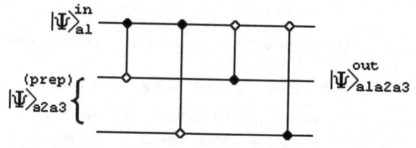

Fig. 22.3. Copying stage of the universal quantum copying machine

$$\rho_{a_1}^{(out)} = \left[| \alpha |^2 (| c_1 |^2 + | c_4 |^2) + | \beta |^2 (| c_2 |^2 + | c_3 |^2)\right] | 0\rangle_{a_1}\langle 0 |$$
$$+ [\alpha\beta^*(c_1 c_4^* + c_4 c_1^*) + \alpha^*\beta(c_2 c_3^* + c_3 c_2^*)] | 0\rangle_{a_1}\langle 1 |$$
$$+ [\alpha\beta^*(c_2 c_3^* + c_3 c_2^*) + \alpha^*\beta(c_1 c_4^* + c_4 c_1^*)] | 1\rangle_{a_1}\langle 0 |$$
$$+ \left[| \alpha |^2 (| c_2 |^2 + | c_3 |^2) + | \beta |^2 (| c_1 |^2 + | c_4 |^2)\right] | 1\rangle_{a_1}\langle 1 |, \quad (22.20)$$

$$\rho_{a_2}^{(out)} = \left[| \alpha |^2 (| c_1 |^2 + | c_2 |^2) + | \beta |^2 (| c_3 |^2 + | c_4 |^2)\right] | 0\rangle_{a_2}\langle 0 |$$
$$+ [\alpha\beta^*(c_1 c_2^* + c_2 c_1^*) + \alpha^*\beta(c_3 c_4^* + c_4 c_3^*)] | 0\rangle_{a_2}\langle 1 |$$
$$+ [\alpha\beta^*(c_3 c_4^* + c_4 c_3^*) + \alpha^*\beta(c_1 c_2^* + c_2 c_1^*)] | 1\rangle_{a_2}\langle 0 |$$
$$+ \left[| \alpha |^2 (| c_3 |^2 + | c_4 |^2) + | \beta |^2 (| c_1 |^2 + | c_2 |^2)\right] | 1\rangle_{a_2}\langle 1 |, \quad (22.21)$$

$$\rho_{a_3}^{(out)} = \left[| \alpha |^2 (| c_1 |^2 + | c_3 |^2) + | \beta |^2 (| c_2 |^2 + | c_4 |^2)\right] | 0\rangle_{a_3}\langle 0 |$$
$$+ [\alpha\beta^*(c_1 c_3^* + c_3 c_1^*) + \alpha^*\beta(c_2 c_4^* + c_4 c_2^*)] | 0\rangle_{a_3}\langle 1 |$$
$$+ [\alpha\beta^*(c_2 c_4^* + c_4 c_2^*) + \alpha^*\beta(c_1 c_3^* + c_3 c_1^*)] | 1\rangle_{a_3}\langle 0 |$$
$$+ \left[| \alpha |^2 (| c_2 |^2 + | c_4 |^2) + | \beta |^2 (| c_1 |^2 + | c_3 |^2)\right] | 1\rangle_{a_3}\langle 1 |. \quad (22.22)$$

It is possible to express the operator $\rho_{a_j}^{(out)}$ in the form:

$$\rho_{a_j}^{(out)} = s_j \rho_{a_j}^{(id)} + \frac{1 - s_j}{2} I,$$

$$= \left(\frac{1 + s_j}{2} | \alpha |^2 + \frac{1 - s_j}{2} | \beta |^2\right) | 0\rangle_{a_j}\langle 0 | + s_j \alpha\beta^* | 0\rangle_{a_j}\langle 1 |$$

$$+ \left(\frac{1 - s_j}{2} | \alpha |^2 + \frac{1 + s_j}{2} | \beta |^2\right) | 1\rangle_{a_j}\langle 1 | + s_j \alpha^*\beta | 1\rangle_{a_j}\langle 0 |, \quad (22.23)$$

where s_j is a scaling factor. One can show that this factor is related to the Fidelity by the relation $s = 2F - 1$.

Now, we may have the following different situations:

Case A (duplicator): Assume that $\rho_{a_1}^{(out)} = \rho_{a_2}^{(out)}$, that is two copies are equal and the third one is different $s_1 = s_2 = s \neq s_3$.

Here we assume the c_i coefficients are real and that α and β are complex. Using (22.20, 22.21, 22.23) we get:

$$c_1 = \sqrt{s}, c_2 = c_4 = \sqrt{\frac{1 - s}{2}}, c_3 = 0, s = \frac{2}{3}, \quad (22.24)$$

and therefore the Fidelity is $F = \frac{5}{6}$ as expected. There are many solutions for the rotation angles. Some of them are:

$$\theta_1 = 0.5535743584, \theta_2 = -2.776728825, \theta_3 = -2.909768850$$

$$\theta_1 = -2.588018296, \theta_2 = 0.3648638288, \theta_3 = -2.909768850$$

$$\theta_1 = 0.5535743584, \theta_2 = 0.3648638288, \theta_3 = 0.2318238040$$

etc.

The preparation state in this case is given by

$$| \psi \rangle_{a_2,a_3}^{(\text{prep})} = \frac{1}{\sqrt{6}} (2 \, | \, 00 \rangle_{a_2,a_3} + | \, 01 \rangle_{a_2,a_3} + | \, 11 \rangle_{a_2,a_3}) \qquad (22.25)$$

Case B (triplicator): All three outputs are equal, $\rho_{a_1}^{(\text{out})} = \rho_{a_2}^{(\text{out})} = \rho_{a_3}^{(\text{out})}$ in which case the solution is

$$c_1 = \sqrt{\frac{1+3s}{4}}, c_2 = c_3 = c_4 = c = \sqrt{\frac{1-s}{4}}, s = \frac{2}{3}, \qquad (22.26)$$

thus one finds again $F = \frac{5}{6}$; however, the above conditions can only be satisfied for the c_i coefficients as well as α and β real, so in a sense these are not universal copies, because in general the Fidelity will depend on the initial state. The preparation state is given by

$$| \psi \rangle_{a_2,a_3}^{(\text{prep})} = \frac{1}{\sqrt{12}} (3 \, | \, 00 \rangle_{a_2,a_3} + | \, 01 \rangle_{a_2,a_3} + | \, 10 \rangle_{a_2,a_3} + | \, 11 \rangle_{a_2,a_3}), \qquad (22.27)$$

and the solutions for the angles corresponding to this preparation are

$$\theta_1 = -2.748893572, \theta_2 = 0.1699184548, \theta_3 = -2.748893573,$$
$$\theta_1 = 0.3926990820, \theta_2 = -2.971674199, \theta_3 = -2.748893573,$$
$$\theta_1 = 0.3926990820, \theta_2 = 0.1699184548, \theta_3 = 0.3926990820,$$

etc.

Finally, the output state is given by (22.18), replacing the corresponding preparation states of the cases A and B.

22.3.3 Output States

Using (22.18) and the input state $| \psi \rangle_{a_1}^{(\text{in})} = \alpha \, | \, 0 \rangle_{a_1} + \beta \, | \, 1 \rangle_{a_1}$, we get different outputs for cases A and B, namely:
(Case A):

$$| \psi \rangle_{a_1,a_2,a_3}^{(\text{out})} = P_{31} P_{21} P_{13} P_{12} \, | \, \psi \rangle_{a_1}^{(in)} \, | \, \psi \rangle_{a_2,a_3}^{(prep)}$$

$$= P_{31} P_{21} P_{13} P_{12} [\sqrt{\frac{2}{3}} \alpha \, | \, 000 \rangle_{a_1,a_2,a_3} + \sqrt{\frac{1}{6}} \alpha \, | \, 001 \rangle_{a_1,a_2,a_3} + \sqrt{\frac{1}{6}} \alpha \, | \, 011 \rangle_{a_1,a_2,a_3}$$

$$+ \sqrt{\frac{2}{3}} \beta \, | \, 100 \rangle_{a_1,a_2,a_3} + \sqrt{\frac{1}{6}} \beta \, | \, 101 \rangle_{a_1,a_2,a_3} + \sqrt{\frac{1}{6}} \beta \, | \, 111 \rangle_{a_1,a_2,a_3}]$$

$$= [\sqrt{\frac{2}{3}} \alpha \, | \, 000 \rangle_{a_1,a_2,a_3} + \sqrt{\frac{1}{6}} \alpha \, | \, 101 \rangle_{a_1,a_2,a_3} + \sqrt{\frac{1}{6}} \alpha \, | \, 011 \rangle_{a_1,a_2,a_3}$$

$$+ \sqrt{\frac{2}{3}} \beta \, | \, 111 \rangle_{a_1,a_2,a_3} + \sqrt{\frac{1}{6}} \beta \, | \, 010 \rangle_{a_1,a_2,a_3} + \sqrt{\frac{1}{6}} \beta \, | \, 100 \rangle_{a_1,a_2,a_3}]. \quad (22.28)$$

(Case B):

$$|\psi\rangle^{(\text{out})}_{a_2,a_3,a_3} = P_{31}P_{21}P_{13}P_{12}\frac{1}{\sqrt{12}}[3\alpha\mid 000\rangle_{a_1,a_2,a_3} + \alpha\mid 001\rangle_{a_1,a_2,a_3}$$

$$+\alpha\mid 010\rangle_{a_1,a_2,a_3} + \alpha\mid 011\rangle_{a_1,a_2,a_3} + 3\beta\mid 100\rangle_{a_1,a_2,a_3}$$

$$+\beta\mid 101\rangle_{a_1,a_2,a_3} + \beta\mid 110\rangle_{a_1,a_2,a_3} + \beta\mid 111\rangle_{a_1,a_2,a_3}]$$

$$= \frac{1}{\sqrt{12}}[3\alpha\mid 000\rangle_{a_1,a_2,a_3} + \alpha\mid 101\rangle_{a_1,a_2,a_3} + \alpha\mid 110\rangle_{a_1,a_2,a_3} + \alpha\mid 011\rangle_{a_1,a_2,a_3}$$

$$+3\beta\mid 111\rangle_{a_1,a_2,a_3} + \beta\mid 010\rangle_{a_1,a_2,a_3} + \beta\mid 001\rangle_{a_1,a_2,a_3} + \beta\mid 100\rangle_{a_1,a_2,a_3}].$$

$$(22.29)$$

22.3.4 Summary and Discussion

The quality of a copy is measured by the following parameters:

a) The fidelity F
b) The distance between the density operator and the ideal one
c) An ideal copying machine should produce copies with no correlation between them. However, this is not true in our case. The two or three copies are entangled.

The fidelity of both the duplicator and the triplicator is $\frac{5}{6}$. The distance $D\left(\rho_{a_j}^{(\text{out})}, \rho_{a_j}^{(\text{id})}\right) = \frac{1}{18}$ in both cases.

Now, we take a look at the entanglement between copies .

For the duplicator if we form $\rho_{a_1,a_2}^{(\text{out})}$ from $\mid\psi\rangle_{a_1,a_2,a_3}^{(\text{out})}$ and tracing over a_3, we readily get:

$$\rho_{a_1,a_2}^{(\text{out})} = \frac{1}{6}\begin{pmatrix} 4\mid\beta\mid^2 & 2\alpha^*\beta & 2\alpha^*\beta & 0 \\ 2\alpha\beta^* & 1 & 1 & 2\alpha^*\beta \\ 2\alpha\beta^* & 1 & 1 & 2\alpha^*\beta \\ 0 & 2\alpha\beta^* & 2\alpha\beta^* & 4\mid\alpha\mid^2 \end{pmatrix}, \quad (22.30)$$

in the basis $\mid 11\rangle_{a_1,a_2}, \mid 10\rangle_{a_1,a_2}, \mid 01\rangle_{a_1,a_2}, \mid 00\rangle_{a_1,a_2}$. The corresponding partial transpose matrix is

$$\rho_{a_1,a_2}^{PT(\text{out})} = \frac{1}{6}\begin{pmatrix} 4\mid\beta\mid^2 & 2\alpha\beta^* & 2\alpha^*\beta & 1 \\ 2\alpha^*\beta & 1 & 0 & 2\alpha^*\beta \\ 2\alpha\beta^* & 0 & 1 & 2\alpha\beta^* \\ 1 & 2\alpha\beta^* & 2\alpha^*\beta & 4\mid\alpha\mid^2 \end{pmatrix}. \quad (22.31)$$

Now we apply the Perez–Horodecki criteria and find the eigenvalues of $\rho_{a_1,a_2}^{PT(\text{out})}$. They are

$$\left[\frac{1}{6}, \frac{1}{6}, \frac{2-\sqrt{5}}{6}, \frac{2+\sqrt{5}}{6}\right],$$

thus one of the eigenvalues is negative, which implies that the state is not separable (entangled state).

The reader can check that the same applies to the triplicator, that is, we also get an entangled state.

22.4 Quantum Processors

22.4.1 Introduction

In this section, we will make further use of the quantum gates to create a stochastic quantum processor. We start by some relevant definitions, followed by a theoretical background necessary for the proposal of a quantum stochastic processor.

As we already discussed in the previous chapter, the quantum gates are unitary transformations acting on one or more qubits in a given sequence. In general, these transformations are implemented by static quantum gate arrays that depend on the particular operation or experiment. The disadvantage of this is that we require a different array for a given operation.

Instead, we may have an array of fixed gates, or processor, that takes as an input not only the data qubits but also the program qubits. This idea was suggested by Nielsen and Chuang [7].

We start defining the Hilbert spaces of the program $\mathcal{H}_P$ and the data $\mathcal{H}_D$. The system is initially in the state $\mid d\rangle_D \otimes \mid P_U\rangle_P$, where $\mid d\rangle_D \in \mathcal{H}_D$ and $\mid P_U\rangle_P \in \mathcal{H}_P$. The dynamics of the programmable gate array is given by a fixed unitary operator G (see Fig. 22.4), that implements a unitary operation U given by the state of the program $\mid P_U\rangle_P$ resulting in the following

$$G[\mid d\rangle_D \otimes \mid P_U\rangle_P] = (U \mid d_1\rangle_D) \otimes \mid R_U\rangle_P. \tag{22.32}$$

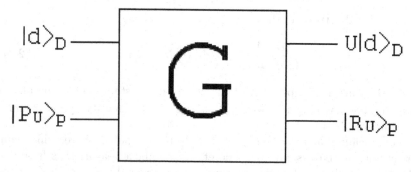

Fig. 22.4. Programmable array of quantum gates. The fixed gate array G takes a data qubit as an input and applies to it a unitary transformation U, previously specified by the program qubit

After the application of G, the state of the data $| \, d\rangle_D$ has been transformed by a unitary operation U to $U \, | \, d\rangle_D$.

$| \, R_U\rangle_P$ is the residual program state that is independent of the data state. This last statement can be shown quite easily. Let us assume that the operator G acts on two different data states $| \, d_1\rangle_D$ y $| \, d_2\rangle_D$, with the same program state $| \, P_U\rangle_P$,

$$G[\, | \, d_1\rangle_D \otimes | \, P_U\rangle_P] = (U \, | \, d_1\rangle_D) \otimes | \, R_U^1\rangle_P,$$

$$G[\, | \, d_2\rangle_D \otimes | \, P_U\rangle_P] = (U \, | \, d_2\rangle_D) \otimes | \, R_U^2\rangle_P. \qquad (22.33)$$

Consider now the inner product

$$\langle d_1 \, | \, d_2\rangle_D = \langle d_1 \, | \, d_2\rangle_D \langle R_U^1 \, | \, R_U^2\rangle_P. \qquad (22.34)$$

If $\langle d_1 \, | \, d_2\rangle_D \neq 0$ then $\langle R_U^1 \, | \, R_U^2\rangle_P. = 1$, and therefore, the residual program states do not depend on the data. The case $\langle d_1 \, | \, d_2\rangle_D = 0$ can be shown in a similar way.

One can also ask a valid question: how many qubits are required for a given operation?

Let us assume that $| \, A\rangle_P$ and $| \, B\rangle_P$ are program states that belong to a common Hilbert space $\mathcal{H}_P$ and that they implement the Unitary operations U_A and U_B, respectively. Let us also assume that these operations are different and distinguishable. For an arbitrary data state $| \, d\rangle_D \in \mathcal{H}_D$ and a gate array G we have

$$G[\, | \, d\rangle_D \otimes | \, A\rangle_P] = (U_A \, | \, d_1\rangle_D) \otimes | \, R_A\rangle_P,$$

$$G[\, | \, d\rangle_D \otimes | \, B\rangle_P] = (U_B \, | \, d_1\rangle_D) \otimes | \, R_B\rangle_P. \qquad (22.35)$$

Assuming that performing the inner product of the above equations, we get

$$\langle B \, | \, A\rangle_P = \langle d \, | \, U_B^\dagger U_A \, | \, d\rangle_D \langle R_B \, | \, R_A\rangle_P. \qquad (22.36)$$

Now, if $\langle B \, | \, A\rangle_P \neq 0$, then

$$\frac{\langle B \, | \, A\rangle_P}{\langle R_B \, | \, R_A\rangle_P} = \langle d \, | \, U_B^\dagger U_A \, | \, d\rangle_D. \qquad (22.37)$$

As the l.h.s. is independent of the state $| \, d\rangle_D$, state that is completely arbitrary, one has $U_B^\dagger U_A = (\exp i\alpha)I$, where I is the identity operator and α an arbitrary phase.

This means that U_A and U_B are equal up to a global phase, which contradicts our initial assumption. Therefore, the above reasoning is false and $\langle B \, | \, A\rangle_P = 0$, that is the two programs are orthogonal.

We can generalize this result to N different operators. Therefore, we conclude that given N different unitary operators $U_1, U_2 \ldots U_N$, these can be implemented by N orthogonal program states $| \, P_1\rangle, | \, P_2\rangle, | \, P_3\rangle \ldots | \, P_N\rangle$, that is, the Hilbert space H_P that implements N unitary operations is of dimension N.

A tragic consequence of all this is that it is not possible, with a finite gate array, to implement an arbitrary unitary operation deterministically.

22.4.2 One Qubit Stochastic Processor

Despite of the bad result from the previous section, we still can do something if we relax the deterministic character of the processor and accept the possibility of a probabilistic processor.

Vidal, Massanes and Cirac ([8, 9, 10]) propose such a processor that implements the rotation of a qubit , given by

$$U_\alpha = \exp\left(i\alpha\frac{\sigma_z}{2}\right), \tag{22.38}$$

for an arbitrary phase $\alpha \in [0, 2\pi]$. This transformation has a very simple interpretation: the rotation of α radians around the z axis, in the *Bloch sphere* of one qubit.

To understand how to implement the operation 22.38 using fixed quantum gates, we define first the program and data states as follows:

$$| \alpha\rangle_P \equiv \frac{1}{\sqrt{2}}(e^{i\alpha/2} | 0\rangle_P + e^{-i\alpha/2} | 1\rangle_P), \tag{22.39}$$

$$| d\rangle_d \equiv (A | 0\rangle_d + B | 1\rangle_d), \tag{22.40}$$

where $|A|^2 + |B|^2 = 1$. We notice that $| \alpha\rangle_P$ is a known state and $| d\rangle_d$ is completely general and normally unknown. One can easily show that the operation described in 22.32 can be realized with a C-NOT gate.

If we describe the C-NOT gate as a G operator

$$G_{C-NOT} =| 0\rangle\langle 0 |_d I_{P+} | 1\rangle\langle 1 |_d \otimes\sigma_x, \tag{22.41}$$

with $\sigma_x = (| 1\rangle\langle 0 | + | 0\rangle\langle 1 |)_P$, then we can identify the data qubit as the control and the program as the target. It is simple to show that

$$G_{C-NOT} | 0\rangle_d | 0\rangle_P =| 0\rangle_d | 0\rangle_P,$$

$$G_{C-NOT} | 0\rangle_d | 1\rangle_P =| 0\rangle_d | 1\rangle_P,$$

$$G_{C-NOT} | 1\rangle_d | 0\rangle_P =| 1\rangle_d | 1\rangle_P,$$

$$G_{C-NOT} | 1\rangle_d | 1\rangle_P =| 1\rangle_d | 0\rangle_P. \tag{22.42}$$

For an initial state $| d\rangle_d | \alpha\rangle_P$, we get

$$C - NOT\,[| d\rangle_d | \alpha\rangle_P] = G_{C-NOT}\,[| d\rangle_d | \alpha\rangle_P]$$

$$= \frac{1}{\sqrt{2}}(Ae^{i\alpha/2} \mid 0\rangle_d + Be^{-i\alpha/2} \mid 1\rangle_d) \otimes \mid 0\rangle_\mathcal{P}$$

$$+ \frac{1}{\sqrt{2}}(Ae^{-i\alpha/2} \mid 0\rangle_d + Be^{i\alpha/2} \mid 1\rangle_d) \otimes \mid 1\rangle_\mathcal{P}$$

$$= \frac{1}{\sqrt{2}} \left[(U_\alpha \mid d\rangle_d) \otimes \mid 0\rangle_P + (U_\alpha^\dagger \mid d\rangle_d) \otimes \mid 1\rangle_P \right],$$

where we used $U_\alpha \mid d\rangle_d = \frac{1}{\sqrt{2}}(Ae^{i\alpha/2} \mid 0\rangle_d + Be^{-i\alpha/2} \mid 1\rangle_d)$.

We notice that if we perform a measurement in the program qubit, using the $\{\mid 0\rangle_{\mathcal{P}_1}, \mid 1\rangle_{\mathcal{P}_1}\}$, the data qubit will collapse to two possible states, $U_\alpha \mid d\rangle_d$ or $U_\alpha^{3\dagger} \mid d\rangle_d$, both with probabilities $p = 1/2$. We say then that we managed to apply the Unitary transformation U with a success probability of $\frac{1}{2}$.

To improve the success probability, one can introduce an extra gate, a Toffoli , as shown in the Fig. 22.5.

Now the processor has two Hilbert spaces, $\mathcal{H}_{\mathcal{P}_1}$ and $\mathcal{H}_{\mathcal{P}_2}$, to store the program. Now we introduce an additional program state $\mid 2\alpha\rangle_{\mathcal{P}_2}$, defined as $\mid 2\alpha\rangle_{P_2} \equiv \frac{1}{\sqrt{2}}(e^{i(2\alpha)/2} \mid 0\rangle_{\mathcal{P}_2} + e^{-i(2\alpha)/2|} \mid 1\rangle_{\mathcal{P}_2})$.

We study now the effect of the Toffoli gate . When the output of the C-NOT gate is $\mid 0\rangle_{\mathcal{P}_1}$, which corresponds to the correct application of U_α to the data, then the output in the first program register ($\mathcal{H}_{\mathcal{P}_1}$) does not change.

On the other hand, if the output of the first program register $\mathcal{H}_{\mathcal{P}_1} \mid 1\rangle_{\mathcal{P}_1}$ indicating that the wrong operator $U_\alpha^\dagger$ was applied (failure), the Toffoli gate acts effectively as a C-NOT gate between the data register $\mathcal{H}_d$ and the second program register $\mathcal{H}_{\mathcal{P}_2}$ in such a way that we correct the failure. If in the second program register, we put $\mid 2\alpha\rangle_{\mathcal{P}_2}$, there is again a probability of $1/2$ of applying the correct transformation $U_{2\alpha}U_\alpha^\dagger = U_\alpha$ to $\mid d\rangle_d$ when measuring both program registers $\mathcal{H}_{\mathcal{P}_1}$ and $\mathcal{H}_{\mathcal{P}_2}$. Thus the success probability has increased to $p = 1/2 + 1/4 = 3/4$. The wrong alternative corresponding to $U_\alpha^{\dagger 3}$

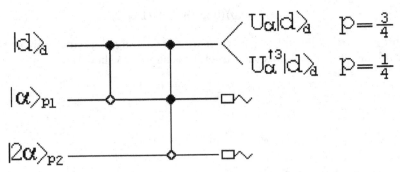

Fig. 22.5. Quantum Stochastic Processor of a qubit. The inputs to the processor are the data register $\mid d\rangle_d$ and the programs $\mid \alpha\rangle_{P_1}$ and $\mid 2\alpha\rangle_{P_2}$. When operating, it will detect the wrong result with probability $\frac{1}{4}$, if in the program register, we get $\mid 1\rangle_{P_1}$ and $\mid 1\rangle_{P_2}$. The correct result $U_\alpha \mid d\rangle_d$ is obtained in all other cases, with a probability $\frac{3}{4}$

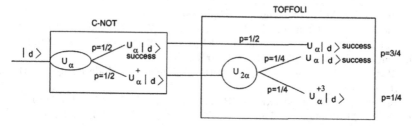

Fig. 22.6. Schematic operation of the quantum stochastic processor

occurs with a probability of $1/4$, when one gets $|1\rangle$ in both programs. The process is explained pictorially in Fig. 22.6.

To summarize, we have proposed a fixed quantum gate array to perform unitary transformations, in this case, the rotation of a qubit around the z-axis, in the Bloch sphere, with a success rate of $p = 3/4$.

One can generalize this scheme with a generalized Toffoli gate with N program registers, as suggested by Vidal et al. ([8]). For this purpose, additional programs $|2^0\alpha\rangle_{\mathcal{P}_1}, |2^1\alpha\rangle_{\mathcal{P}_2}, |2^2\alpha\rangle_{\mathcal{P}_3}\ldots |2^{N-1}\alpha\rangle_{\mathcal{P}_N}$ are required, getting a success probability of $p = 1 - 2^{-N}$.

For recent discussions on probabilistic programmable quantum processors, see ([9, 10, 11, 12]).

Problems

22.1. Show that the Fidelity of both the duplicator and the triplicator is $\frac{5}{6}$.

22.2. Show that the output state for a triplicator is entangled.

22.3. Show that the output state of the processor, after the C-NOT gate and the Toffoli is:

$$|\Omega\rangle = \frac{1}{2}\begin{bmatrix} (\exp(3i\alpha) \mid 0\rangle_1\langle 0 \mid + \exp(i\alpha) \mid 1\rangle_1\langle 1 \mid) \mid d\rangle_d \mid 00\rangle_{23} \\ +(\exp(-i\alpha) \mid 0\rangle_1\langle 0 \mid + \exp(-3i\alpha) \mid 1\rangle_1\langle 1 \mid) \mid d\rangle_d \mid 01\rangle_{23} \\ +(\exp(i\alpha) \mid 0\rangle_1\langle 0 \mid + \exp(-i\alpha) \mid 1\rangle_1\langle 1 \mid) \mid d\rangle_d \mid 10\rangle_{23} \\ +(\exp(-3i\alpha) \mid 0\rangle_1\langle 0 \mid + \exp(3i\alpha) \mid 1\rangle_1\langle 1 \mid) \mid d\rangle_d \mid 11\rangle_{23} \end{bmatrix}.$$

22.4. From the problem 22.3, find the probability of success of the processor.

References

1. Wootters, W.K., Zurek, W.H.: Nature (London), **299**, 802 (1982)
2. Buzek, V., Hillery, M.: Phys. Rev. A, **54**, 1844 (1996)
3. Schumacher, B.: Phys. Rev. A, **51**, 2738 (1995)
4. Scarani, V., Iblisdir, S., Gisin, N.: Rev. Mod. Phys., **77**, 1225 (2005)
5. Buzek, V., Braunstein, S.L., Hillery, M., Bruss, D.: Phys. Rev. A, **56**, 3446 (1997)

6. Hernandez, M.L.: Lic. Thesis, PUC (2005)
7. Nielsen, M.A., Chuang, I.L.: Phys. Rev. Lett., **79**, 321 (1997)
8. Vidal, G., Masanes, L., Cirac, J.I.: Phys. Rev. Lett., **88**, 047905 (2002)
9. Blackburn, P.: Ms Thesis, PUC (2006)
10. Blackburn, P., Orszag, M.: International Journal of Modern.Physics. B, **20**, 1679 (2006)
11. Brazier, A., Buzek, V., Knight, P.L.: Phys. Rev. A, **71**, 032306 (2005)
12. Hillery, M., Ziman, M., Buzek, V.: Phys. Rev. A, **69**, 042311 (2004)

A. Operator Relations

A.1 Theorem 1

Let A and B be two non-commuting operators, then [1]

$$\exp(\alpha A)B\exp(-\alpha A) = B + \alpha\,[A, B] + \frac{\alpha^2}{2!}\,[A, [A, B]] + \dots. \tag{A.1}$$

Proof.

Let

$$f_1(\alpha) = \exp(\alpha A)B\exp(-\alpha A)\,, \tag{A.2}$$

then, one can expand f_1 in Taylor series about the origin. We first evaluate the derivatives

$$f_1'(\alpha) = \exp(\alpha A)(AB - BA)\exp(-\alpha A)\,,$$

so

$$f_1'(0) = [A, B]\,. \tag{A.3}$$

Similarly

$$f_1''(\alpha) = \exp(\alpha A)(A\,[A, B] - [A, B]\,A)\exp(-\alpha A)\,,$$

so that

$$f_1''(0) = [A, [A, B]]\,. \tag{A.4}$$

Now, we write the Taylor's expansion

$$f_1(\alpha) = f_1(0) + \alpha f_1'(0) + \frac{\alpha^2}{2!}f_1''(0) + \dots \tag{A.5}$$

or

$$\exp(\alpha A)B\exp(-\alpha A) = B + \alpha\,[A, B] + \frac{\alpha^2}{2!}\,[A, [A, B]] + \dots \tag{A.6}$$

A particular case is when $[A, B] = c$, where c is a c-number, then

$$\exp(\alpha A)B\exp(-\alpha A) = B + \alpha c\,, \tag{A.7}$$

in which case $\exp(\alpha A)$ acts as a displacement operator.

A.2 Theorem 2: The Baker–Campbell–Haussdorf Relation

Let A and B be two non-commuting operators such that

$$[A, [A, B]] = [B, [A, B]] = 0 , \tag{A.8}$$

then

$$\exp\left[\alpha(A + B)\right] = \exp \alpha A \exp \alpha B \exp\left[-\frac{\alpha^2}{2}[A, B]\right] \tag{A.9}$$

$$= \exp \alpha B \exp \alpha A \exp\left[\frac{\alpha^2}{2}[A, B]\right] .$$

Proof.
Define

$$f_2(\alpha) \equiv \exp \alpha A \exp \alpha B . \tag{A.10}$$

Then

$$\frac{\mathrm{d}f_2(\alpha)}{\mathrm{d}\alpha} = [A + \exp(\alpha A)B \exp -(\alpha A)]f_2(\alpha) \tag{A.11}$$

$$= (A + B + \alpha [A, B])f_2(\alpha) ,$$

where in the last step, we used (A.6).
Also, from the definition of $f_2(\alpha)$, we can write

$$\frac{\mathrm{d}f_2(\alpha)}{\mathrm{d}\alpha} = \exp(\alpha A)A \exp \alpha B + \exp(\alpha A)\exp(\alpha B)B \tag{A.12}$$

$$= \exp \alpha A \exp \alpha B [\exp(-\alpha B)A \exp \alpha B + B]$$

$$= f_2(\alpha)(A + B + \alpha [A, B]) .$$

By comparing (A.11) and (A.12), we can see that $f_2(\alpha)$ commutes with $(A + B + \alpha [A, B])$, thus one can integrate as a c-number differential equation, getting

$$f_2(\alpha) = \exp\left[(A + B)\alpha + \frac{\alpha^2}{2}[A, B]\right] = \exp \alpha(A+B) \exp \frac{\alpha^2}{2}[A, B] , \tag{A.13}$$

thus obtaining the desired result.
Another application of the Theorem 1 is taking

$$A = aa^\dagger, \tag{A.14}$$

$$B = a \text{ or } a^\dagger .$$

As

$$[n, a] = -a \tag{A.15}$$

and the higher order commutators also give a with alternating signs, thus

$$\exp(\alpha n)a\exp(-\alpha n) = a - \alpha a + \frac{\alpha^2}{2}a + ... = \exp(-\alpha)a . \tag{A.16}$$

Similarly

$$\exp(\alpha n)a^\dagger \exp(-\alpha n) = \exp(\alpha)a^\dagger . \tag{A.17}$$

A.3 Theorem 3: Similarity Transformation

$$\exp(\alpha A)f(B)\exp-(\alpha A) = f(\exp(\alpha A)(B)\exp-(\alpha A)) . \tag{A.18}$$

Proof.

We start with the following identity

$$[\exp(\alpha A)(B)\exp-(\alpha A)]^n = \exp(\alpha A)B\exp(-\alpha A)\exp(\alpha A)B\exp(-\alpha A)...$$
$$= \exp(\alpha A)B^n \exp(-\alpha A) .$$

Then, for any function $f(B)$ that can be expanded in power series, the Theorem 3 follows.

As an interesting application, let us calculate

$$\exp\left(-\alpha a^\dagger + \alpha^* a\right) f(a, a^\dagger) \exp\left(\alpha a^\dagger - \alpha^* a\right)$$
$$= f[\exp\left(-\alpha a^\dagger + \alpha^* a\right) a \exp\left(\alpha a^\dagger - \alpha^* a\right) ,$$
$$\exp\left(-\alpha a^\dagger + \alpha^* a\right) a^\dagger \exp\left(\alpha a^\dagger - \alpha^* a\right)]$$
$$= f(a + \alpha, a^\dagger + \alpha^*) .$$

Also

$$\exp\left(-\alpha a^\dagger\right) f(a, a^\dagger) \exp\left(\alpha a^\dagger\right) = f(a + \alpha, a^\dagger) , \tag{A.19}$$

$$\exp\left(\alpha^* a\right) f(a, a^\dagger) \exp\left(-\alpha^* a\right) = f(a, a^\dagger + \alpha^*) , \tag{A.20}$$

$$\exp\left(\alpha n\right) f(a, a^\dagger) \exp\left(-\alpha n\right) = f[a \exp(-\alpha), a^\dagger \exp(\alpha)] \tag{A.21}$$

Other properties:

One can easily show that

$$[a, a^{\dagger l}] = l a^{\dagger l-1} = \frac{da^{\dagger l}}{da^\dagger}, \tag{A.22}$$

$$[a^\dagger, a^l] = -l a^{l-1} = -\frac{da^l}{da} .$$

A more general version of the above relations is for a function $f(a, a)$ which may be expanded in power series of a and $a^\dagger$

$$[a, f(a, a^\dagger)] = \frac{\partial f(a, a^\dagger)}{\partial a^\dagger} , \tag{A.23}$$

$$[a^\dagger, f(a, a^\dagger)] = -\frac{\partial f(a, a^\dagger)}{\partial a} . \tag{A.24}$$

Reference

1. Louisell, W.H.: Quantum Statistical Properties of Radiation. John Wiley, New York (1973)

B. The Method of Characteristics

We have a first-order partial differential equation: [1]

$$Pp + Qq = R ,$$
(B.1)

where $P = P(x, y, z), Q = Q(x, y, z), R = R(x, y, z)$, and

$$p \equiv \frac{\partial z}{\partial x}, q \equiv \frac{\partial z}{\partial y} ,$$
(B.2)

and we wish to find a solution of (B.1), of the form

$$z = f(x, y) .$$
(B.3)

The general solution of (B.1) is

$$F(u, v) = 0 ,$$
(B.4)

where F is an arbitrary function, and

$$u(x, y, z) = c_1 ,$$
$$v(x, y, z) = c_2 ,$$
(B.5)

is a solution of the equations

$$\frac{\mathrm{d}x}{P} = \frac{\mathrm{d}y}{Q} = \frac{\mathrm{d}z}{R} .$$
(B.6)

Proof.
If (B.5) are solutions of (B.6), then the equations

$$\frac{\partial u}{\partial x}\mathrm{d}x + \frac{\partial u}{\partial y}\mathrm{d}y + \frac{\partial u}{\partial z}\mathrm{d}z = 0 ,$$
(B.7)

and

$$\frac{\mathrm{d}x}{P} = \frac{\mathrm{d}y}{Q} = \frac{\mathrm{d}z}{R} ,$$

must be compatible; thus, we must have

$$Pu_x + Qu_y + Ru_z = 0 \,, \tag{B.8}$$

and similarly for v

$$Pv_x + Qv_y + Rv_z = 0 \,. \tag{B.9}$$

On the other hand, if x and y are independent variables and $z = z(x,y)$, then from (B.5), we get

$$u_x + u_z \frac{\partial z}{\partial x} = 0 \,, \tag{B.10}$$

$$u_y + u_z \frac{\partial z}{\partial y} = 0 \,,$$

and substituting (B.10) into (B.8) we get

$$\left[-P\frac{\partial z}{\partial x} - Q\frac{\partial z}{\partial y} + R \right] \frac{\partial u}{\partial z} = 0 \,,$$

and (B.1) is satisfied.

The second part of the proof is to show that the general solution of (B.1) is

$$F(u, v) = 0 \,. \tag{B.11}$$

From (B.11), one writes

$$\frac{\partial F}{\partial x} = \frac{\partial F}{\partial u} \left(\frac{\partial u}{\partial x} + \frac{\partial u}{\partial z}\frac{\partial z}{\partial x} \right) + \frac{\partial F}{\partial v} \left(\frac{\partial v}{\partial x} + \frac{\partial v}{\partial z}\frac{\partial z}{\partial x} \right) = 0 \,, \tag{B.12}$$

$$\frac{\partial F}{\partial y} = \frac{\partial F}{\partial u} \left(\frac{\partial u}{\partial y} + \frac{\partial u}{\partial z}\frac{\partial z}{\partial y} \right) + \frac{\partial F}{\partial v} \left(\frac{\partial v}{\partial y} + \frac{\partial v}{\partial z}\frac{\partial z}{\partial y} \right) = 0 \,. \tag{B.13}$$

We finally notice that (B.13) is satisfied considering (B.10).

Example.

Find the general solution of the equation

$$x^2 \frac{\partial z}{\partial x} + y^2 \frac{\partial z}{\partial y} = (x + y)z \,. \tag{B.14}$$

In this case

$$P = x^2 \,, \tag{B.15}$$

$$Q = y^2 \,,$$

$$R = (x + y)z \,,$$

and we have to find the solution of

$$\frac{dx}{x^2} = \frac{dy}{y^2} = \frac{dz}{(x + y)z} \,. \tag{B.16}$$

Integrating, first

$$\frac{\mathrm{d}x}{x^2} = \frac{\mathrm{d}y}{y^2} \ ,$$

we get

$$x^{-1} + y^{-1} = c_1' \ . \tag{B.17}$$

On the other hand,

$$\frac{\mathrm{d}x - \mathrm{d}y}{x^2 - y^2} = \frac{(\frac{x^2}{y^2} - 1)\mathrm{d}y}{x^2 - y^2} = \frac{\mathrm{d}y}{y^2} = \frac{\mathrm{d}z}{(x+y)z} \ ,$$

from where we get

$$\frac{x - y}{z} = c_2 = v \ . \tag{B.18}$$

Combining (B.17) and (B.18), we get

$$\frac{xy}{z} = c_1 = u \ , \tag{B.19}$$

so the general solution can be put as

$$F\left(\frac{xy}{z}, \frac{x-y}{z}\right) = 0 \ , \tag{B.20}$$

or if we write (B.20) in the equivalent form

$$u = g(v) \ , \tag{B.21}$$

then the solution is

$$\frac{xy}{z} = g\left(\frac{x-y}{z}\right) \ . \tag{B.22}$$

Reference

1. Sneddon, I.: Elements of Partial Differential Equations. (Mc-Graw Hill, New York (1957)

C. Proof

In this Appendix, we show the equation

$$\left[\langle\sum_{j\neq k}\delta(t-t_j)\delta(t'-t_k)\rangle_S - R^2\right]\rho_{aa}^2 = -pR\delta(t-t') . \qquad \text{(C.1)}$$

For regular pumping, one can put $t_j = t_0 + j\tau$, where τ is the constant time interval between the atoms and t_0 some arbitrary time origin [1].

In this case, there are no pumping fluctuations, and therefore, there are no correlations between the products of delta functions, that is

$$\sum_{j,k}\langle\delta(t-t_j)\delta(t'-t_k)\rangle_S = \sum_{j}\langle\delta(t-t_j)\rangle_S \sum_{k}\langle\delta(t'-t_k)\rangle_S \qquad \text{(C.2)}$$

$$= R^2 .$$

Now, we split the l.h.s. of the above equation in two parts

$$\sum_{j\neq k}\langle\delta(t-t_j)\delta(t'-t_k)\rangle_S + \sum_{j=k}\langle\delta(t-t_j)\delta(t'-t_k)\rangle_S = R^2 , \qquad \text{(C.3)}$$

$$\sum_{j\neq k}\langle\delta(t-t_j)\delta(t'-t_k)\rangle_S + R\delta(t-t') = R^2 ,$$

thus proving the relation

$$\left[\langle\sum_{j\neq k}\delta(t-t_j)\delta(t'-t_k)\rangle_S - R^2\right]\rho_{aa}^2 = -pR\delta(t-t')$$

for $p = 1$.

In the Poissonian case, t_j is totally uncorrelated from $t_k (j\neq k)$, so

$$\sum_{j\neq k}\langle\delta(t-t_j)\delta(t'-t_k)\rangle_S = \sum_{j}\langle\delta(t-t_j)\rangle_S \sum_{k}\langle\delta(t'-t_k)\rangle_S = R^2 , \qquad \text{(C.4)}$$

which proves (C.1) for $p = 0$.

Notice that in the above result, we are missing an atom in the second summation, so the above result is approximate, the approximation being very good when $R \gg 1$. (The error is of the order of R compared to R^2.)

A more general proof is found in the reference [2].

References

1. An excellent discussion on this point, as well and on noise supression in quantum optical systems is found in: Davidovich, L.: Rev. Mod. Phys., **68**, 127 (1996)
2. Benkert, C., Scully, M.O., Bergou, J., Davidovich, L., Hillery, M., Orszag, M.: Phys. Rev. A, **41**, 2756 (1990)

D. Stochastic Processes in a Nutshell

D.1 Introduction

Classical Mechanics gives a deterministic view of the dynamical variables of a system. This of course is true, when one is not in a chaotic regime.

On the other hand, in many cases, the system under study is only described by the time evolution of probability distributions.

To show these ideas with an example, we take a look at the random walk in one dimension, by now, a classical problem [3].

A person moves in a line, taking random steps forward or backward, with equal probability, at fixed time intervals τ.

Calling the position $x_n = na$, then the probability that it occupies the site x_n at time t is $P(x_n \mid t)$ and obeys the equation

$$P(x_n \mid t + \tau) = \frac{1}{2} P(x_{n-1} \mid t) + \frac{1}{2} P(x_{n+1} \mid t) . \tag{D.1}$$

Now, we go to the continuum limit, letting τ and a become small, but with finite $\frac{a^2}{\tau}$. Then

$$P(x \mid t + \tau) = P(x \mid t) + \tau \frac{\partial}{\partial t} P(x \mid t) + \dots \tag{D.2}$$

$$P(x_{n\pm 1} \mid t) = P(x \pm a \mid t) = P(x \mid t) \pm a \frac{\partial}{\partial x} P(x \mid t)$$

$$+ \frac{a^2}{2} \frac{\partial^2}{\partial x^2} P(x \mid t) + \dots ,$$

and inserting the above expansions in (D.1), we get

$$\tau \frac{\partial}{\partial t} P(x \mid t) + \mathcal{O}(\tau^2) = \frac{a^2}{2} \frac{\partial^2}{\partial x^2} P(x \mid t) + \mathcal{O}(a^4) + \dots \tag{D.3}$$

Now, letting $\tau, a \to 0$ with

$$D \equiv \frac{a^2}{\tau} , \tag{D.4}$$

D being the diffusion coefficient, we get **a diffusion or Fokker–Planck Equation:**

$$\frac{\partial}{\partial t} P(x \mid t) = \frac{D}{2} \frac{\partial^2}{\partial x^2} P(x \mid t) . \tag{D.5}$$

D.2 Probability Concepts

Let us call ω an event and let A describe a set of events , thus

$$\omega \in A , \tag{D.6}$$

meaning that the event ω belongs to the set of events A [2] .

Also, we call Ω the set of all the events and Φ the set of no events.

We now introduce the probability of A, $P(A)$, satisfying the following axioms

i) $P(A) \geq 0$ for all A.

ii) $P(\Omega) = 1$.

iii) If $A_i(i = 1, 2, 3...)$ is a countable collection of non-overlapping sets, such that

$$A_i \cap A_j = \Phi, i \neq j , \tag{D.7}$$

then

$$P(\cup_i A_i) = \sum_i P(A_i) . \tag{D.8}$$

Now, we are ready to define the joint and conditional probabilities.

Joint probability

$$P(A \cap B) = P \{\omega \in A \text{ and } \omega \in B\} . \tag{D.9}$$

Conditional probability

$$P(A \mid B) = \frac{P(A \cap B)}{P(B)} , \tag{D.10}$$

which satisfies the intuitive idea of a conditional probability that $\omega \in A$ (given that we know that $\omega \in B$) is given by the joint probability of A and B divided by the probability of B.

Now, suppose we have a collection of sets B_i, such that

$$B_i \cap B_j = \Phi , \tag{D.11}$$

$$\cup_i(A \cap B_i) = A \cap (\cup_i B_i) = A . \tag{D.12}$$

Now, by the axiom iii

$$\sum_i P(A \cap B_i) = P(\cup_i(A \cap B_i)) = P(A) , \tag{D.13}$$

thus

$$\sum_i P(A, B_i) = \sum_i P(A \mid B_i)P(B_i) = P(A) , \tag{D.14}$$

or, put it in words, if we sum the joint probability over the mutually exclusive events B_i, it eliminates that variable. These ideas will be useful later to derive the Chapman–Kolmogorov Equation.

D.3 Stochastic Processes

We have a time-dependent random variable $X(t)$ and measure the values $x_1, x_2, x_3...$ at times $t_1, t_2, t_3...$, then the joint probability densities

$$P(x_1, t_1; x_2, t_2; ...)$$

describe completely the system, which is referred to as a stochastic process.

One can also define the conditional probability densities as

$$P(x_1, t_1; x_2, t_2; ... \mid y_1, \tau_1; y_2, \tau_2; ...) = \tag{D.15}$$
$$P(x_1, t_1; x_2, t_2; ...y_1, \tau_1; y_2, \tau_2; ..)/P(y_1, \tau_1; y_2, \tau_2; ..) \,,$$

where the time sequence increases as

$$t_1 \geq t_2 \geq ... \geq \tau_1 \geq \tau_2...$$

Some simple examples:

a) Complete independence.

In this case $X(t)$ is completely independent of past and future, or

$$P(x_1, t_1; x_2, t_2; ...) = \sqcap_i P(x_i, t_i) \,. \tag{D.16}$$

b) The next simplest case is the Markov Process, where the conditional probability is entirely determined by the knowledge of the most recent condition, that is

$$P(x_1, t_1; x_2, t_2; ... \mid y_1, \tau_1; y_2, \tau_2; ...) = P(x_1, t_1; x_2, t_2; ... \mid y_1, \tau_1) \,. \tag{D.17}$$

It is simple to show, that for the Markovian case, an arbitrary joint probability can be written as

$$P(x_1, t_1; x_2, t_2; ...x_n, t_n) = \sqcap_{i=1}^{n-1} P(x_i, t_i \mid x_{i-1}, t_{i-1})P(x_n, t_n) \,. \tag{D.18}$$

D.3.1 The Chapman–Kolmogorov Equation

As we saw in the previous section, summing over all mutually exclusive variables, eliminates that variable, in other words

$$\sum_B P(A \cap B \cap C...) = P(A \cap C...) \,. \tag{D.19}$$

Now, we apply this idea to a stochastic process

$$P(x,t \mid x_0,t_0) = \int dy P(x,t;y,s \mid x_0,t_0) \qquad (\text{D.20})$$

$$= \int dy P(x,t \mid y,s;x_0,t_0)P(y,s \mid x_0,t_0) .$$

Next, we apply the Markov condition, getting the Chapman–Kolmogorov Equation

$$P(x,t \mid x_0,t_0) = \int dy P(x,t \mid y,s)P(y,s \mid x_0,t_0) . \qquad (\text{D.21})$$

In the above analysis, t_0 is any initial time for which $x(t_0) = x_0$, and s is an intermediate time $t_0 \leq s \leq t$, and $x(s) = y$.

At this point, we observe that $P(x,t \mid x_0,t_0)$ is a probability density, satisfying the initial condition

$$P(x,t \mid x_0,t_0) \mid_{t=t_0} = \delta(t - t_0) , \qquad (\text{D.22})$$

and the normalization condition

$$\int dx P(x,t \mid x_0,t_0) = 1 . \qquad (\text{D.23})$$

Now, going back to (D.21), we write $t = s + \Delta t$, and expand in Δt

$$P(x,s+\Delta t \mid x_0,t_0) = \int dy \left[P(x,s \mid y,s) + \Delta t \frac{\partial P(xt \mid y,s)}{\partial t} \mid_{t=s} \right] P(y,s \mid x_0,t_0) ,$$

or

$$P(x,s+\Delta t \mid x_0,t_0) = P(x,s \mid x_0,t_0) + \Delta t \int dy W(x \mid y)P(y,s \mid x_0,t_0) , \qquad (\text{D.24})$$

where $W(x \mid y)$ is the transition rate, defined as

$$W(x \mid y) = \frac{\partial P(x,s \mid y,s)}{\partial t} \mid_{t=s} . \qquad (\text{D.25})$$

Letting $\Delta t \to 0$, (D.24) becomes

$$\frac{\partial P(x,t \mid x_0,t_0)}{\partial t} = \int dy W(x \mid y)P(y,t \mid x_0,t_0) . \qquad (\text{D.26})$$

This is the **forward Chapman–Kolmogorov equation.**

By integrating (D.26), one can easily verify that

$$\int dx W(x \mid y) = 0 . \qquad (\text{D.27})$$

The transition probability can be split into two parts, one that does not change plus the change, that is

$$W(x \mid y) = W_0(x)\delta(x - y) + W_1(x \mid y) , \qquad \text{(D.28)}$$

and integrating the above equation in x and using (D.27), we get

$$W_0(y) = -\int W_1(x \mid y)dx ,$$

so the forward Chapman–Kolmogorov equation now reads as

$$\frac{\partial P(x,t \mid x_0, t_0)}{\partial t} = \int dy W_1(x \mid y)P(y,t \mid x_0, t_0) \qquad \text{(D.29)}$$

$$-\int dy W_1(y \mid x)P(x,t \mid x_0, t_0) ,$$

which has the form of a rate equation.

If the random variable X can take discrete values, the forward Chapman–Kolmogorov equation can be written as

$$\frac{\partial P(x_i, t)}{\partial t} = \sum_j \left[W_{ij}P(x_j, t) - W_{ji}P(x_i, t) \right] . \qquad \text{(D.30)}$$

This equation is known as the **Master Equation**.

Many stochastic processes are of a special type called 'birth and death process' or one-step process [4]. They correspond to

$$W_{ij} = r_j \delta_{i,j-1} + g_j \delta_{i,j+1}, (i \neq j) \qquad \text{(D.31)}$$

which permits jumps to adjacent sites.

Also, for the diagonal part

$$W_n = -(r_n + g_n) , \qquad \text{(D.32)}$$

so the master equation reads

$$\dot{P}_n = r_{n+1}P_{n+1} + g_{n-1}P_{n-1} - (r_n + g_n)P_n , \qquad \text{(D.33)}$$

where r_n represents the probability per unit time to jump from $n \to n-1$, and g_n the probability per unit time to go from $n \to n+1$.

Typically, one-step processes occur in atomic transition via one photon (emission and absorption), nuclear excitation and de-excitation, fission, etc.

An interesting example is the Poisson process, defined as

$$r_n = 0 , \qquad \text{(D.34)}$$

$$g_n = q .$$

$$P_n(0) = \delta_{n,0},$$

and the Master Equation is

$$\dot{P}_n = q(P_{n-1} - P_n) . \tag{D.35}$$

This is a **one-sided random walk**.
To solve it, we use the characteristic function:

$$G(s,t) = \langle \exp ins \rangle = \sum_n P_n(t) \exp ins , \tag{D.36}$$

with boundary condition $G(s,0) = 1$.
Multiplying the Master Equation by $\exp ins$ and summing over n, we get

$$\sum_n \exp(ins)\, \dot{P}_n = q \sum_n [P_{n-1}\exp(ins) - P_n\exp(ins)]$$

or

$$\frac{\partial G(s,t)}{\partial t} = q(\exp(is) - 1)G(s,t) . \tag{D.37}$$

It is simple to verify that the solution of (D.37) is

$$G(s,t) = \exp\{tq\,[\exp(is) - 1]\} \tag{D.38}$$
$$= \exp(-tq) \sum_n \frac{(\exp is)^n (tq)^n}{n!} ,$$

thus comparing with (D.36), we finally get

$$P_n(t) = \exp(-tq)\frac{(tq)^n}{n!} , \tag{D.39}$$

which is a Poisson distribution with $\langle n \rangle = tq$.

D.4 The Fokker–Planck Equation

Sometimes, instead of discrete jumps, one chooses to describe the random process as a continuous one.
If we take, for example, in the Chapman–Kolmogorov equation [3]:

$$\Phi(w \mid x) \equiv W(x + w \mid x) , \tag{D.40}$$

then

$$\frac{\partial P(x,t \mid x_0,t_0)}{\partial t} = \int dw\,\Phi(w \mid x - w)P(x - w, t \mid x_0,t_0) \tag{D.41}$$

$$= \int \exp\left(-w\frac{\partial}{\partial x}\right) \left[\Phi(w \mid x)P(x,t \mid x_0,t_0)\right] dw$$

$$= \int \left[1 - w\frac{\partial}{\partial x} + \frac{1}{2}w^2\frac{\partial^2}{\partial x^2} + ..\right] \left[\Phi(w \mid x)P(x,t \mid x_0,t_0)\right] dw ,$$

and because $\int dw\Phi(w\mid x) = 0$, we get

$$\frac{\partial P(x,t\mid x_0,t_0)}{\partial t} = \sum_{n=1}^{\infty} \frac{(-1)^n}{n!} \frac{\partial^n}{\partial x^n} \left[Q_n(x)P(x,t\mid x_0,t_0) \right] , \tag{D.42}$$

with

$$Q_n(x) = \int w^n \Phi(w\mid x)dw . \tag{D.43}$$

Many times, the above equation is truncated, keeping only the first two terms, getting the Fokker–Planck Equation.

In one dimension, with $Q_1 = A, Q_2 = B$, we get

$$\frac{\partial P(x,t\mid x_0,t_0)}{\partial t} = -\frac{\partial}{\partial x} \left[A(x,t)P(x,t\mid x_0,t_0) \right] \tag{D.44}$$

$$+\frac{1}{2}\frac{\partial^2}{\partial x^2} \left[B(x,t)P(x,t\mid x_0,t_0) \right] .$$

A simple generalization to more variables leads to the Fokker Planck equation

$$\frac{\partial P(\mathbf{x},t\mid \mathbf{x}_0,t_0)}{\partial t} = -\sum_i \frac{\partial}{\partial x_i} \left[\mathbf{A}_i(\mathbf{x},t)P(\mathbf{x},t\mid \mathbf{x}_0,t_0) \right] \tag{D.45}$$

$$+\frac{1}{2}\sum_{i,j} \frac{\partial^2}{\partial x_i \partial x_j} \left[\mathbf{B}_{ij}(\mathbf{x},t)P(\mathbf{x},t\mid \mathbf{x}_0,t_0) \right] ,$$

where $\mathbf{A}$ is the drift vector and $\mathbf{B}$ the diffusion matrix. This equation can also be written as

$$\frac{\partial P(\mathbf{x},t\mid \mathbf{x}_0,t_0)}{\partial t} + \sum_i \frac{\partial}{\partial x_i} \mathbf{J}_i(\mathbf{x},t) = 0 , \tag{D.46}$$

$$\mathbf{J}_i(\mathbf{x},t) = \left[\mathbf{A}_i(\mathbf{x},t)P(\mathbf{x},t\mid \mathbf{x}_0,t_0) \right] \tag{D.47}$$

$$-\frac{1}{2}\sum_j \frac{\partial}{\partial x_j} \left[\mathbf{B}_{ij}(\mathbf{x},t)P(\mathbf{x},t\mid \mathbf{x}_0,t_0) \right] .$$

$\mathbf{J}_i(\mathbf{x},t)$ is interpreted as a probability current.

Let us take a one-dimensional example.

D.4.1 The Wiener Process

We take the articular case $A = 0$, $B = 1$, so the Fokker–Planck now reads [2]

$$\frac{\partial P(w,t\mid w_0,t_0)}{\partial t} = \frac{1}{2}\frac{\partial^2}{\partial w^2} \left[P(w,t\mid w_0,t_0) \right] . \tag{D.48}$$

Once more, we use the characteristic function

$$\phi(s,t) = \int dw \exp(isw) P(w,t \mid w_0, t_0) . \tag{D.49}$$

The differential equation for ϕ is

$$\frac{\partial \phi}{\partial t} = -s^2 \phi . \tag{D.50}$$

We also notice that as $P(w,t \mid w_0, t_0) \mid_{t=t_0} = \delta(w - w_0)$, so $\phi(s, t_0) = \exp isw_0$, and the solution is

$$\phi(s,t) = \exp\left[isw_0 - \frac{1}{2}s^2(t - t_0) \right] , \tag{D.51}$$

which is a Gaussian, whose inverse transform is also a Gaussian

$$P(w,t \mid w_0, t_0) = \frac{1}{\sqrt{2\pi(t - t_0)}} \exp - \frac{(w - w_0)^2}{2(t - t_0)} . \tag{D.52}$$

The first two moments are

$$\langle W \rangle = w_0 , \tag{D.53}$$
$$\langle (\Delta W)^2 \rangle = t - t_0 .$$

This distribution spreads in time and corresponds precisely to Einstein's model for Brownian motion.

An important characteristic of Wiener's process is the independence of the increments, which is interesting for stochastic integration purposes.

We saw that, in general, for Markov Processes, one has

$$P(w_n, t_n; w_{n-1}, t_{n-1}; \ldots w_0, t_0) = \sqcap_{i=0}^{n-1} P(w_{i+1}, t_{i+1} \mid w_i, t_i) P(w_0, t_0)$$

$$= \sqcap_{i=0}^{n-1} \left\{ [2\pi(t_{i+1} - t_i)]^{-\frac{1}{2}} \exp\left[-\frac{(w_{i+1} - w_i)^2}{2(t_{i+1} - t_i)} \right] \right\} P(w_0, t_0) . \tag{D.54}$$

Now we define the Wiener increments as

$$\Delta W_i = W(t_i) - W(t_{i-1}) , \tag{D.55}$$
$$\Delta t_i - t_i - t_{i-1} ,$$

so the joint probability density for the increments is

$$P(\Delta w_n; \Delta w_{n-1}; \ldots \Delta w_1; w_0)$$

$$= \sqcap_{i=1}^{n} \left\{ [2\pi \Delta t_i]^{-\frac{1}{2}} \exp\left[-\frac{(\Delta w_i)^2}{2(\Delta t_i)} \right] \right\} P(w_0, t_0) , \tag{D.56}$$

thus they are statistically independent.

If we define the mean and autocorrelation functions as

$$\langle \mathbf{W}(t) \mid \mathbf{W}_0, t_0 \rangle = \int d\mathbf{w} P(\mathbf{w}, t \mid \mathbf{w}_0, t_0) \mathbf{w} , \qquad (D.57)$$

$$\langle \mathbf{W}(t) \mathbf{W}(t_0)^T \mid \mathbf{W}_0, t_0 \rangle = \int d\mathbf{w} d\mathbf{w}_0 P(\mathbf{w}, t; \mathbf{w}_0, t_0) \mathbf{w} \mathbf{w}_0^T$$

$$= \int d\mathbf{w}_0 \langle \mathbf{W}(t) \mid \mathbf{W}_0, t_0 \rangle \mathbf{w}_0^T P(\mathbf{w}_0, t_0) .$$

For the Wiener process

$$\langle W(t) W(\mathbf{s}) \mid W_0, t_0 \rangle = \langle [W(t) - W(s)] W(\mathbf{s}) \mid W_0, t_0 \rangle + \langle W(s)^2 \rangle , \quad (D.58)$$

and due to the independence of the increments, the first term is zero and

$$\langle W(t) W(\mathbf{s}) \mid W_0, t_0 \rangle = w_0^2 + \min(t - t_0, s - t_0) . \qquad (D.59)$$

D.4.2 General Properties of the Fokker–Planck Equation

The general Fokker–Plank equation reads

$$\frac{\partial P(\mathbf{x}, t \mid \mathbf{x}_0, t_0)}{\partial t} = -\sum_i \frac{\partial}{\partial x_i} [\mathbf{A}_i(\mathbf{x}, t) P(\mathbf{x}, t \mid \mathbf{x}_0, t_0)]$$

$$+ \frac{1}{2} \sum_{i,j} \frac{\partial^2}{\partial x_i \partial x_j} [\mathbf{B}_{ij}(\mathbf{x}, t) P(\mathbf{x}, t \mid \mathbf{x}_0, t_0)] . \quad (D.60)$$

As we mentioned before, the first term in the r.h.s is the drift term, which will rule the deterministic motion, and the second one is the diffusion term, which will cause the probability to broaden. This different role of the two terms can be easily seen if we calculate $\langle x_i \rangle$ and $\langle x_i x_j \rangle$. One can easily show that

$$\frac{d\langle x_i \rangle}{dt} = \langle \mathbf{A}_i \rangle , \qquad (D.61)$$

$$\frac{d\langle x_i x_j \rangle}{dt} = \langle x_i \mathbf{A}_j \rangle + \langle x_j \mathbf{A}_i \rangle + \frac{1}{2} \langle \mathbf{B}_{ij} + \mathbf{B}_{ji} \rangle .$$

D.4.3 Steady-State Solution

Very often in optics and other areas of physics, one is not really interested in the time-dependent solution of the Fokker–Planck equation , but rather in the steady state. Thus, we set the time derivative to zero and get

$$\sum_i \frac{\partial}{\partial x_i} \left[-\mathbf{A}_i(\mathbf{x},t)P(\mathbf{x},t \mid \mathbf{x}_0,t_0) + \frac{1}{2} \sum_j \frac{\partial}{\partial x_j} \left[\mathbf{B}_{ij}(\mathbf{x},t)P(\mathbf{x},t \mid \mathbf{x}_0,t_0) \right] \right] = 0 ,$$

$$(D.62)$$

and if the constant current is set to zero (detailed balance) , one gets

$$\mathbf{A}_i(\mathbf{x},t)P(\mathbf{x},t \mid \mathbf{x}_0,t_0) = \frac{1}{2} \sum_j \frac{\partial}{\partial x_j} \left[\mathbf{B}_{ij}(\mathbf{x},t)P(\mathbf{x},t \mid \mathbf{x}_0,t_0) \right] \qquad (D.63)$$

or

$$2\mathbf{A}_i - \sum_j \frac{\partial \mathbf{B}_{ij}}{dx_j} = \sum_j \mathbf{B}_{ij} \frac{1}{P(\mathbf{x},t \mid \mathbf{x}_0,t_0)} \frac{\partial P(\mathbf{x},t \mid \mathbf{x}_0,t_0)}{\partial x_j}$$

$$= \sum_j \mathbf{B}_{ij} \frac{\partial In\left[P(\mathbf{x},t \mid \mathbf{x}_0,t_0)\right]}{\partial x_j} ,$$

and defining a Potential function $V(\mathbf{x})$ by $P(\mathbf{x},t \mid \mathbf{x}_0,t) = N \exp(-V(\mathbf{x}))$, we get for V

$$-\frac{\partial V(\mathbf{x})}{\partial x_i} = 2 \sum_j \mathbf{B}_{ij}^{-1} \mathbf{A}_j - \sum_{j,k} \mathbf{B}_{ij}^{-1} \frac{\partial \mathbf{B}_{jk}}{\partial x_k} . \qquad (D.64)$$

Integrating (D.64), we get for the probability distribution

$$P_{SS}(\mathbf{x}) = N \exp \left[\int \sum_{i,j} 2\mathbf{B}_{ij}^{-1} \mathbf{A}_j(\mathbf{x})dx_i - \int \sum_{i,j,k} \mathbf{B}_{ij}^{-1} \frac{\partial \mathbf{B}_{jk}}{\partial x_k} dx_i \right] . \quad (D.65)$$

In particular, for $\mathbf{B}_{ij} = D\delta_{ij}$, we get

$$P_{SS}(\mathbf{x}) = N \exp \int \frac{2}{D} \mathbf{A}(\mathbf{x})dx \qquad (D.66)$$

D.5 Stochastic Differential Equations

D.5.1 Introduction

One way of treating the motion of a Brownian particle, or any other problem with a random force, is via a Langevin or Stochastic differential equation

$$\dot{V} = -\gamma V + L(t) , \qquad (D.67)$$

where, in the case of a Brownian particle, the r.h.s. is the force of the fluid over the particle and is made up of two components:

a) The damping force $-\gamma V$

b) A rapidly varying force $L(t)$, independent of the particle's velocity that accounts for the collisions of the water molecules with the Brownian particle, whose average is zero. Thus

$$\langle L(t) \rangle = 0 \tag{D.68}$$
$$\langle L(t)L(t') \rangle = D\delta(t - t') .$$

$\langle L(t)L(t') \rangle$ is referred to as the two time correlation function.

If one defines the spectrum as the Fourier transform of the two time correlation function

$$S(\omega) = \int_{-\infty}^{+\infty} d\tau \, \exp(i\omega\tau)\langle L(t + \tau)L(t) \rangle , \tag{D.69}$$

we immediately notice that, because the Fourier transform of a delta function is a constant, $L(t)$ has a flat spectrum or it correspond to **white noise.**

Let us assume that the initial velocity of the Brownian particle is deterministic and given by $V(0) = V_0$, then for $t > 0$, for each sample path

$$V(t) = V_0 \exp(-\gamma t) + \exp(-\gamma t) \int_0^t \exp(\gamma t')L(t')dt' . \tag{D.70}$$

Using the properties of L, we can calculate $\langle V \rangle$ and $\langle V^2 \rangle$

$$\langle V(t) \mid V_0, t_0 \rangle = V_0 \exp(-\gamma t) , \tag{D.71}$$
$$\langle V^2(t) \mid V_0, t_0 \rangle = V_0^2 \exp(-2\gamma t) +$$
$$\exp(-2\gamma t) \int_0^t dt' \int_0^t dt'' \exp\gamma(t' + t'')\langle L(t')L(t'') \rangle$$

$$\langle V^2(t) \mid V_0, t_0 \rangle = V_0^2 \exp(-2\gamma t) + \frac{D}{2\gamma}[1 - \exp(-2\gamma t)] . \tag{D.72}$$

When $t \to \infty$

$$\langle V^2(t) \mid V_0, t_0 \rangle = \frac{D}{2\gamma} , \tag{D.73}$$

On the other hand, for short times

$$\langle (\Delta V)^2(t_0 + \Delta t) \mid V_0, t_0 \rangle = D\Delta t + \mathcal{O}(\Delta t)^2 .. \tag{D.74}$$

We also notice, that in this case, the drift and diffusion coefficients are

$$A = \frac{\langle \Delta V \rangle}{\Delta t} = \frac{(\langle V - V_0 \rangle) \mid_{t=t_0 + \Delta t}}{\Delta t} = -\gamma V , \tag{D.75}$$

$$B = \frac{\langle (\Delta V)^2 \rangle}{\Delta t} = D , \tag{D.76}$$

so that the corresponding Fokker–Planck equation is

$$\frac{\partial P(V,t)}{\partial t} = \gamma \frac{\partial}{\partial V}(VP) + \frac{D}{2}\frac{\partial^2 P}{\partial V^2} . \tag{D.77}$$

The above equation describes the so-called **Ornstein–Uhlenbeck process**, corresponding to a linear drift and a constant diffusion term.

We now calculate the power spectrum of V. So we first need the two time correlation function

$$\langle V(t)V(t') \rangle = V_0^2 \exp\left[-\gamma(t+t')\right]$$

$$+ \exp\left[-\gamma(t+t')\right] \int_0^t dt'' \int_0^{t'} dt''' \exp\left[\gamma(t''+t''')\right] \langle L(t'')L(t''') \rangle$$

$$= V_0^2 \exp\left[-\gamma(t+t')\right] + \exp\left[-\gamma(t+t')\right] D \int_0^{t'} dt''' \exp 2\gamma t'''$$

$$\langle V(t)V(t') \rangle = V_0^2 \exp\left[-\gamma(t+t')\right] + \exp\left[-\gamma(t+t')\right] \frac{\Gamma}{2\gamma}\left[\exp(2\gamma t') - 1\right] . \tag{D.78}$$

In steady state, for $t,t' \to \infty$ but with $t - t' = \tau$, we get

$$\langle V(t+\tau)V(t) \rangle = \frac{D}{2\gamma} \exp(-\gamma \mid \tau \mid) . \tag{D.79}$$

Finally, taking the Fourier transform, we get the power spectrum of V

$$\phi_V(\omega) = \frac{1}{2\pi} \int \exp(i\omega\tau) \langle V(t+\tau)V(t) \rangle d\tau \tag{D.80}$$

$$= \frac{1}{2\pi} \frac{D}{\omega^2 + \gamma^2} .$$

D.5.2 Ito Versus Stratonovich

A more general type of Langevin equation can be written as

$$\frac{dx}{dt} = a(x,t) + b(x,t)L(t) , \tag{D.81}$$

where the previous D factor can be absorbed in b, so that

$$\langle L(t)L(t') \rangle = \delta(t - t') , \tag{D.82}$$

$$\langle L \rangle = 0 .$$

Now we define

$$W(t) = \int_0^t L(t')dt' , \qquad (D.83)$$

assumed to be continuous, so that

$$\langle W(t + \Delta t) - W_0(t) \mid W_0, t \rangle = \langle \int_t^{t+\Delta t} dsL(s) \rangle = 0 , \qquad (D.84)$$

$$\langle [W(t + \Delta t) - W_0(t)]^2 \mid W_0, t \rangle = \qquad (D.85)$$
$$\langle \int_t^{t+\Delta t} ds_1 \int_t^{t+\Delta t} ds_2 L(s_1)L(s_2) \rangle = \int_t^{t+\Delta t} ds_1 \int_t^{t+\Delta t} ds_2 \delta(s_1 - s_2) = \Delta t ,$$

therefore, one could write a Fokker–Planck equation for W with

$$A = 0, B = 1 ,$$

which correspond to a Wiener process , and $Ldt = dW$ becomes a Wiener increment.

The stochastic differential equation (D.81) is not fully defined unless one specifies how to integrate it. Normally, this would not be a problem, and the rules of ordinary calculus apply. However, here we must be careful because we are dealing with a rapidly varying function of time $L(t)$.

Thus, we define the integral the mean square limit of a Riemann–Stieltjes sum

$$\int_{t_0}^t f(t')dW(t') = ms \lim_{n \to \infty} \sum_{i=1}^n f(\tau_i)\left[W(t_i) - W(t_{i-1})\right] , \qquad (D.86)$$

where $t_{i-1} \le \tau_i \le t_i$, and we have divided the time interval from $t_0 \to t$ in n intermediate times $t_1 t_2 ... t_n$.

One can verify that *it does matter which* $f(\tau_i)$ *we choose.*
Two popular choices are:

a) Ito with $\tau_i = t_{i-1}$.
b) Stratonovich: $f(\tau_i) = \frac{f(t_i)+f(t_{i-1})}{2}$.

From the above assumptions, one learns how to calculate things with Ito and Stratonovich.

For Stratonovich, we have for example

$$S \int_{t_0}^t W(t')dW(t')$$

$$= ms \lim_{n \to \infty} \sum_{i=1}^n \left[\frac{W(t_i) + W(t_{i-1})}{2}\right] [W(t_i) - W(t_{i-1})]$$

$$= \frac{1}{2} ms \lim_{n \to \infty} \sum_{i=1}^n [W^2(t_i) - W^2(t_{i-1})] = \frac{1}{2}[W^2(t) - W^2(t_0)] ,$$

which obeys the rules of ordinary calculus.

On the other hand, for Ito

$$I \int_{t_0}^{t} W(t') dW(t')$$

$$= ms \lim_{n \to \infty} \sum_{i=1}^{n} [W(t_{i-1})] [W(t_i) - W(t_{i-1})]$$

$$= ms \lim_{n \to \infty} \sum_{i=1}^{n} [W(t_{i-1}) \Delta W(t_i)]$$

$$= ms \lim_{n \to \infty} \frac{1}{2} \sum_{i=1}^{n} \left\{ [W(t_{i-1}) + \Delta W(t_i)]^2 - W(t_{i-1})^2 - \Delta W(t_i)^2 \right\}$$

$$= \frac{1}{2} [W(t)^2 - W(t_0)^2] - ms \lim_{n \to \infty} \frac{1}{2} \sum_{i=1}^{n} \Delta W(t_i)^2 ,$$

and because

$$ms \lim_{n \to \infty} \frac{1}{2} \sum_{i=1}^{n} \Delta W(t_i)^2 = t - t_0 ,$$

we finally get

$$I \int_{t_0}^{t} W(t') dW(t') = \frac{1}{2} [W(t)^2 - W(t_0)^2 - (t - t_0)] . \tag{D.87}$$

Finally, for the Ito integration, one can prove that

$$dW(t)^2 = dt, \tag{D.88}$$
$$dW(t)^{2+N} = 0, N = 1, 2, 3..$$

The details and proof of the above properties are found in Gardiner's book [2].

From these properties, we can see that $dW \sim \sqrt{dt}$, and we have to keep terms up to $(dW)^2$, which differs from the ordinary calculus.

D.5.3 Ito's Formula

Consider a function $f[x(t)]$. We will derive the basic formula for Ito's calculus:

$$df[x(t)] = f[x(t) + dx] - f[x(t)]$$

$$= f'[x(t)] dx + \frac{1}{2} f''[x(t)] dx^2 + ...$$

$$= f'[x(t)][a(x,t)dt + b(x,t)dW]$$
$$+\frac{1}{2}f''[x(t)][b(x,t)dW^2 + ...]$$

and using (D.88), we get

$$df[x(t)] = \left\{a(x,t)f'[x(t)] + \frac{1}{2}b(x,t)f''[x(t)]\right\}dt$$

$$+b(x,t)f'[x(t)]dW. \tag{D.89}$$

The above formula can be easily generalized for many dimensions. Now, we take the average of Ito's formula

$$\frac{d\langle f(x)\rangle}{dt} = \int dx\,\partial_t P(x,t)f(x)$$

$$= \int dx\left[a\partial_x f + \frac{b}{2}\partial_x^2 f\right]P(x,t),$$

and integrating by parts and discarding the surface terms, we get

$$\int dx f(x)\partial_t P(x,t) = \int dx f(x)\left[-\partial_x aP + \frac{1}{2}\partial_x^2 bP\right],$$

thus getting the Ito–Fokker–Planck equation

$$\partial_t P(x,t \mid x_0,t_0) = -\partial_x[a(x,t)P(x,t \mid x_0,t_0)] \tag{D.90}$$

$$+\frac{1}{2}\partial_x^2[b(x,t)P(x,t \mid x_0,t_0)] \tag{D.91}$$

Similarly, for many variables, if one has an Ito stochastic differential equation

$$(I)dx = \mathbf{a}(\mathbf{x},t)dt + \mathbf{b}(\mathbf{x},t)d\mathbf{W}, \tag{D.92}$$

where $d\mathbf{W}$ is an n-component Wiener process, then the corresponding Ito's Fokker–Planck equation is:

$$\partial_t P(\mathbf{x},t \mid \mathbf{x}_0,t_0) = -\sum_i \partial_i[a_i(\mathbf{x},t)P(\mathbf{x},t \mid \mathbf{x}_0,t_0)] \tag{D.93}$$

$$+\frac{1}{2}\sum_{i,j}\partial_i\partial_j\left[\mathbf{b}\mathbf{b}^T(\mathbf{x},t)\right]_{ij}P(\mathbf{x},t \mid \mathbf{x}_0,t_0). \tag{D.94}$$

Thus, from our previous notation

$$\mathbf{B} = \mathbf{b}\mathbf{b}^T \tag{D.95}$$

Similarly, for Stratonovich

$$(S)\mathrm{dx} = \mathbf{a}^S(\mathbf{x}, t)\mathrm{d}t + \mathbf{b}^S(\mathbf{x}, t)\mathrm{d}\mathbf{W}, \qquad (D.96)$$

we get a Stratonovich–Fokker–Planck equation

$$\partial_t P(\mathbf{x}, t \mid \mathbf{x}_0, t_0) = -\sum_i \partial_i \left[a_i^S(\mathbf{x}, t) P(\mathbf{x}, t \mid \mathbf{x}_0, t_0) \right]$$

$$+ \frac{1}{2} \sum_{i,j,k} \partial_i \left[\mathbf{b}_{ik}^S \partial_j \mathbf{b}_{jk}^{ST}(\mathbf{x}, t) \right] P(\mathbf{x}, t \mid \mathbf{x}_0, t_0) . \qquad (D.97)$$

By simple comparison between the two Fokker–Planck equations, we get

$$a_i^S = a_i - \frac{1}{2} \sum_{j,k} \mathbf{b}_{kj} \partial_k \mathbf{b}_{ij}^T, \qquad (D.98)$$

$$\mathbf{b}_{ik}^S = \mathbf{b}_{ik} .$$

This last relation tells us that if we have a given Fokker–Planck equation, it corresponds to a Langevin equation to be integrated a la Ito, with $\mathbf{a}$ and $\mathbf{b}$ drift and diffusion coefficients, and to a Langevin equation to be integrated a la Stratonovich with $\mathbf{a}^S$ and $\mathbf{b}^S$ drift and diffusion coefficients, and the relation between the Ito and the Stratonovich coefficients is given by (D.98).

D.6 Approximate Methods

Non-linear Langevin equations are difficult to solve exactly.

We present here the Ω-expansion of **Van Kampen** [5], where Ω is the size or number of particles of our system.

We consider a variable X that is proportional to the particle number, and define

$$x = \frac{X}{\Omega} . \qquad (D.99)$$

The key point in Van Kampen's expansion is that we can separate [1]:

$$x(t) = x_0(t) + \sqrt{\epsilon} y(t) , \qquad (D.100)$$

where $x_0(t)$ is the deterministic part, $y(t)$ represents the fluctuations and $\epsilon = \frac{1}{\Omega}$.

This decomposition is based on the Central Limit Theorem that says that for large Ω, the fluctuations of X around its mean value go as Ω.

Of course, this expansion fails, as we shall see, near an instability point.

We also assume that in the stochastic equation, the small parameter $\sqrt{\epsilon}$ is the noise strength, so we write

$$\mathrm{d}x = a(x)\mathrm{d}t + \sqrt{\epsilon}\mathrm{d}W(t) , \qquad (D.101)$$

and

$$x(t) = x_0(t) + \sqrt{\epsilon}x_1(t) + \epsilon x_2(t) + \dots \tag{D.102}$$

Differentiating x and expanding $a(x)$ around x_0, we get

$$dx_0(t) + \sqrt{\epsilon}dx_1(t) + \epsilon dx_2(t) + \dots$$

$$= a(x_0)dt + a'(x_0)(x - x_0)dt + \frac{1}{2}a''(x_0)(x - x_0)^2dt + \dots + \sqrt{\epsilon}dW(t)$$

$$= a(x_0)dt + a'(x_0)\left[\sqrt{\epsilon}x_1(t) + \epsilon x_2(t) + \dots\right]dt +$$

$$\frac{1}{2}a''(x_0)\left[\epsilon x_1^2 + \dots\right]dt + +\sqrt{\epsilon}dW(t) \,,$$

and by comparing different orders of ϵ we get

$$dx_0(t) = a(x_0)dt, \tag{D.103}$$

$$dx_1(t) = a'(x_0)x_1(t)dt + dW(t) \,, \tag{D.104}$$

$$dx_2(t) = a'(x_0)x_2(t)dt + \frac{1}{2}a''(x_0)x_1^2dt \,, \tag{D.105}$$

and so on.

The initial conditions for $x_1, x_2, x_3 \dots$ are

$$x_1(0) = 0, x_2(0) = 0, \text{ etc} : \tag{D.106}$$

Now, we take a non-trivial example.

A particle, in one dimension, under the action of a double-well potential

$$a(x) = -\frac{dV}{dx} \,, \tag{D.107}$$

with

$$V = -\frac{\gamma}{2}x^2 + \frac{g}{4}x^4 \,. \tag{D.108}$$

The stochastic differential equation, in this case is

$$dx(t) = (\gamma x - gx^3)dt + \sqrt{\epsilon}dW(t) \,. \tag{D.109}$$

The shape of the potential is described in the Fig. D.1

We will consider the case $\gamma > 0$. Near the equilibrium positions, the drift is practically zero, and the noise term in the stochastic equation is quite important.

On the other hand, very far from the equilibrium positions, the motion is dominated by a large drift and is practically a deterministic one.

For $\gamma > 0$, $x = 0$ is an unstable equilibrium position and $x = \pm\sqrt{\frac{\gamma}{g}}$ are stable ones.

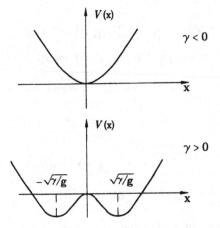

Fig. D.1. Double-well potential for the cases $\gamma < 0$ (upper curve) and $\gamma > 0$ (lower curve)

Applying our method in this example, we get

$$dx_0 = (\gamma x_0 - g x_0^3)dt, x_0(0) = h \, , \tag{D.110}$$

$$dx_1 = (\gamma - 3g x_0^2)x_1 dt + dW, x_1(0) = 0 \, , \tag{D.111}$$

$$dx_2 = \left[(\gamma - 3g x_0^2)x_2 - 3g x_0 x_1^2\right] dt, x_2(0) = 0 \, , \tag{D.112}$$

The solution for the deterministic motion is

$$x_0(t) = \frac{h \exp(\gamma t)}{\sqrt{1 + \frac{g}{\gamma}h^2 \left[\exp(2\gamma t) - 1\right]}} \, , \tag{D.113}$$

so that if we choose the unstable equilibrium point, that is the initial condition $h = 0$, the we get $x_0(t) = 0$, and for the stable equilibrium points, $h = \pm\sqrt{\frac{\gamma}{g}}$, we get $x_0(t) = \pm\sqrt{\frac{\gamma}{g}}$, as it should.

For the first case: $h = x_0(t) = 0$, we get

$$dx_1 = \gamma x_1 dt + dW(t) \, , \tag{D.114}$$

$$dx_2 = \gamma x_2 dt \, , \tag{D.115}$$

and the solution to the above equations are

$$x_1(t) = \int_o^\infty \exp\left[\gamma(t - t')\right] dW(t') \, , \tag{D.116}$$

$$x_2(t) = 0.$$

In the stable case $x_0 = h = \pm\sqrt{\frac{\gamma}{g}}$, we get

$$dx_1(t) = -2\gamma x_1 dt + dW(t), \tag{D.117}$$

$$dx_2(t) = \left(-2\gamma x_2 \mp 3g\sqrt{\frac{\gamma}{g}}x_1^2\right)dt ,$$

and the solutions are

$$x_1(t) = \int_0^t dW(t')\exp\left[-2\gamma(t-t')\right] \tag{D.118}$$

$$x_2(t) = \mp 3g\sqrt{\frac{\gamma}{g}}\int_0^t dt'x_1^2(t')\exp\left[-2\gamma(t-t')\right] .$$

Now, we notice that in all cases $\langle x_1 \rangle = 0$, so

$$\langle x^2(t)\rangle = x_0^2 + \epsilon\left(\langle x_1^2\rangle + 2x_0\langle x_2\rangle\right) + ... \tag{D.119}$$

and in the two cases, we can write

$$\langle x^2(t)\rangle_{\text{uns}} = \frac{\epsilon}{2\gamma}\left[\exp(2\gamma t) - 1\right] , \tag{D.120}$$

$$\langle x^2(t)\rangle_{\text{stable}} = \frac{\gamma}{g} + \frac{\epsilon}{4\gamma}\left[1 - \exp(-4\gamma t)\right] - \frac{3\epsilon}{4\gamma}\left[1 - \exp(-2\gamma t)\right]^2 + ...$$

In the limit $t \to \infty$, $\langle x^2(\infty)\rangle_{\text{uns}}$ diverges, while $\langle x^2(t)\rangle_{\text{stable}} \to \frac{\gamma}{g} - \frac{\epsilon}{2\gamma}$.

As we can see, the perturbative expansion gives the correct answer when starting from stable equilibrium points but it diverges when starting from an unstable equilibrium. In this last case, the perturbative expansion is no longer valid.

References

1. Tombesi P.: In: Gomez, B., Moore, S.M., Rodriguez-Vargas, A.M., Rueda, A. (eds) Stochastic Processes Applied to Physics and Other Related Fields. World Scientific, Singapore (1983)
2. Gardiner, C.W.: Handbook of Stochastic Methods. Springer Verlag, Berlin (1983)
3. Stenholm, S.: In: Meystre, P., Scully, M.O. (eds) Quantum Optics, Experimental Gravitation and Measurement Theory. Plenum Press, NY (1983)
4. Van Kampen, N.G.: Stochastic Processes in Physics and Chemistry. North-Holland, Amsterdam (1981).
5. Van Kampen, N.G.: Can. J. Phys., **39**, 551 (1961)

E. Derivation of the Homodyne Stochastic Schrödinger Differential Equation

Here we present the detailed derivation of the Homodyne Schrödinger differential equation. We start from the expansion given by (16.36), which in the two-jump situation, and neglecting the commutators between the jump operators and the no-jump evolution, can be expressed as

$$\rho(\Delta t) = \sum_{m_1,m_2=0}^{\infty} \frac{(\Delta t)^{m_1+m_2}}{m_1!m_2!} S(\Delta t) J_2^{m_2} J_1^{m_1} \rho(0) . \qquad \text{(E.1)}$$

The probability of m_1 and m_2 quantum jumps of the respective types is given by

$$P_{m_1,m_2}(\Delta t) = \frac{(\Delta t)^{m_1+m_2}}{m_1!m_2!} Tr\left[S(\Delta t) J_2^{m_2} J_1^{m_1} \rho(0)\right] . \qquad \text{(E.2)}$$

The master equation of the field, corresponding to a lossy cavity at temperature T, may be written as

$$\frac{d\rho}{dt} = (J_1 + J_2)\rho - \frac{\gamma}{2}\rho[a^\dagger a(1 + 2\langle n\rangle_{\text{th}}) + 2\varepsilon(1 + \langle n\rangle_{\text{th}})a^\dagger$$

$$+2\varepsilon\langle n\rangle_{\text{th}}a + \langle n\rangle_{\text{th}} + \varepsilon^2(1 + 2\langle n\rangle_{\text{th}})]$$

$$-\frac{\gamma}{2}[a^\dagger a(1 + 2\langle n\rangle_{\text{th}}) + 2\varepsilon(1 + \langle n\rangle_{\text{th}})a + 2\varepsilon\langle n\rangle_{\text{th}}a^\dagger$$

$$+\langle n\rangle_{\text{th}} + \varepsilon^2(1 + 2\langle n\rangle_{\text{th}})]\rho \qquad \text{(E.3)}$$

Therefore, according to the discussion given in Chap. 16, one possible way of writing $S(\Delta t)$ is

$$S(\Delta t)\rho = N(\Delta t)\rho N^\dagger(\Delta t) , \qquad \text{(E.4)}$$

with

$$N(\Delta t) = \exp\left\{-\frac{\gamma(\Delta t)}{2}[a^\dagger a(1 + 2\langle n\rangle_{\text{th}}) + 2\varepsilon(1 + \langle n\rangle_{\text{th}})a^\dagger\right.$$

$$\left.+2\varepsilon\langle n\rangle_{\text{th}}a + \langle n\rangle_{\text{th}} + \varepsilon^2(1 + 2\langle n\rangle_{\text{th}})]\right\} . \qquad \text{(E.5)}$$

Using (E.2) and (E.5), we can write

$$P_{m_1,m_2}(\Delta t) = \left[\frac{\exp(\mu_1)(\mu_1)^{m_1}}{m_1!}\right]\left[\frac{\exp(\mu_2)(\mu_2)^{m_2}}{m_2!}\right]$$

$$Tr[\exp(\beta)\left(1+\frac{a^\dagger}{\varepsilon}\right)^{m_2}\left(1+\frac{a}{\varepsilon}\right)^{m_1}\rho\left(1+\frac{a^\dagger}{\varepsilon}\right)^{m_1}$$

$$\left(1+\frac{a}{\varepsilon}\right)^{m_2}\exp(\beta^\dagger)] \tag{E.6}$$

'with

$$\mu_1 = \gamma(\Delta t)\varepsilon^2(1+\langle n\rangle_{\text{th}}) , \tag{E.7}$$

$$\mu_2 = \gamma(\Delta t)\varepsilon^2\langle n\rangle_{\text{th}} ,$$

$$\beta = -\frac{\gamma(\Delta t)}{2}\{a^\dagger a(1+2\langle n\rangle_{\text{th}})$$

$$+2[\varepsilon(1+\langle n\rangle_{\text{th}})a + \varepsilon\langle n\rangle_{\text{th}}a^\dagger] + \langle n\rangle_{\text{th}}\} .$$

From (E.6), we can calculate $\langle m_i\rangle$ and $\sigma_i^2 = \langle m_i^2\rangle - \langle m_i\rangle^2$ up to order $\left(\frac{1}{\varepsilon}\right)^{\frac{3}{2}}$. The result is

$$\langle m_i\rangle = \mu_i(1+\frac{2}{\varepsilon}\langle X\rangle) ,$$

$$\sigma_i^2 = \langle \mu_i\rangle . \tag{E.8}$$

Now, we turn to the final step of this calculation, which yields the time evolution of the state vector. After repeated jump and no-jump events, the unnormalized wavefunction for the field can be written as

$$|\widetilde{\psi}\rangle_f(\Delta t) = N(\Delta t - t_m)C_2N(t_m - t_{m-1})C_1N...|\psi\rangle_f(0)$$

or, except for an overall phase factor,

$$|\widetilde{\psi}\rangle_f(\Delta t) = N(\Delta t)C_2^{m_2}C_1^{m_1}\psi\rangle_f(0) , \tag{E.9}$$

where the symbol $\sim$ indicates that the vector is not normalized.

Using (E.5) and (16.51), one can write, up to a normalization constant

$$|\widetilde{\psi}\rangle_f(\Delta t) = \exp\left(-\frac{\gamma(\Delta t)}{2}\{a^\dagger a(1+\langle n\rangle_{\text{th}})\right.$$

$$\left.+2[\varepsilon(1+\langle n\rangle_{\text{th}})a^\dagger + 2\varepsilon\langle n\rangle_{\text{th}}a]\}\right)$$

$$\left(1+\frac{a^\dagger}{\varepsilon}\right)^{m_2}\left(1+\frac{a}{\varepsilon}\right)^{m_1}|\psi\rangle_f(0) \tag{E.10}$$

or expanding up to $\varepsilon^{-\frac{3}{2}}$

$$| \tilde{\psi} \rangle_f(\Delta t) = \left\{ 1 - \frac{\gamma(\Delta t)}{2} [a^\dagger a(1 + \langle n \rangle_{th}) + aa^\dagger \langle n \rangle_{th}] \right.$$

$$\left. -\gamma(\Delta t)\varepsilon[a(1 + \langle n \rangle_{th}) + a^\dagger \langle n \rangle_{th})] \right\}$$

$$\times \left[1 + \frac{1}{\varepsilon}(m_1 a + m_2 a^\dagger) \right] | \psi \rangle_f(0) \tag{E.11}$$

We are interested in the limit $\varepsilon \to \infty$. In deriving (E.11), we considered ε large, $\gamma(\Delta t) \sim \varepsilon^{-\frac{3}{2}}$ and $m_1, m_2, \mu_1, \mu_2 \sim \varepsilon^{\frac{1}{2}}$. Now we consider two random numbers with non-zero average m_1, m_2

$$m_1 = \langle m_1 \rangle + \frac{\sigma_1}{\sqrt{\Delta t}} \Delta W_1 \,,$$

$$m_2 = \langle m_2 \rangle + \frac{\sigma_2}{\sqrt{\Delta t}} \Delta W_2 \,, \tag{E.12}$$

which satisfy

$$\langle (\Delta W_1)^2 \rangle = \langle (\Delta W_2)^2 \rangle = \Delta t \,. \tag{E.13}$$

We notice that ΔW_i are two independent Wiener Processes.

Finally, (E.11) can be written as

$$\Delta^{m_1, m_2} | \tilde{\psi} \rangle_f(\Delta t) = | \tilde{\psi} \rangle_f(\Delta t) - | \tilde{\psi} \rangle_f(0)$$

$$= \left\{ \left[-\frac{\gamma}{2}(a^\dagger a(1 + \langle n \rangle_{th}) - \frac{\gamma}{2}aa^\dagger \langle n \rangle_{th} \right. \right.$$

$$+2\gamma\langle X \rangle a(1 + \langle n \rangle_{th})$$

$$\left. +a^\dagger \langle n \rangle_{th} \right] \Delta t + a^\dagger \sqrt{\gamma \langle n \rangle_{th}} \Delta W_2$$

$$\left. +a\sqrt{\gamma(1 + \langle n \rangle_{th})} \Delta W_1 \right\} | \tilde{\psi} \rangle_f(0) \,,$$

which is the desired result.

F. Fluctuations

We want to calculate $d\langle(\Delta a^\dagger a)^2\rangle$ and $Md\langle(\Delta a^\dagger a)^2\rangle$.

We do it first in a simple case T=0, $O = a^\dagger a; C = \sqrt{\gamma}a; H = \hbar\omega a^\dagger a$.

$$
\begin{aligned}
d\,\langle(\Delta a^\dagger a)^2\rangle = {} & \gamma\delta t\{-\langle a^\dagger a a^\dagger a a^\dagger a\rangle + 2\langle a^\dagger a a^\dagger a\rangle\langle a^\dagger a\rangle \\
& - 2\langle a^\dagger a\rangle\langle a^\dagger a\rangle\langle a^\dagger a\rangle + \langle a^\dagger a a^\dagger a\rangle\langle a^\dagger a\rangle\} \\
& - \langle a^\dagger a a^\dagger a\rangle\delta N + \langle a^\dagger a\rangle\langle a^\dagger a\rangle\delta N \\
& + \frac{\langle a^\dagger a^\dagger a a^\dagger a a\rangle\langle a^\dagger a\rangle - \langle a^\dagger a^\dagger a a\rangle\langle a^\dagger a^\dagger a a\rangle}{\langle a^\dagger a\rangle\langle a^\dagger a\rangle}\delta N\ ,
\end{aligned}
\tag{F.1}
$$

or

$$
\begin{aligned}
d\,\langle(\Delta a^\dagger a)^2\rangle = {} & -\gamma\delta t\langle(\Delta a^\dagger a)(\Delta a^\dagger a)(\Delta a^\dagger a)\rangle \\
& - \langle(\Delta a^\dagger a)^2\rangle\delta N \\
& + \frac{\langle a^\dagger a^\dagger a a^\dagger a a\rangle\langle a^\dagger a\rangle - \langle a^\dagger a^\dagger a a\rangle\langle a^\dagger a^\dagger a a\rangle}{\langle a^\dagger a\rangle\langle a^\dagger a\rangle}\delta N
\end{aligned}
\tag{F.2}
$$

Now, we apply the above results to the more interesting case $T > 0$, $O = a^\dagger a; C_1 = \sqrt{(\langle n\rangle_{\mathrm{th}} + 1)\gamma}a, C_2 = \sqrt{\gamma\langle n\rangle_{\mathrm{th}}}a^\dagger; H = \hbar\omega a^\dagger a$:

$$
\begin{aligned}
d\,\langle(\Delta a^\dagger a)^2\rangle = {} & -\gamma(\langle n\rangle_{\mathrm{th}} + 1)\langle(\Delta a^\dagger a))(\Delta a^\dagger a)(\Delta a^\dagger a)\rangle dt \\
& - \langle(\Delta a^\dagger a)^2\rangle\delta N_1 \\
& + \frac{(\langle a^\dagger a a^\dagger a a^\dagger a\rangle\langle a^\dagger a\rangle - \langle a^\dagger a a^\dagger a\rangle\langle a^\dagger a a^\dagger a\rangle)\delta N_1}{\langle a\dagger a\rangle\langle a^\dagger a\rangle} \\
& + \gamma\langle n\rangle_{\mathrm{th}}dt[-\langle a a^\dagger a a^\dagger a a^\dagger\rangle + 2\langle a a^\dagger a a^\dagger\rangle - \langle a a^\dagger\rangle \\
& + 2\langle a a^\dagger a a^\dagger\rangle\langle a^\dagger a\rangle - 2\langle a a^\dagger\rangle\langle a^\dagger a\rangle - \langle a a^\dagger\rangle\langle a^\dagger a\rangle\langle a^\dagger a\rangle \\
& + \langle a^\dagger a a^\dagger a\rangle\langle a a^\dagger\rangle - \langle a^\dagger a\rangle\langle a^\dagger a\rangle\langle a a^\dagger\rangle] \\
& - \langle(\Delta a^\dagger a)^2\rangle\delta N_2 \\
& + \frac{(\langle a a^\dagger a a^\dagger a a^\dagger\rangle\langle a a^\dagger\rangle - \langle a a^\dagger a a^\dagger\rangle\langle a a^\dagger a a^\dagger\rangle)\delta N_2}{\langle a a\dagger\rangle\langle a a^\dagger\rangle}\ .
\end{aligned}
\tag{F.3}
$$

In the above expression, neither the deterministic nor the stochastic term is definitely non-increasing. But in the mean, it does decrease

$$M\frac{\mathrm{d}\langle(\Delta a^\dagger a)^2\rangle}{\mathrm{d}t} = -\gamma(\langle n\rangle_{\mathrm{th}} + 1)\frac{\langle(\Delta a^\dagger a)a^\dagger a\rangle\langle a^\dagger a(\Delta a^\dagger a)\rangle}{\langle a\dagger a\rangle}$$

$$- \gamma\langle n\rangle_{\mathrm{th}}\frac{\langle(\Delta aa^\dagger)aa^\dagger\rangle\langle aa^\dagger(\Delta aa^\dagger)\rangle}{\langle aa\dagger\rangle} \le 0\,. \tag{F.4}$$

G. The No-Cloning Theorem [1]

We assume that we have a device able to duplicate an arbitrary quantum state. That is, if the system is initially in the state $| \psi \rangle$

$$| \psi \rangle \otimes | \alpha \rangle \rightarrow U(| \psi \rangle \otimes | \alpha \rangle) = | \phi \rangle \otimes | \alpha \rangle \otimes | \alpha \rangle ,$$

where $| \phi \rangle$ is the state of the system after performing its copying. Similarly, for a different input state, we would have

$$| \psi \rangle \otimes | \beta \rangle \rightarrow U(| \psi \rangle \otimes | \beta \rangle) = | \phi' \rangle \otimes | \beta \rangle \otimes | \beta \rangle .$$

Taking the inner product of these two states, we get

$$\langle \psi | \psi \rangle \langle \alpha | \beta \rangle = \langle \phi | \phi' \rangle \langle \alpha | \beta \rangle \langle \alpha | \beta \rangle .$$

In the above equation, $\langle \psi | \psi \rangle = 1$ and $0 < |< \alpha | \beta >| < 1$, so we conclude that $\langle \phi | \phi' \rangle \langle \alpha | \beta \rangle = 1$, which is impossible, because $| \langle \phi | \phi' \rangle | \leq 1$.

Thus, the system represented by $| \psi \rangle$ cannot exist.

Reference

1. Peres, A. Quantum Theory; Concepts and Methods. Kluwer, The Netherlands (1995)

H. The Universal Quantum Cloning Machine [1]

We will develop the Universal copying machine, as proposed by Buzek et al. The following transformation is proposed

$$| 0\rangle_a \, | Q\rangle_x \to | 0\rangle_a \, | 0\rangle_b \, | Q_0\rangle_x + (| 0\rangle_a \, | 1\rangle_b + | 1\rangle_a \, | 0\rangle_b) \, | Y_0\rangle_x , \qquad \text{(H.1)}$$

$$| 1\rangle_a \, | Q\rangle_x \to | 1\rangle_a \, | 1\rangle_b \, | Q_1\rangle_x + (| 0\rangle_a \, | 1\rangle_b + | 1\rangle_a \, | 0\rangle_b) \, | Y_1\rangle_x . \qquad \text{(H.2)}$$

As there are several free parameters, we can impose some conditions, namely

$$_x\langle Q_i \mid Q_i\rangle_x + 2 {}_x\langle Y_i \mid Y_i\rangle_x = 1, i = 0, 1 ,$$

$$_x\langle Y_0 \mid Y_1\rangle_x = {}_x\langle Y_1 \mid Y_0\rangle_x = 0 ,$$

$$_x\langle Q_i \mid Y_i\rangle_x = 0, i = 0, 1 ,$$

$$_x\langle Q_0 \mid Q_1\rangle_x = 0 , \qquad \text{(H.3)}$$

where the Qs and Ys are states of the copying machine.

With the above assumptions, one can write $\rho_{ab}^{(\text{out})}$, describing the modes a and b after the copying of a pure state $| \psi\rangle = \alpha \, | 0\rangle + \beta \, | 1\rangle$ as

$$\begin{aligned}
\rho_{ab}^{(\text{out})} = {} & \alpha^2 \, | 00\rangle\langle 00 \, |_x \, \langle Q_0 \mid Q_0\rangle_x + \sqrt{2}\alpha\beta \, | 00\rangle\langle + \, |_x \, \langle Y_1 \mid Q_0\rangle_x \\
& + \sqrt{2}\alpha\beta \, | +\rangle\langle 00 \, |_x \, \langle Q_0 \mid Y_1\rangle_x \\
& + \left[2\alpha_x^2\langle Y_0 \mid Y_0\rangle_x + 2\beta_x^2\langle Y_1 \mid Y_1\rangle_x \right] \, | +\rangle\langle + \, | \\
& + \sqrt{2}\alpha\beta \, | +\rangle\langle 11 \, |_x \, \langle Q_1 \mid Y_0\rangle_x + \sqrt{2}\alpha\beta \, | 11\rangle\langle + \, |_x \, \langle Y_0 \mid Q_1\rangle_x \\
& + \beta^2 \, | 11\rangle\langle 11 \, |_x \, \langle Q_1 \mid Q_1\rangle_x .
\end{aligned} \qquad \text{(H.4)}$$

Now, if we trace over the b mode, we get the density matrix for the a-mode

$$\begin{aligned}
\rho_a^{(\text{out})} = {} & | 0\rangle_a\langle 0 \, | \, (\alpha^2 + \beta_x^2\langle Y_1 \mid Y_1\rangle_x - \alpha_x^2\langle Y_0 \mid Y_0\rangle_x) \\
& + | 0\rangle_a\langle 1 \, | \, \alpha\beta \, ({}_x\langle Q_1 \mid Y_0\rangle_x + {}_x\langle Y_1 \mid Q_0\rangle_x) \\
& + \alpha\beta \, | 1\rangle_a\langle 0 \, | \, ({}_x\langle Q_0 \mid Y_1\rangle_x + {}_x\langle Y_0 \mid Q_1\rangle_x) \\
& + | 1\rangle_a\langle 1 \, | \, (\beta^2 + \alpha_x^2\langle Y_0 \mid Y_0\rangle_x - \beta_x^2\langle Y_1 \mid Y_1\rangle_x) .
\end{aligned} \qquad \text{(H.5)}$$

The density operator $\rho_b^{(\text{out})} = \rho_a^{(\text{out})}$, in other words the two output modes are equal, but different to the input state. To quantify the difference, we use the 'distance' (22.7), giving as a result

$$D_\alpha = 2x^2(4\alpha^4 - 4\alpha^2 + 1) + 2\alpha^2(1 - \alpha^2)(e - 1)^2 , \tag{H.6}$$

with

$$_x\langle Y_0 \mid Y_0\rangle_x =_x \langle Y_1 \mid Y_1\rangle_x \equiv x , \tag{H.7}$$

$$_x\langle Y_0 \mid Q_1\rangle_x =_x \langle Q_0 \mid Y_1\rangle_x =_x \langle Q_1 \mid Y_0\rangle_x =_x \langle Y_1 \mid Q_0\rangle_x \equiv \frac{e}{2} , \tag{H.8}$$

which are the two free parameters, with $0 \le x \le \frac{1}{2}, 0 \le e \le \frac{1}{\sqrt{2}}$.

The first requirement is that the distance D_α be independent of the input, that is of α. So we impose

$$\frac{\partial D_\alpha}{\partial(\alpha^2)} = 0 , \tag{H.9}$$

which gives us a relation between the parameters

$$e = 1 - 2x ,$$

so that D_α becomes input independent $D_\alpha = 2x^2$.

We also require a condition on the two-mode density. The distance between the density operator and its ideal version should be input independent, that is $D_{ab} = Tr(\rho_{ab}^{(\text{out})} - \rho_{ab}^{(\text{id})})^2$ satisfies

$$\frac{\partial D_{ab}}{\partial(\alpha^2)} = 0 . \tag{H.10}$$

After some algebra, one gets

$$D_{ab} = (f_{11})^2 + 2(f_{12})^2 + 2(f_{13})^2$$

$$(f_{22})^2 + 2(f_{23})^2 + (f_{33})^2 ,$$

with $f_{11} = \alpha^4 - \alpha^2(1 - 2x)$, $f_{12} = \sqrt{2}\alpha\beta(\alpha^2 - \frac{1}{2}(1 - 2x))$, $f_{13} = (\alpha\beta)^2$, $f_{22} = 2((\alpha\beta)^2 - x)$, $f_{23} = \sqrt{2}\alpha\beta(\beta^2 - \frac{1}{2}(1 - 2x))$, $f_{33} = \beta^4 - \beta^2(1 - 2x)$.

Now, the (H.9) can be solved, giving $x = \frac{2}{9}$.

If we write $\rho_a^{(\text{out})}$ in the basis $\mid \psi\rangle = \alpha \mid 0\rangle + \beta \mid 1\rangle$ and $\mid \psi_\perp\rangle = \alpha \mid 0\rangle - \beta \mid 1\rangle$, we readily get

$$\rho_{a_1}^{(\text{out})} = \frac{5}{6} \mid \psi\rangle_{a_1}\langle\psi \mid + \frac{1}{6} \mid \psi_\perp\rangle_{a_1}\langle\psi_\perp \mid .$$

Reference

1. Buzek, V., Hillery, M.: Phys. Rev. A, **54**, 1844 (1996)

I. Hints to Solve the Problems

Chapter1

1.1 Use (1.11) and (1.12).
1.2 Calculate $\langle n^2 \rangle$
1.3 Verify the solution using (1.3)

Chapter2

2.1 Use (2.36)

Chapter3

3.1 Iterate (3.27) many times.
3.2 See Appendix A
3.3 See Appendix A
3.4 Follow the text from (3.39) to (3.45)
3.5 Use (3.44) and:

$$\frac{\partial}{\partial z}\delta_{11}^T(\rho) = \frac{\partial}{\partial z}\delta(\rho) + \frac{i}{(2\pi)^3}\int_{-\infty}^{\infty}\frac{k_3 k_1^2}{k^2}\exp(i\mathbf{k}\cdot\rho)d\mathbf{k}$$

$$\frac{\partial}{\partial x}\delta_{13}^T(\rho) = \frac{i}{(2\pi)^3}\int_{-\infty}^{\infty}\frac{k_3 k_1^2}{k^2}\exp(i\mathbf{k}\cdot\rho)d\mathbf{k}.$$

Chapter4

4.1 Define

$$a^\dagger \mid \beta\rangle = \beta \mid \beta\rangle,$$

and follow the same procedure as in (4.2) to (4.6).
4.2 Use

$$\mid \alpha\rangle = \exp(-\frac{\alpha\alpha^*}{2})\exp(\alpha a^\dagger) \mid 0\rangle,$$

and write

$$\mid \alpha\rangle\langle \alpha \mid = \exp(-\alpha\alpha^*)\exp(\alpha a^\dagger) \mid 0\rangle\langle 0 \mid \exp(\alpha^* a).$$

4.3 Use (4.6)

4.4 Convert the sums into integrals

4.5 Use (4.31) and (4.32)

4.6 Start from (4.2)

4.7 To prove the last property, use the second one for continuous spectrum.

4.8 Use the results of Problem (4.7)

Chapter5

5.1 Use a procedure similar to the one leading to (5.27)

5.2 Use the results of Problem (5.1)

5.3 See Reference [1]

5.4 See Reference [1]

5.5 Use the results of Problem (5.1)

Chapter6

6.1 Calculate $\langle n^2 \rangle$ as we did for $\langle n \rangle$ in (6.95)

Chapter7

7.1 Use (4.16)

7.2 Use (3.19)

7.3 Use the commutation relation

$$[a, a^{\dagger n}] = n a^{\dagger (n-1)}.$$

7.4 Use (A.23) and (A.24)

7.5 First show that

$$a^l F^{(n)}(a^\dagger, a) = \mathcal{N}(a + \frac{\partial}{\partial a^\dagger})^l F^{(n)}(a^\dagger, a),$$

$$F^{(n)}(a^\dagger, a) a^{\dagger l} = \mathcal{N}(a^\dagger + \frac{\partial}{\partial a})^l F^{(n)}(a^\dagger, a).$$

Chapter8

8.1 Find the eigenvalues and eigenvectors of H_n.

8.2 Use (8.46) and (8.38)

8.3 Approximate (8.55)

Chapter9

9.2 Verify the definitions, using the results of Problem (9.1)

9.3 Use the rules given by (9.49)

9.4 Use the rules given by (9.49)

9.5 Use the rules given by (9.49)

9.6 Find $\langle a^2 \rangle$ and $\langle a^{\dagger 2} \rangle$ from an equation similar to (9.21)

Chapter10

10.1 Use (10.59), (10.60), (10.61)
10.2 Calculate the Fourier Transform of the result of Problem (10.1)
10.3 Use (10.84), (10.85), (10.86).

Chapter11

11.1 Use (11.5) and (11.6)
11.2 Use (11.8)
11.3 Start from (11.22) and approximate the trigonometric functions.
11.4 Start from (11.23)

Chapter12

12.1 Use the Generalized Einstein relations.
12.2 Use a procedure similar to (12.35–12.38)
12.3 See Reference [1]
12.4 Take ε^2 and differentiate with respect to time and use (12.65)

Chapter13

13.1 Star from (13.12)
13.2 Use (13.24) and follow the rules given by (9.49). Then one gets a Fokker-Planck equation in terms of α_1 and α_2. To go to polar coordinates, define

$$\alpha_1 = \rho_1 \exp(i\theta_1); \alpha_2 = \rho_2 \exp(i\theta_2),$$

then, one has

$$\frac{\partial}{\partial \alpha_1} = \frac{1}{2}\exp(-i\theta_1)\frac{\partial}{\partial \rho_1} + \frac{1}{2i}\frac{\exp(-i\theta_1)}{\rho_1}\frac{\partial}{\partial \theta_1},$$

$$\frac{\partial}{\partial \alpha_2} = \frac{1}{2}\exp(-i\theta_2)\frac{\partial}{\partial \rho_2} + \frac{1}{2i}\frac{\exp(-i\theta_2)}{\rho_2}\frac{\partial}{\partial \theta_2},$$

$$\frac{\partial}{\partial \theta_1} = \frac{1}{2}\frac{\partial}{\partial \mu}+,$$

$$\frac{\partial}{\partial \theta_2} = \frac{1}{2}\frac{\partial}{\partial \mu} - \frac{\partial}{\partial \theta},$$

where

$$\mu = \frac{\theta_1 + \theta_2}{2},$$

$$\theta = \frac{\theta_1 - \theta_2}{2}.$$

13.3 Use the results of the Problem (13.2)

Chapter14

14.1 Start from (14.62)
14.2 Start from (14.66)
14.3 Integrate (14.38) over ω.
14.4 Part (b). Use the quadratic part of the formula for $\frac{1}{\tau}$.
14.5 Use the results of the problem (14.4) for the case $\omega \gg \omega_j$

Chapter15

15.1 Use (15.5), (15.6) and (15.7)
15.2 Use (15.7)
15.3 Use (15.5)
15.4 See Reference [5]
15.5 Use (15.49)
15.6 See Reference [17]

Chapter16

16.1 See Reference [new29]
16.2 See Reference [11]

Chapter17

17.1 Use (17.7)
17.2 Verify that $[H, c] = 0$.
17.4 See Reference [18]

Chapter18

18.1 Use (18.49) and (18.50)
18.2 Use (A.16) and (A.17)
18.3 See Appendix A
18.4 See Reference [1]
18.5 See Reference [1]
18.6 See Reference [1]

Chapter19

19.1 See Reference [19]
19.2 See Reference [19]

Chapter 20

20.1 Use (20.34), (20.41) to verify (20.42)
20.2 See References [9], [10], [11], [12].

Chapter 21

21.1 Apply the definition of K.
21.2 Check the signs of the eigenvalues of the partially transposed density matrix.
21.3 Apply the NPT criterion.

Chapter 22

22.2 Follow a procedure similar to the duplicator
22.3 Apply the two gates to the input data and programs.

Index